NONLINEAR EFFECTS IN OPTICAL FIBERS

NONLINEAR EFFECTS IN OPTICAL FIBERS

MÁRIO F. S. FERREIRA

OPTICAL SOCIETY OF AMERICA

A JOHN WILEY & SONS, INC., PUBLICATION

Published by John Wiley & Sons, Inc., Hoboken, New Jersey
Published simultaneously in Canada

For general information on our other products and services or for technical support, please
contact our Customer Care Department within the United States at (800) 762-2974, outside the
United States at (317) 572-3993 or fax (317) 572-4002.

Wiley also publishes its books in a variety of electronic formats. Some content that appears in print
may not be available in electronic formats. For more information about Wiley products, visit our
web site at www.wiley.com.

Library of Congress Cataloging-in-Publication Data:

Ferreira, Mário F. S.
 Nonlinear effects in optical fibers / Mario F. S. Ferreira.
  p. cm.
 ISBN 978-0-470-46466-3 (hardback)
 1. Fiber optics. 2. Nonlinear optics. I. Title.
 QC448.F45 2010
 621.36'92–dc22

                               2010036857

Printed in Singapore

oBook ISBN: 978-1-118-00339-8
ePDF ISBN: 978-1-118-00337-4
ePub ISBN: 978-1-118-00338-1

10 9 8 7 6 5 4 3 2 1

CONTENTS

Preface   **xi**

1   Introduction   **1**

References / 5

2   Electromagnetic Wave Propagation   **9**

2.1   Wave Equation for Linear Media / 9
2.2   Electromagnetic Waves / 11
2.3   Energy Density and Flow / 13
2.4   Phase Velocity and Group Velocity / 14
2.5   Reflection and Transmission of Waves / 16
    2.5.1   Snell's Laws / 16
    2.5.2   Fresnel Equations / 17
2.6   The Harmonic Oscillator Model / 21
2.7   The Refractive Index / 23
2.8   The Limit of Geometrical Optics / 24
Problems / 26
References / 27

3   Optical Fibers   **29**

3.1   Geometrical Optics Description / 30
    3.1.1   Planar Waveguides / 30
    3.1.2   Step-Index Fibers / 33
    3.1.3   Graded-Index Fibers / 36

3.2   Wave Propagation in Fibers / 39
    3.2.1   Fiber Modes / 39
    3.2.2   Single-Mode Fibers / 42
3.3   Fiber Attenuation / 44
3.4   Modulation and Transfer of Information / 45
3.5   Chromatic Dispersion in Single-Mode Fibers / 46
    3.5.1   Unchirped Input Pulses / 48
    3.5.2   Chirped Input Pulses / 52
    3.5.3   Dispersion Compensation / 53
3.6   Polarization Mode Dispersion / 54
    3.6.1   Fiber Birefringence and the Intrinsic PMD / 55
    3.6.2   PMD in Long Fiber Spans / 57
Problems / 60
References / 61

4   The Nonlinear Schrödinger Equation   **63**

4.1   The Nonlinear Polarization / 63
    4.1.1   The Nonlinear Wave Equation / 66
4.2   The Nonlinear Refractive Index / 66
4.3   Importance of Nonlinear Effects in Fibers / 68
4.4   Derivation of the Nonlinear Schrödinger Equation / 70
    4.4.1   Propagation in the Absence of Dispersion and Nonlinearity / 73
    4.4.2   Effect of Dispersion Only / 73
    4.4.3   Effect of Nonlinearity Only / 74
    4.4.4   Normalized Form of NLSE / 74
4.5   Soliton Solutions / 75
    4.5.1   The Fundamental Soliton / 76
    4.5.2   Solutions of the Inverse Scattering Theory / 77
    4.5.3   Dark Solitons / 80
4.6   Numerical Solution of the NLSE / 81
Problems / 83
References / 84

5   Nonlinear Phase Modulation   **85**

5.1   Self-Phase Modulation / 86
    5.1.1   SPM-Induced Phase Shift / 86
    5.1.2   The Variational Approach / 89
    5.1.3   Impact on Communication Systems / 93
    5.1.4   Modulation Instability / 94

5.2 Cross-Phase Modulation / 97

    5.2.1 XPM-Induced Phase Shift / 97

    5.2.2 Impact on Optical Communication Systems / 100

    5.2.3 Modulation Instability / 103

    5.2.4 XPM-Paired Solitons / 105

Problems / 106

References / 107

6 Four-Wave Mixing   **111**

6.1 Wave Mixing / 112

6.2 Mathematical Description / 114

6.3 Phase Matching / 115

6.4 Impact and Control of FWM / 118

6.5 Fiber Parametric Amplifiers / 123

    6.5.1 FOPA Gain and Bandwidth / 123

6.6 Parametric Oscillators / 128

6.7 Nonlinear Phase Conjugation with FWM / 131

6.8 Squeezing and Photon Pair Sources / 133

Problems / 135

References / 135

7 Intrachannel Nonlinear Effects   **139**

7.1 Mathematical Description / 140

7.2 Intrachannel XPM / 142

7.3 Intrachannel FWM / 147

7.4 Control of Intrachannel Nonlinear Effects / 149

Problems / 153

References / 153

8 Soliton Lightwave Systems   **155**

8.1 Soliton Properties / 156

    8.1.1 Soliton Interaction / 157

8.2 Perturbation of Solitons / 159

    8.2.1 Perturbation Theory / 160

    8.2.2 Fiber Losses / 160

8.3 Path-Averaged Solitons / 162

    8.3.1 Lumped Amplification / 163

    8.3.2 Distributed Amplification / 164

    8.3.3 Timing Jitter / 166

8.4   Soliton Transmission Control / 168

    8.4.1   Fixed-Frequency Filters / 169

    8.4.2   Sliding-Frequency Filters / 170

    8.4.3   Synchronous Modulators / 173

    8.4.4   Amplifier with Nonlinear Gain / 174

8.5   Dissipative Solitons / 176

    8.5.1   Analytical Results of the CGLE / 176

    8.5.2   Numerical Solutions of the CGLE / 180

8.6   Dispersion-Managed Solitons / 183

    8.6.1   The True DM Soliton / 183

    8.6.2   The Variational Approach to DM Solitons / 185

8.7   WDM Soliton Systems / 189

Problems / 192

References / 193

9   **Other Applications of Optical Solitons**   **199**

9.1   Soliton Fiber Lasers / 199

    9.1.1   The First Soliton Laser / 200

    9.1.2   Figure-Eight Fiber Laser / 201

    9.1.3   Nonlinear Loop Mirrors / 201

    9.1.4   Stretched-Pulse Fiber Lasers / 202

    9.1.5   Modeling Fiber Soliton Lasers / 203

9.2   Pulse Compression / 204

    9.2.1   Grating-Fiber Compressors / 204

    9.2.2   Soliton-Effect Compressors / 207

    9.2.3   Compression of Fundamental Solitons / 210

9.3   Fiber Bragg Gratings / 213

    9.3.1   Pulse Compression Using Fiber Gratings / 214

    9.3.2   Fiber Bragg Solitons / 216

Problems / 220

References / 220

10   **Polarization Effects**   **225**

10.1   Coupled Nonlinear Schrödinger Equations / 226

10.2   Nonlinear Phase Shift / 227

10.3   Solitons in Fibers with Constant Birefringence / 229

10.4   Solitons in Fibers with Randomly Varying Birefringence / 234

10.5   PMD-Induced Soliton Pulse Broadening / 236

10.6   Dispersion-Managed Solitons and PMD / 240

Problems / 242

References / 242

11   Stimulated Raman Scattering   **245**

11.1   Raman Scattering in the Harmonic Oscillator Model / 246

11.2   Raman Gain / 250

11.3   Raman Threshold / 252

11.4   Impact of Raman Scattering on Communication
       Systems / 255

11.5   Raman Amplification / 258

11.6   Raman Fiber Lasers / 264

Problems / 269

References / 270

12   Stimulated Brillouin Scattering   **273**

12.1   Light Scattering at Acoustic Waves / 274

12.2   The Coupled Equations for Stimulated Brillouin
       Scattering / 277

12.3   Brillouin Gain and Bandwidth / 278

12.4   Threshold of Stimulated Brillouin Scattering / 280

12.5   SBS in Active Fibers / 282

12.6   Impact of SBS on Communication Systems / 284

12.7   Fiber Brillouin Amplifiers / 286

       12.7.1   Amplifier Gain / 287

       12.7.2   Amplifier Noise / 289

       12.7.3   Other Applications of the SBS Gain / 290

12.8   SBS Slow Light / 293

12.9   Fiber Brillouin Lasers / 296

Problems / 300

References / 301

13   Highly Nonlinear and Microstructured Fibers   **305**

13.1   The Nonlinear Parameter in Silica Fibers / 306

13.2   Microstructured Fibers / 309

13.3   Non-Silica Fibers / 314

13.4   Soliton Self-Frequency Shift / 317

13.5   Four-Wave Mixing / 320

13.6   Supercontinuum Generation / 323

    13.6.1   Basic Physics of Supercontinuum Generation / 323

    13.6.2   Modeling the Supercontinuum / 330

Problems / 332

References / 332

14   Optical Signal Processing   **339**

14.1   Nonlinear Sources for WDM Systems / 340

14.2   Optical Regeneration / 343

14.3   Optical Pulse Train Generation / 349

14.4   Wavelength Conversion / 350

    14.4.1   Wavelength Conversion with FWM / 351

    14.4.2   Wavelength Conversion with XPM / 354

14.5   All-Optical Switching / 358

    14.5.1   XPM-Induced Optical Switching / 359

    14.5.2   Optical Switching Using FWM / 361

Problems / 363

References / 364

Index   **369**

PREFACE

The first generation of fiber-optic communication systems was introduced early in the 1980s and operated at modest values of both the bit rate and the link length. In such circumstances, the nonlinear effects were found to be irrelevant. However, the situation changed dramatically during the 1990s with the advent and commercialization of wideband optical amplifiers, wavelength division multiplexing, and high-speed optoelectronic devices. By the end of that decade, the capacity of lightwave systems had already exceeded 1 Tb/s, as a result of the combination of larger number of WDM channels and increased channel data rates, together with denser channel spacings. Significant performance improvements were achieved in the following years, which paved the way for today's systems with rates approaching 100 Gb/s per channel (wavelength) and wavelength counts of 80–100. On the other hand, higher channel powers are being used in long-haul landline and submarine links in order to increase the distances between amplifiers or repeaters. As a result of all these advances, the nonlinear effects in optical fibers became of paramount importance, since they adversely affect the system performance.

Paradoxically, the same nonlinear phenomena that have several important limitations also offer the promise of addressing the bandwidth bottleneck for signal processing for future ultrahigh-speed optical networks. Electronic devices are not suitable for such systems, due to their cost, complexity, and practical speed limits. Nonlinear optical signal processing, making use of the third-order optical nonlinearity in single-mode fibers, appears as a key and promising technology for improving the transparency and increasing the capacity of future full "photonic networks."

Starting in 1996, new types of fibers, known as photonic crystal fibers, holey fibers, or microstructured fibers, were developed. These fibers have a relatively narrow core, surrounded by a cladding that contains an array of embedded air holes. Structural changes in such fibers profoundly affect their dispersive and nonlinear properties. The efficiency of the nonlinear effects can be further increased if some highly nonlinear materials are used to make the fibers, instead of silica. Using such highly nonlinear fibers, the required fiber length for nonlinear processing could be reduced to the order of centimeters, instead of the several kilometers long conventional silica fibers. All these advances have led to considerable growth in the field of nonlinear fiber optics during the last decade.

This book provides an introduction to the fascinating world of nonlinear phenomena occurring inside the optical fibers. Though the main emphasis is placed on the physical background of the different nonlinear effects, the technical aspects associated with their impact on optical communication systems, as well as their potential applications, particularly for signal processing, pulse generation, and amplification, are also discussed. An attempt has been made to include the latest and most significant research results in this area. Moreover, several problems are included at the end of each chapter. These aspects contribute to make this book of potential interest to senior undergraduate and graduate students enrolled in M.S. and Ph.D. degree programs, engineers and technicians involved with the fiber-optics industry, and researchers working in the field of nonlinear fiber optics.

I am deeply grateful to the many students and colleagues with whom I have interacted over the years. All of them have contributed to this book either directly or indirectly. In particular, I thank especially Sofia Latas for providing many figures, as well as Margarida Facão, Armando Pinto, and Nelson Muga for several discussions concerning different parts of this text. Last but not least, I thank my family for understanding why I needed to remain working during many of our weekends and vacations.

MÁRIO F. S. FERREIRA

Aveiro, Portugal
February 2011

1

INTRODUCTION

The propagation of light in optical fibers is based on the phenomenon of total internal reflection, which is known since 1854, when John Tyndall demonstrated the transmission of light along a stream of water emerging from a hole in the side of a tank [1]. Glass fibers were fabricated since the 1920s, but their use remained restricted to medical applications until the 1960s. The use of such fibers for optical communications was not practical, due to their high losses (~1000 dB/km). However, Kao and Hockham [2] suggested in 1966 that optical fibers could be used in communication systems if their losses were reduced below 20 dB/km. Following an intense activity on the purification of fused silica, such goal was achieved in 1970 by Corning Glass Works, in the United States. Further technological progress allowed the reduction of fiber loss to 0.2 dB/km near the 1550 nm spectral region by 1979 [3]. This achievement led to a revolution in the field of optical fiber communications.

Besides the loss, the fiber dispersion constitutes actually another main problem affecting the performance of an optical communication system. An example of this is the mechanism of *group velocity dispersion* (GVD), which arises as the frequency components of the signal pulse propagate with different velocities, determining the broadening of the pulse. Dispersive pulse broadening and loss both increase in direct proportion to the length of the link. Traditionally, repeater stations have been used at appropriate intervals over long links for detecting, electrically amplifying, filtering, and then regenerating the optical signal. However, such repeaters are complicated and become expensive to use in large quantities. Fiber amplifiers appear in most cases as an attractive alternative to the electronic repeaters. A single amplifier is able to boost the power in multiple wavelengths simultaneously, whereas a separate electronic repeater is needed for each wavelength. This simple fact made feasible the development and deployment of dense wavelength division multiplexed (DWDM) systems, which have revolutionized network communication systems since the 1990s.

The expansion of fiber networks to encompass larger areas coupled with the use of longer distances between amplifiers or repeaters means that higher optical power levels are needed. In addition, the ever-increasing bit rates imply the use of shorter pulses having higher intensities. Both these changes increase the likelihood of various nonlinear processes in the fibers. In fact, the nonlinearities of fused silica, from which optical fibers are made, are weak compared to those of many other materials. However, nonlinear effects can be readily observed in optical fibers due to both their rather small field cross sections, which results in high field intensities, and the long interaction lengths provided by them, which significantly enhances the efficiency of the nonlinear processes. These nonlinear processes can impose significant limitations in high-capacity fiber transmission systems.

It seems paradoxical that the same nonlinear phenomena that impose several important limitations also offer the promise of addressing the bandwidth bottleneck for signal processing for future ultrahigh-speed optical networks. In fact, electronic devices are not suitable for such systems, due to their cost, complexity, and practical speed limits. All-optical signal processing appears, therefore, as a key and promising technology for improving the transparency and increasing the capacity of future full "photonic networks" [4].

Nonlinear optical signal processing appears as a potential solution to this demand. In particular, the third-order $\chi^{(3)}$ optical nonlinearity in silica-based single-mode fibers offers a significant promise in this regard [5]. This happens not only because the third-order nonlinearity is nearly instantaneous—having a response time typically <10 fs—but also because it is responsible for a wide range of phenomena, which can be used to construct a great variety of all-optical signal processing devices.

Silica fiber nonlinearities can be classified into two main categories: stimulated scattering effects (Raman and Brillouin) and effects arising from the nonlinear index of refraction. Stimulated scattering is manifested as intensity-dependent gain or loss, while the nonlinear index gives rise to an intensity-dependent phase of the optic field. The first experimental demonstration of fiber nonlinearities was Erich Ippen's CS_2-core fiber Raman laser in early 1970 [6]. Subsequently, Smith's theoretical paper on stimulated Raman and Brillouin scattering in silica fibers [7] and the first experimental demonstration of stimulated Raman scattering in a single-mode fiber by Stolen et al. [8] were two landmarks in this field.

Stimulated Raman scattering (SRS) results from the interaction between the photons and the molecules of the medium and leads to the transfer of the light intensity from the shorter to the longer wavelengths. The SRS gain in silica has a wide bandwidth on the order of 12 THz ($\sim$100 nm at 1.5 µm) due to its amorphous nature. Thus, SRS can lead to the crosstalk between different WDM channels, becoming the most detrimental of the scattering effects in such systems.

Besides the negative aspect pointed out above, the Raman effect can also find several positive applications. One of the readily apparent advantages of Raman gain in glass fibers was the possibility of constructing wideband amplifiers and tunable oscillators [9]. Indeed, the first SRS work also demonstrated a Raman oscillator using mirrors to provide feedback in a 190 cm fiber [8]. However, the goal of a tunable continuous-wave (CW) fiber Raman laser would have to wait for longer low-loss single-mode fibers. It was not until 1983 that studies of Raman amplification from laboratories around the world began to appear. By the end of the 1980s, the signal-to-noise advantages of Raman amplification appeared to be well understood [10].

However, the lack of efficient high-power fiber-coupled pump lasers prevented the practical use of Raman amplification by that time. After 1988, the lightwave world concentrated on the erbium fiber amplifier, and it was not until 1997 that system experiments using Raman amplifiers started to appear. Following those early demonstrations, the use of Raman amplification in transmission systems has become quite common.

Brillouin scattering originates from the interaction between the pump light and acoustic waves generated in the fiber. In this way, a strong wave traveling in one direction provides narrowband gain, with a linewidth on the order of 20 MHz, for light propagating in the opposite direction. Stimulated Brillouin scattering (SBS) in fibers was observed for the first time in 1972 by Ippen and Stolen [11], who used a pulsed narrowband xenon laser operating at 535.3 nm.

The peak of the Brillouin gain coefficient is over 100 times greater than the Raman gain peak, which makes SBS the dominant nonlinear process in silica fibers under some circumstances. This is particularly the case in fiber transmission systems using narrow-linewidth lasers. SBS can be detrimental to such systems in a number of ways: by originating a severe signal attenuation, by causing multiple frequency shifts in some cases, and by introducing a high-intensity backward coupling into the transmission optics. However, Brillouin gain can also find some useful applications, namely, as an inline fiber amplifier [12,13], for channel selection in a closely spaced wavelength-multiplexed network [14,15], temperature and strain sensing [16,17], all-optical slow-light control [18,19], optical storage [20], and so on.

The intensity-dependent refractive index of silica gives rise to three effects: self-phase modulation (SPM), cross-phase modulation (XPM), and four-wave mixing (FWM). The SPM effect corresponds to a spectral broadening of the pulse determined by its own power temporal variation. The first observation of this phenomenon in silica fibers occurred in a 1975 experiment by Lin and Stolen [21]. Earlier, Hasegawa and Tappert [22] suggested the existence of fiber solitons, resulting from a balance between SPM and anomalous GVD. Such solitons were indeed observed experimentally by Mollenauer et al. [23] in 1980 and subsequently led to a number of advances in the generation and control of ultrashort pulses [24,25]. The advent of fiber amplifiers fueled research on optical solitons and eventually led to new types of solitons, such as dispersion-managed and dissipative solitons [26–33].

The XPM effect is similar to the SPM effect but the spectral broadening of the pulses is now due to the influence of other pulses propagating at the same time in the fiber. This effect becomes especially important in WDM systems, where a large number of pulses with different carrier wavelengths are usually transmitted in one fiber. Since the bit pattern in the different channels is completely random, the cancellation of this effect through an intelligent system design will be impossible in practice. The XPM appears indeed as the fundamental effect that determines the maximum capacity of optical transmission systems. However, XPM can also be used with advantage in several applications in the area of nonlinear optical processing [34–37].

Due to the FWM effect, beating between two channels at their difference frequency modulates the signal phase at that frequency, generating new frequencies as sidebands. The occurrence of the FWM phenomenon in optical fibers was observed for the first time by Stolen et al. [38] using a 9 cm long multimode fiber pumped by a double-pulsed YAG laser at 532 nm. In a WDM system, if the channels are equally spaced, the new components generated by FWM fall at the original channel frequencies, giving rise to crosstalk [39–41]. In contrast to SPM and XPM, which become more significant for high bit rate systems, the FWM does not depend on the bit rate. The efficiency of this phenomenon depends strongly on phase matching conditions, as well as on the channel spacing, chromatic dispersion, and fiber length.

Besides its obvious application in generating new frequencies over a broad spectral range, FWM can also be used to amplify signals over a broad band around the fiber zero-dispersion wavelength. Moreover, FWM can be used for nonlinear phase conjugation, frequency conversion, optical switching, generation of squeezed states of light, and as a source of entangled photon pairs [42–53].

Starting in 1996, new types of fibers, known as tapered fibers, photonic crystal fibers, and microstructured fibers, were developed [54–58]. Structural changes in such fibers profoundly affect their dispersive and nonlinear properties. As a result, new phenomena were observed, such as the supercontinuum generation, in which the optical spectrum of ultrashort pulses is broadened by a factor of more than 200 over a length of only 1 m or less [59–62]. The efficiency of the nonlinear effects can be further increased using fibers made of materials with a nonlinear refractive index higher than that of the silica glass, namely, lead silicate, tellurite, bismuth glasses, and chalcogenide glasses [63–66]. Using such highly nonlinear fibers (HNLFs), the required fiber length for nonlinear processing can be reduced to the order of centimeters, instead of the several kilometers long conventional fibers.

This book is intended to provide a comprehensive account of the various nonlinear effects occurring in optical fibers. An overview is given of the impact of these effects on communication systems, as well as of their potential in different applications, particularly for signal processing, pulse generation, and amplification. This book can be roughly divided into five parts. The first part, consisting of Chapters 2–4, presents the basic concepts and equations that will be used in the rest of the book. Chapter 2 provides a review of the fundamental concepts and properties related to light propagation in linear dielectric media. The harmonic oscillator model is used to describe the interaction between an optical wave and the matter. Chapter 3 discusses the basic linear properties of optical fibers in the perspective of their use in communication systems, a special attention being paid to the phenomena of chromatic dispersion and polarization mode dispersion. A brief introduction to nonlinear optics, the derivation of the nonlinear Schrödinger equation, and a discussion of its soliton solutions are presented in Chapter 4.

The second part, consisting of Chapters 5–7, is dedicated to the description of nonlinear effects arising from the intensity-dependent refractive index of optical fibers. Chapter 5 describes the phenomena of self-phase modulation and cross-phase modulation, as well as their impact on communication systems. Chapter 6 deals with the four-wave mixing process, including some important applications of this phenomenon, such as parametric amplification, parametric oscillation, optical phase conjugation, and the generation of squeezed states of light. While both XPM and

FWM appear as *interchannel* nonlinear effects, the nonlinear interaction among the pulses of the same channel is discussed in Chapter 7 in which two *intrachannel* effects are considered: the intrachannel cross-phase modulation (IXPM) and the intrachannel four-wave mixing (IFWM). Both IXPM and IFWM can occur only when the pulses overlap in time, at least partly, during their propagation, as happens in dispersion-managed transmission systems.

The third part, consisting of Chapters 8–10, is dedicated to the topic of optical fiber solitons and their applications. Chapter 8 deals with the use of optical solitons in communication systems, considering both constant dispersion and dispersion-managed fiber links. Other applications and phenomena involving optical solitons are discussed in Chapter 9. The polarization effects on soliton propagation, considering the cases of both constant and randomly varying birefringence, are discussed in Chapter 10.

The fourth part, consisting of Chapters 11 and 12, presents a discussion of resonant fiber nonlinear effects. Chapter 11 is dedicated to the stimulated Raman scattering effect, whereas Chapter 12 deals with the stimulated Brillouin scattering effect. The similarities and main differences between these two effects, the limitations that they impose on communication systems, and some important applications are discussed in both chapters.

The fifth and last part, consisting of Chapters 13 and 14, is dedicated to the description of the more relevant types of highly nonlinear fibers, together with some of their actual applications in nonlinear optical signal processing. Chapter 13 describes silica-based conventional highly nonlinear fibers, microstructured fibers, and fibers made of highly nonlinear materials, as well as some novel nonlinear phenomena that can be observed with them. Chapter 14 highlights the importance of highly nonlinear fibers to realize different functions in the area of optical signal processing, namely, multiwavelength sources, pulse generation, all-optical regeneration, wavelength conversion, and optical switching.

REFERENCES

1. J. Tyndall, *Proc. R. Inst.* **1**, 446 (1854).
2. K. C. Kao and G. A. Hockham, *Proc. IEE* **113**, 1151 (1966).
3. T. Miya, Y. Terunuma, F. Hosaka, and T. Miyoshita, *Electron. Lett.* **15**, 106 (1979).
4. M. Saruwatari, *IEEE J. Sel. Top. Quantum Electron.* **6**, 1363 (2000).
5. T. Okuno, M. Onishi, T. Kashiwada, S. Ishikawa, and M. Nishimura, *IEEE J. Sel. Top. Quantum Electron.* **5**, 1385 (1999).
6. E. P. Ippen, *Appl. Phys. Lett.* **16**, 303 (1970).
7. R. G. Smith, *Appl. Opt.* **11**, 2489 (1972).
8. R. H. Stolen, E. P. Ippen, and A. R. Tynes, *Appl. Phys. Lett.* **20**, 62 (1972).
9. E. P. Ippen, C. K. N. Patel, and R. H. Stolen, U.S. Patent 3,705,992, 1971.
10. Y. Aoki, *J. Lightwave Technol.* **6**, 1225 (1988).
11. E. P. Ippen and R. H. Stolen, *Appl. Phys. Lett.* **21**, 539 (1972).

12. M. F. Ferreira, J. F. Rocha, and J. L. Pinto, *Opt. Quantum Electron.* **26**, 35 (1994).

13. R. W. Tkach and A. R. Chraplyvy, *Opt. Quantum Electron.* **21**, S105 (1989).

14. D. W. Smith, C. G. Atkins, D. Cotter, and R. Wyatt, *Electron. Lett.* **22**, 556 (1986).

15. A. R. Chraplyvy and R. W. Tkach, *Electron. Lett.* **22**, 1084 (1986).

16. M. Nikles, L. Thevenaz, and P. A. Robert, *Opt. Lett.* **21**, 738 (1996).

17. K. Hotate and M. Tanaka, *IEEE Photon. Technol. Lett.* **14**, 179 (2002).

18. K. Y. Song, M. G. Herraez, and L. Thevenaz, *Opt. Express* **13**, 82 (2005).

19. Z. Zhu, A. M. Dawes, D. J. Gauthier, L. Zhang, and A. E. Willner, *J. Lightwave Technol.* **25**, 201 (2007).

20. Z. Zhu, D. J. Gauthier, and R. W. Boyd, *Science* **318** 1748 (2007).

21. C. Lin and R. H. Stolen, *Appl. Phys. Lett.* **28**, 216 (1976).

22. A. Hasegawa and F. Tappert, *Appl. Phys. Lett.* **23**, 142 (1973).

23. L. F. Mollenauer, R. H. Stolen, and J. P. Gordon, *Phys. Rev. Lett.* **45**, 1095 (1980).

24. L. F. Mollenauer and R. H. Stolen, *Opt. Lett.* **9**, 13 (1984).

25. J. D. Kafka and T. Baer, *Opt. Lett.* **12**, 181 (1987).

26. N. Akhmediev and A. Ankiewicz, *Dissipative Solitons*, Springer, Berlin, 2005.

27. A. A. Ankiewicz, N. N. Akhmediev, and N. Devine, *Opt. Fiber Technol.* **13**, 91 (2007).

28. M. F. Ferreira and S. V. Latas, in *Optical Fibers Research Advances*, Nova Science Publishers, 2008, Chapter 10.

29. N. J. Smith, F. M. Knox, N. J. Doran, K. J. Blow, and I. Bennion, *Electron. Lett.* **32**, 54 (1996).

30. W. Forysiak, F. M. Knox, and N. J. Doran, *Opt. Lett.* **19**, 174 (1994).

31. A. Hasegawa, S. Kumar, and Y. Kodama, *Opt. Lett.* **22**, 39 (1996).

32. M. Suzuki, I. Morita, N. Edagawa, S. Yamamoto, H. Toga, and S. Akiba, *Electron. Lett.* **31**, 2027 (1995).

33. A. Hasegawa (Ed.), *New Trends in Optical Soliton Transmission Systems*, Kluwer, Dordrecht, The Netherlands, 1998.

34. T. Yamamoto, E. Yoshida, and M. Nakazawa, *Electron. Lett.* **34**, 1013 (1998).

35. B. E. Olsson, P. Öhlén, L. Rau, and D. J. Blumenthal, *IEEE Photon. Technol. Lett.* **12**, 846 (2000).

36. J. Li, B. E. Olsson, M. Karlsson, and P. A. Andrekson, *IEEE Photon. Technol. Lett.* **15**, 770 (2003).

37. J. H. Lee, T. Tanemura, K. Kikuchi, T. Nagashima, T. Hasegawa, S. Ohara, and N. Sugimoto, *Opt. Lett.* **39**, 1267 (2005).

38. R. H. Stolen, J. E. Bjorkhholm, and A. Ashkin, *Appl. Phys. Lett.* **24**, 308 (1974).

39. N. Shibata, R. P. Braun, and R. G. Waarts, *IEEE J. Quantum Electron.* **23**, 1205 (1987).

40. F. Forghieri, R. W. Tkach, and A. R. Chraplyvy, *J. Lightwave Technol.* **13**, 889 (1995).

41. H. Suzuki, S. Ohteru, and N. Takachio, *IEEE Photon. Technol. Lett.* **11**, 1677 (1999).

42. J. Hansryd and P. A. Andrekson, *IEEE Photon. Technol. Lett.* **13**, 194 (2001).

43. M. E. Marhic, N. Kagi, T.-K. Chiang, and L. G. Kazovsky, *Opt. Lett.* **21**, 573 (1996).

44. M. E. Marhic, K. K. Y. Wong, L. G. Kazovsky, and T. E. Tsai, *Opt. Lett.* **2**, 1439 (2002).

45. T. Torounnidis and P. Andrekson, *IEEE Photon. Lett.* **19**, 650 (2007).

46. J. E. Sharping, M. A. Foster, A. L. Gaeta, J. Lasri, O. Lyngnes, and K. Vogel, *Opt. Express* **15**, 1474 (2007).

47. J. E. Sharping, J. R. Sanborn, M. A. Foster, D. Broaddus, and A. L. Gaeta, *Opt. Express* **16**, 18050 (2008).

48. J. E. Sharping, *J. Lightwave Technol.* **26**, 2184 (2008).

49. J. Pina, B. Abueva, and G. Goedde, *Opt. Commun.* **176**, 397 (2000).

50. M. D. Levenson, R. M. Shelby, A. Aspect, M. Reid, and D. F. Walls, *Phys. Rev. A* **32**, 1550 (1985).

51. G. J. Milburn, M. D. Levenson, R. M. Shelby, S. H. Perlmutter, R. G. DeVoe, and D. F. Walls, *J. Opt. Soc. Am. B* **4**, 1476 (1987).

52. J. E. Sharping, et al., *Opt. Lett.* **26**, 367 (2001).

53. W. H. Reeves, D. V. Skryabin, F. Biancalana, J. C. Knight, P. St. J. Russell, F. G. Omenetto, A. Efimov, and A. J. Taylor, *Nature* **424**, 511 (2003).

54. L. M. Tong, J. Y. Lou, and E. Mazur, *Opt. Express* **12**, 1025 (2004).

55. C. M. Cordeiro, W. J. Wadsworth, T. A. Birks, and P. St. J. Russel, *Opt. Lett.* **30**, 1980 (2005).

56. J. C. Knight, T. A. Birks, P. St. J. Russel, and D. M. Atkin, *Opt. Lett.* **21**, 1547 (1996).

57. J. Broeng, D. Mogilevstev, S. E. Barkou, and A. Bjarklev, *Opt. Fiber Technol.* **5**, 305 (1999).

58. P. St. J. Russell, *Science* **299**, 358 (2003).

59. S. Coen, A. H. Chau, R. Leonhardt, J. D. Harvey, J. C. Knight, W. J. Wadsworth, and P. St. J. Russel, *J. Opt. Soc. Am. B* **19**, 753 (2002).

60. A. Kudlinski, A. K. George, J. C. Knight, J. C. Travers, A. B. Rulkov, S. V. Popov, and J. R. Taylor, *Opt. Express* **14**, 5715 (2006).

61. A. V. Gorbach and D. V. Skryabin, *Nat. Photon.* **1**, 653 (2007).

62. A. Kudlinski and A. Mussot, *Opt. Lett.* **33**, 2407 (2008).

63. M. Asobe, T. Kanamori, and K. Kubodera, *IEEE Photon. Technol. Lett.* **4**, 362 (1992).

64. K. Kikuchi, K. Taira, and N. Sugimoto, *Electron. Lett.* **38**, 166 (2002).

65. J. H. Lee, K. Kikuchi, T. Nagashima, et al., *Opt. Express* **13**, 3144 (2005).

66. V. G. Ta'eed, N. J. Baker, L. Fu, et al., *Opt. Express* **15**, 9205 (2007).

2

ELECTROMAGNETIC WAVE PROPAGATION

Light is an electromagnetic phenomenon consisting of electric and magnetic fields that are solutions of Maxwell's equations. These equations provide the mathematical foundation used to model and evaluate the flow of electromagnetic energy in all situations, of which optical fibers constitute a particular case. The purpose of this chapter is to review the fundamental concepts and properties related to light propagation in linear dielectric media. From Maxwell's equations, we will derive the linear wave equation and discuss the main properties of electromagnetic waves. Moreover, the harmonic oscillator model will be used to describe the interaction between an optical wave and the matter. Using such a model, the susceptibility, refractive index, and attenuation of an optical material are discussed. Additional information concerning the subject of this chapter can be found in many textbooks [1–6].

2.1 WAVE EQUATION FOR LINEAR MEDIA

The mathematical foundation for the description of electromagnetic wave propagation in a dielectric medium is provided by the *Maxwell's equations*. These equations are named after James Maxwell (1831–1879) and can be written as follows:

$$\nabla \cdot \mathbf{D} = \rho \tag{2.1}$$

$$\nabla \cdot \mathbf{B} = 0 \tag{2.2}$$

$$\nabla \times \mathbf{E} = -\frac{\partial}{\partial t}\mathbf{B} \tag{2.3}$$

$$\nabla \times \mathbf{H} = \mathbf{J} + \frac{\partial}{\partial t}\mathbf{D} \tag{2.4}$$

where $\mathbf{E}$ and $\mathbf{H}$ are the electric and magnetic field vectors, respectively, $\mathbf{D}$ and $\mathbf{B}$ are the corresponding electric and magnetic flux densities, $\mathbf{J}$ is the current density vector, and ρ is the charge density. The electric flux density and the electric field are related in the form

$$\mathbf{D} = \varepsilon\mathbf{E} = \varepsilon_0\mathbf{E} + \mathbf{P} \tag{2.5}$$

where ε is the permittivity of the medium, ε_0 is the vacuum permittivity, and $\mathbf{P}$ is the induced electric polarization. On the other hand, the relation between the magnetic flux density and the magnetic field is given by

$$\mathbf{B} = \mu\mathbf{H} = \mu_0\mathbf{H} + \mathbf{M} \tag{2.6}$$

where μ is the permeability of the medium, μ_0 is the vacuum permeability, and $\mathbf{M}$ is the induced magnetic polarization. Since silica, from which optical fibers are made, is a nonmagnetic material, we set $\mathbf{M} = 0$ in the following. The constants μ_0 and ε_0 have the following values:

$$\mu_0 = 4\pi \times 10^{-7}\ \text{H/m} \tag{2.7}$$

$$\varepsilon_0 = \frac{10^{-9}}{36\pi}\ \text{F/m} \tag{2.8}$$

Equations (2.1) and (2.2) correspond to Gauss's law for the electric field and Gauss's law for the magnetic field, respectively, while Eq. (2.3) is Faraday's law of induction and Eq. (2.4) is Ampere's circuital law.

The electric and magnetic fields can be considered as two aspects of a sole physical phenomenon: the electromagnetic field. In the following, we analyze the main characteristics of such a field. We confine our analysis to isotropic, homogeneous, and sourceless materials, so that $\mathbf{J} = 0$ and $\rho = 0$.

The curl of Eq. (2.3) gives

$$\nabla \times (\nabla \times \mathbf{E}) = \nabla \times \left(-\frac{\partial}{\partial t}\mathbf{B}\right) = -\frac{\partial}{\partial t}(\nabla \times \mathbf{B}) \tag{2.9}$$

The left-hand side of Eq. (2.9) can be simplified using the following vector identity:

$$\nabla \times (\nabla \times \mathbf{E}) = \nabla(\nabla \cdot \mathbf{E}) - \nabla \cdot \nabla\mathbf{E} \tag{2.10}$$

When $\rho = 0$, we have from Eqs. (2.1) and (2.5) that $\nabla \cdot \mathbf{E} = 0$. In such a case, Eq. (2.10) gives the result

$$\nabla \times (\nabla \times \mathbf{E}) = -\nabla \cdot \nabla\mathbf{E} = -\nabla^2\mathbf{E} \tag{2.11}$$

and Eq. (2.9) becomes

$$-\nabla^2\mathbf{E} = -\frac{\partial}{\partial t}(\nabla \times \mathbf{B}) \tag{2.12}$$

Using Eqs. (2.4)–(2.6), Eq. (2.12) can be written in the form

$$\nabla^2 \mathbf{E} - \mu\varepsilon \frac{\partial^2}{\partial t^2}\mathbf{E} = 0 \qquad (2.13)$$

A similar procedure can be used to obtain an equation for $\mathbf{B}$:

$$\nabla^2 \mathbf{B} - \mu\varepsilon \frac{\partial^2}{\partial t^2}\mathbf{B} = 0 \qquad (2.14)$$

Equations (2.13) and (2.14) are wave equations, with the wave's velocity v given by

$$v = \frac{1}{\sqrt{\mu\varepsilon}} \qquad (2.15)$$

The connection of the velocity of light with the electric and magnetic properties of a material was one of the most important results of Maxwell's theory. Considering Eqs. (2.7) and (2.8), the following value for the velocity of light in the vacuum is obtained:

$$v = c = 2.997924562 \times 10^8 \text{ m/s} \qquad (2.16)$$

In a material, the velocity of light is less than c. The *index of refraction, n,* of the material is defined as the ratio of the speed of light in the vacuum, c, to its speed in the material, v:

$$n = \frac{c}{v} = \sqrt{\frac{\varepsilon\mu}{\varepsilon_0\mu_0}} \qquad (2.17)$$

The index of refraction given by Eq. (2.17) corresponds to the real part of the complex refractive index, which will be discussed in Section 2.7. In the case of nonmagnetic materials, we have $\mu \approx \mu_0$ and the index of refraction is determined solely by the permittivity of the medium ε, which depends on the frequency of the incident electromagnetic wave.

2.2 ELECTROMAGNETIC WAVES

Equations (2.13) and (2.14) have the following solutions in the form of harmonic plane waves:

$$\mathbf{E} = \text{Re}\{\mathbf{E}_0 \, e^{i(\mathbf{k} \cdot \mathbf{r} - \omega t + \phi)}\} \qquad (2.18)$$

$$\mathbf{B} = \text{Re}\{\mathbf{B}_0 \, e^{i(\mathbf{k} \cdot \mathbf{r} - \omega t + \phi)}\} \qquad (2.19)$$

where $\mathbf{E}_0$ and $\mathbf{B}_0$ are constant vectors giving the direction and amplitude of oscillations, ω is the angular frequency, $\mathbf{k}$ is the wave vector, and Re indicates the real part. Hereafter we will drop the "Re", but it will be understood that the physical fields are given by the real part of the complex field appearing in our equations. Considering Eqs. (2.1) (with $\rho = 0$), (2.5), and (2.18), we have

$$\nabla \cdot \mathbf{E} = i\mathbf{k} \cdot \mathbf{E} = 0 \qquad (2.20)$$

In a similar way, using Eqs. (2.2) and (2.19), we obtain

$$\nabla \cdot \mathbf{B} = i\mathbf{k} \cdot \mathbf{B} = 0 \qquad (2.21)$$

Equations (2.20) and (2.21) show that both $\mathbf{E}$ and $\mathbf{B}$ must be perpendicular to the direction of propagation, which is given by $\mathbf{k}$.

Assuming that the electric and magnetic fields are given by Eqs. (2.18) and (2.19), Eq. (2.3) becomes

$$i\mathbf{k} \times \mathbf{E} = i\omega \mathbf{B} \qquad (2.22)$$

or

$$\mathbf{B} = \frac{1}{\omega}(\mathbf{k} \times \mathbf{E}) = \frac{1}{vk}(\mathbf{k} \times \mathbf{E}) \qquad (2.23)$$

Thus,

$$\mathbf{B} = \frac{1}{v}(\hat{\mathbf{s}} \times \mathbf{E}) \qquad (2.24)$$

where $\hat{\mathbf{s}} = \mathbf{k}/k$ is the unit vector in the propagation direction. Equation (2.24) contains three important aspects concerning the electromagnetic waves:

1. $\mathbf{B}$ is perpendicular to $\mathbf{E}$
2. $\mathbf{B}$ is in phase with $\mathbf{E}$
3. the magnitudes of $\mathbf{B}$ and $\mathbf{E}$ are related as $B = E/v$

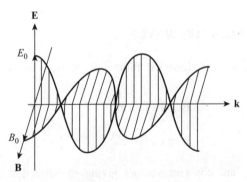

Figure 2.1 Propagation of a plane electromagnetic wave.

Figure 2.1 represents the propagation of a plane electromagnetic wave in a direction indicated by the wave vector **k**.

2.3 ENERGY DENSITY AND FLOW

Any text on electromagnetic theory demonstrates that the energy density associated with an electromagnetic wave is given by

$$U = \frac{1}{2}[\mathbf{D} \cdot \mathbf{E} + \mathbf{B} \cdot \mathbf{H}] \tag{2.25}$$

Using the constitutive relations given by Eqs. (2.5) and (2.6), we obtain

$$U = \frac{1}{2}\left[\varepsilon|\mathbf{E}^2| + \frac{|\mathbf{B}|^2}{\mu}\right] = \frac{1}{2}\left[\varepsilon + \frac{1}{\mu v^2}\right]|\mathbf{E}|^2 = \varepsilon|\mathbf{E}|^2 \tag{2.26}$$

In free space, we have

$$U = \varepsilon_0|\mathbf{E}|^2 = \frac{|\mathbf{B}|^2}{\mu_0} \tag{2.27}$$

The presence of both an electric and a magnetic field at the same point in space results in a flow of the field energy. The energy flux density is described by the *Poynting vector*, **S**, defined as

$$\mathbf{S} = \mathbf{E} \times \mathbf{H} = \frac{1}{\mu}\mathbf{E} \times \mathbf{B} \tag{2.28}$$

The energy flux density in a given direction, indicated by the unit vector $\hat{\mathbf{u}}$, is given by the scalar product $\hat{\mathbf{u}} \cdot \mathbf{S}$.

We will use a plane wave to determine some of the properties of the Poynting vector. Since **S** involves terms quadratic in **E**, it is necessary to use the real form of **E**:

$$\mathbf{E} = \mathbf{E}_0 \cos\phi, \quad \phi = \mathbf{k} \cdot \mathbf{r} - \omega t + \varphi \tag{2.29}$$

Also, from Eq. (2.23),

$$\mathbf{B} = \mathbf{B}_0 \cos\phi = \frac{1}{vk}(\mathbf{k} \times \mathbf{E}_0)\cos\phi \tag{2.30}$$

The Poynting vector becomes

$$\mathbf{S} = \frac{1}{\mu}\mathbf{E}_0 \times \frac{1}{vk}(\mathbf{k} \times \mathbf{E}_0)\cos^2\phi = \frac{1}{\mu v}|\mathbf{E}_0|^2\hat{\mathbf{s}}\cos^2\phi \tag{2.31}$$

Since the frequencies associated with light are very high (10^{14}–10^{15} Hz), we normally do not detect the magnitude of $\mathbf{S}$, but rather its temporal average over a time T determined by the response time of the detector used. Considering that the time average of $\cos^2 \phi$ over many periods is 1/2, the time-averaged value of the magnitude of the Poynting vector is given by

$$I \equiv \langle |\mathbf{S}| \rangle = \frac{1}{2\mu v} |\mathbf{E}_0|^2 \tag{2.32}$$

$I = \langle |\mathbf{S}| \rangle$ is called the *flux density* and has units of W/m^2.

The energy density is given from Eq. (2.26) by

$$U = \varepsilon |\mathbf{E}_0|^2 \cos^2 \phi \tag{2.33}$$

with a time average

$$\langle U \rangle = \frac{\varepsilon}{2} |\mathbf{E}_0|^2 \tag{2.34}$$

We may use the definition of the wave velocity, given by Eq. (2.15), to relate the density of flux I to the average energy density $\langle U \rangle$ in the form

$$I = v \langle U \rangle \tag{2.35}$$

This corresponds to a general result:

Energy flux density = (energy density) × (propagation speed)

2.4 PHASE VELOCITY AND GROUP VELOCITY

Since the refractive index of the medium is frequency dependent, the phase velocity of a wave is in general also a function of the frequency. This fact has important implications when the propagating waves are composed of several frequencies, as is the case in applications using the modulation of light. The velocity of the carrier and the velocity of the modulation will be in general different.

Let us consider the simple situation of a propagating plane wave containing only two frequencies. The total real electric field of such a wave can be written as the sum of fields of the two waves, which we assume to propagate in the z-direction and to have the same amplitude E_{01}:

$$E(z,t) = E_{01}[\cos(k_1 z - \omega_1 t) + \cos(k_2 z - \omega_2 t)] = 2E_{01}\cos(k_m z - \omega_m t)\cos(k_a z - \omega_a t) \tag{2.36}$$

where

$$\omega_a = \frac{1}{2}(\omega_1 + \omega_2), \quad \omega_m = \frac{1}{2}(\omega_1 - \omega_2) \tag{2.37}$$

and

$$k_a = \frac{1}{2}(k_1 + k_2), \quad k_m = \frac{1}{2}(k_1 - k_2) \qquad (2.38)$$

The quantities ω_a and k_a are the average angular frequency and the average propagation constant, respectively, whereas the quantities ω_m and k_m are designated the *modulation frequency* and the *modulation propagation constant*, respectively.

The total field can be regarded as a traveling (carrier) wave of frequency ω_a having a time-varying or modulated amplitude $E_0(z, t)$ such that

$$E(z, t) = E_0(z, t)\cos(k_a z - \omega_a t) \qquad (2.39)$$

where

$$E_0(z, t) = 2E_{01} \cos(k_m z - \omega_m t) \qquad (2.40)$$

The phase velocity of the carrier wave can be obtained from its phase $\varphi = (k_a z - \omega_a t)$ using the relation

$$v = -\frac{(\partial \varphi / \partial t)_z}{(\partial \varphi / \partial z)_t} \qquad (2.41)$$

which gives the result

$$v = \frac{\omega_a}{k_a} \qquad (2.42)$$

Concerning the propagation of the modulation envelope, the rate at which it advances is known as the *group velocity*, v_g. The group velocity is obtained from Eq. (2.41), considering the phase of the envelope $(k_m z - \omega_m t)$, and is given by

$$v_g = \frac{\omega_m}{k_m} = \frac{\omega_1 - \omega_2}{k_1 - k_2} = \frac{\Delta\omega}{\Delta k} \qquad (2.43)$$

The function describing the dependence of ω on k, $\omega = \omega(k)$, is called dispersion relation. When the frequency range $\Delta\omega$ is small, the ratio $\Delta\omega/\Delta k$ tends to the derivative of the dispersion relation and the group velocity becomes

$$v_g = \frac{d\omega}{dk} \qquad (2.44)$$

Since $\omega = kv$, Eq. (2.44) yields

$$v_g = v + k\frac{dv}{dk} \qquad (2.45)$$

The modulation or signal propagates with a velocity v_g that may be greater than, equal to, or less than the phase velocity of the carrier, v. In nondispersive media, v does not depend on k, and $v_g = v$. In dispersive media, in which $n(k)$ is known, $\omega = kc/n$ and the group velocity can be written in the form

$$v_g = \frac{c}{n} - \frac{kc}{n^2}\frac{dn}{dk} \qquad (2.46)$$

The group velocity can be considered as the propagation velocity of a "group" of waves with frequencies distributed over an infinitesimally small bandwidth centered on ω_a. In the presence of a broad frequency spectrum, the slope of the curve $\omega(k)$ may change over the range of the spectrum. As a consequence, different spectral components propagate at different group velocities, leading to signal distortion. This problem is called *group velocity dispersion* and will be discussed in Chapter 3 in the context of optical fibers.

2.5   REFLECTION AND TRANSMISSION OF WAVES

The phenomenon of reflection and transmission of plane waves at interfaces between dielectrics is useful in exploiting and understanding the behavior of light in dielectric waveguides. Of interest are not only the relations among the angles of incidence, reflection, and refraction, but also the fractions of optical power that are reflected and transmitted at the boundaries, as well as the phase shifts that occur on reflection.

2.5.1   Snell's Laws

Let us consider a monochromatic plane wave incident on a boundary between two media with refractive indices n_1 and n_2 (Fig. 2.2). The incident wave is given by

$$\mathbf{E}_i = \mathbf{E}_{0i} \exp\{i(\mathbf{k}_i \cdot \mathbf{r} - \omega t)\} \qquad (2.47)$$

and can be decomposed into two waves with the same frequency—a reflected wave, $\mathbf{E}_r$, and a transmitted wave, $\mathbf{E}_t$—given by

$$\mathbf{E}_r = \mathbf{E}_{0r} \exp\{i(\mathbf{k}_r \cdot \mathbf{r} - \omega t)\} \qquad (2.48)$$

$$\mathbf{E}_t = \mathbf{E}_{0t} \exp\{i(\mathbf{k}_t \cdot \mathbf{r} - \omega t)\} \qquad (2.49)$$

A relation among the three waves, valid for all points on the interface and for any instant of time, can be verified only if their phases are the same. This condition gives

$$\mathbf{k}_i \cdot \mathbf{r} = \mathbf{k}_r \cdot \mathbf{r} = \mathbf{k}_t \cdot \mathbf{r} \qquad (2.50)$$

From Eq. (2.50), we conclude that

$$\mathbf{k}_r - \mathbf{k}_i = b_1 \hat{\mathbf{N}} \qquad (2.51)$$

$$\mathbf{k}_t - \mathbf{k}_i = b_2 \hat{\mathbf{N}} \qquad (2.52)$$

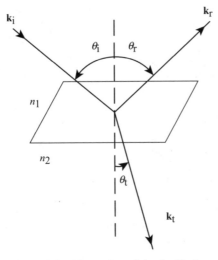

Figure 2.2   Illustration of the Snell's law.

where $\hat{\mathbf{N}}$ is a vector perpendicular to the interface, whereas b_1 and b_2 are scalars. Considering that $\mathbf{k}_i = k_0 n_1 \hat{\mathbf{s}}_i$, $\mathbf{k}_r = k_0 n_1 \hat{\mathbf{s}}_r$, and $\mathbf{k}_t = k_0 n_2 \hat{\mathbf{s}}_t$, the previous relations can be written in the form

$$n_1(\hat{\mathbf{s}}_r - \hat{\mathbf{s}}_i) = a_1 \hat{\mathbf{N}} \tag{2.53}$$

$$n_2 \hat{\mathbf{s}}_t - n_1 \hat{\mathbf{s}}_i = a_2 \hat{\mathbf{N}} \tag{2.54}$$

where a_1 and a_2 are scalars. The projection of these relations on the interface gives

$$\theta_i = \theta_r \tag{2.55}$$

$$n_1 \sin \theta_i = n_2 \sin \theta_t \tag{2.56}$$

Equations (2.55) and (2.56) are known as Snell's laws of reflection and refraction, respectively.

Referring to Fig. 2.2, where $n_2 > n_1$ is assumed, and noting that the geometry is the same if the ray direction is reversed, let us consider what happens if a ray inside the refractive medium meets the surface at a large angle of incidence θ_t, such that $\sin \theta_t > n_1/n_2$. In such a case, Eq. (2.56) would give $\sin \theta_i > 1$, which means that there can be no ray above the surface and that all incident light is reflected. This phenomenon is known as *total internal reflection* and constitutes the basic mechanism that explains the propagation of light along optical fibers.

2.5.2 Fresnel Equations

A wave encountering a boundary between dielectric media with different refractive indices will be not only refracted but also partly reflected. The ratios of the amplitudes

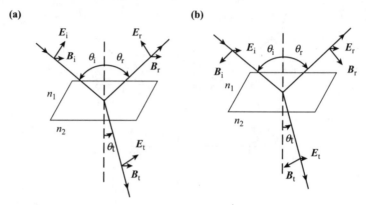

Figure 2.3 Illustration of the incident, reflected, and refracted rays for two cases of plane polarization. The directions of the vector fields are shown for **E**-vector (a) in the plane of incidence and (b) perpendicular to it.

of the reflected and transmitted waves to that of the incident wave are known as *reflection* and *transmission coefficients, r* and *t*, respectively. For an electromagnetic wave, the *Fresnel equations* express the way in which these coefficients depend on the angles of incidence and refraction, and on the polarization of the wave, given by the direction of the electric field.

In Fig. 2.3, the incident, reflected, and refracted rays are shown for two cases of plane polarization, when the **E**-vector is (a) in the plane of incidence and (b) perpendicular to it. The vectors **E** and **B** are related to the wave vector as given in Eq. (2.23).

The components of the **E, H,** and **B** fields parallel to the boundary must be the same on both sides of the boundary. Considering the case of Fig. 2.3a, these boundary conditions give

$$B_i + B_r = B_t \tag{2.57}$$

$$E_i \cos \theta_i - E_r \cos \theta_r = E_t \cos \theta_t \tag{2.58}$$

Substituting in Eq. (2.57) $B = (n/c)E$ and combining with Eq. (2.58), we obtain the following results for the reflection and transmission coefficients:

$$r_{\parallel} = \left(\frac{E_r}{E_i}\right)_{\parallel} = \frac{n_2 \cos \theta_i - n_1 \cos \theta_t}{n_2 \cos \theta_i + n_1 \cos \theta_t} \tag{2.59}$$

$$t_{\parallel} = \left(\frac{E_t}{E_i}\right)_{\parallel} = \frac{2n_1 \cos \theta_i}{n_2 \cos \theta_i + n_1 \cos \theta_t} \tag{2.60}$$

where the subscript $\parallel$ refers to the case in which the **E** field is parallel to the plane of incidence. Using Snell's law of refraction, we can rewrite Eqs. (2.59) and (2.60) in terms of angles only:

$$r_{\parallel} = \frac{\tan(\theta_i - \theta_t)}{\tan(\theta_i + \theta_t)} \tag{2.61}$$

$$t_{\parallel} = \frac{2 \sin \theta_t \cos \theta_i}{\sin(\theta_i + \theta_t)\cos(\theta_i - \theta_t)} \tag{2.62}$$

A similar analysis for the case of Fig. 2.3b, in which the electric field vector is perpendicular to the plane of incidence, gives

$$r_{\perp} = \left(\frac{E_r}{E_i}\right)_{\perp} = \frac{n_1 \cos \theta_i - n_2 \cos \theta_t}{n_1 \cos \theta_i + n_2 \cos \theta_t} \tag{2.63}$$

$$t_{\perp} = \left(\frac{E_t}{E_i}\right)_{\perp} = \frac{2n_1 \cos \theta_i}{n_1 \cos \theta_i + n_2 \cos \theta_t} \tag{2.64}$$

Using Snell's law of refraction, the above relations can be written in the form

$$r_{\perp} = -\frac{\sin(\theta_i - \theta_t)}{\sin(\theta_i + \theta_t)} \tag{2.65}$$

$$t_{\perp} = \frac{2 \sin \theta_t \cos \theta_i}{\sin(\theta_i + \theta_t)} \tag{2.66}$$

Figure 2.4 illustrates the variation of the amplitude reflection and transmission coefficients with the angle of incidence for a boundary between air ($n_1 \approx 1$) and glass ($n_2 = 1.5$).

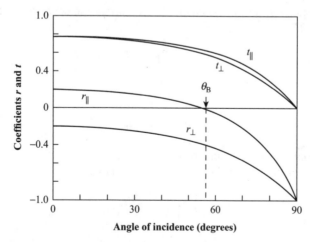

Figure 2.4   Variation of the amplitude reflection and transmission coefficients with the angle of incidence for a boundary between air ($n_1 \approx 1$) and glass (degrees $n_2 = 1.5$).

It can be verified from Eq. (2.61) that $r_{\parallel}$ goes through zero when $\theta_i + \theta_t = \pi/2$. The angle of incidence $\theta_i = \theta_B$ corresponding to this situation is known as the *Brewster angle*. Light reflected at the Brewster angle becomes completely linearly polarized, with the electric vector normal to the plane of incidence. On the other hand, a glass plate at the Brewster angle is completely transparent for light with the electric vector parallel to the plane of incidence.

The *reflectance R*, which is defined as the ratio of the reflected power to the incident power, is given by

$$R = \frac{P_r}{P_i} = r^2 \tag{2.67}$$

On the other hand, the *transmittance T* is defined as the ratio of the transmitted power to the incident power:

$$T = \frac{P_t}{P_i} = \left(\frac{n_2 \cos \theta_t}{n_1 \cos \theta_i}\right) t^2 \tag{2.68}$$

For normal incidence, we have $\theta_i \approx \theta_t \approx 0$ and $\cos \theta_i \approx \cos \theta_t \approx 1$. In this case, the reflection and transmission coefficients are independent of polarization, becoming

$$r \equiv r_{\parallel} = -r_{\perp} = \frac{n_2 - n_1}{n_1 + n_2} \tag{2.69}$$

$$t = t_{\parallel} = t_{\perp} = \frac{2n_1}{n_1 + n_2} \tag{2.70}$$

It can be verified that the transmission coefficient is always positive, whereas the sign of the reflection coefficient depends on the relative magnitude of the refractive indices n_1 and n_2.

The reflectance R and transmittance T for normal incidence are obtained from Eqs. (2.67)–(2.70):

$$R = r^2 = \left(\frac{n_2 - n_1}{n_1 + n_2}\right)^2 \tag{2.71}$$

$$T = \frac{n_2}{n_1} t^2 = \frac{4n_1 n_2}{(n_1 + n_2)^2} \tag{2.72}$$

We verify from the above results that $R + T = 1$.

In the case of the interface between air ($n_1 = 1$) and glass ($n_2 = 1.5$), we have

$$r = \frac{1.5 - 1}{1.5 + 1} = 0.2 \tag{2.73}$$

$$t = \frac{2}{1.5 + 1} = 0.8 \tag{2.74}$$

The reflectance loss of 4% at normal incidence for a typical air/glass surface becomes a serious problem in the multicomponent lenses of several optical instruments. The losses can, however, be significantly reduced by coating the surface with a transparent layer with adequate values of refractive index and thickness.

2.6 THE HARMONIC OSCILLATOR MODEL

When an optical wave propagates in a given medium, the relatively heavy nucleus of the atoms of the medium cannot follow its rather high frequency and only the elastically bound valence electrons are excited. The electrons oscillate with the frequency of the optical field around their equilibrium state and, as a consequence, emit waves by their own. Figure 2.5 illustrates this oscillator; the light valence electron is represented by a mass m, whereas the forces between the atom and the valence electron are represented by a spring of constant k.

The equation of motion of this oscillator can be written in the form

$$ m\frac{d^2\mathbf{X}(t)}{dt^2} + \gamma_d\frac{d\mathbf{X}(t)}{dt} + k\mathbf{X}(t) = q\mathbf{E}(t) \tag{2.75} $$

where $\mathbf{X}$ is the displacement coordinate, q is charge of the electron, γ_d is a damping coefficient, and $\mathbf{E}(t) = \mathbf{E}_0\,e^{-i\omega t}$ is the external electrical field. Since the displacement coordinate $\mathbf{X}$ changes periodically with the frequency ω of the electric field, it can be written as $\mathbf{X}(t) = \mathbf{X}_0\,e^{-i\omega t}$. Substituting $\mathbf{X}(t)$ and $\mathbf{E}(t)$ in Eq. (2.75) and introducing the resonance frequency of the oscillator $\omega_0 = \sqrt{k/m}$, we obtain the result

$$ \mathbf{X}_0 = \frac{q\mathbf{E}_0}{m}\frac{1}{(\omega_0^2-\omega^2)-i\omega\gamma_m} \tag{2.76} $$

where $\gamma_m = \gamma_d/m$.

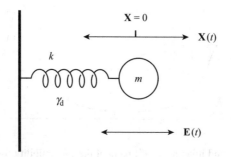

Figure 2.5 Schematic illustration of the harmonic oscillator.

The dipole moment of the oscillator is given by

$$\mathbf{p}(t) = q\mathbf{X}(t) = q\mathbf{X}_0\, e^{-i\omega t} \tag{2.77}$$

Each atom or oscillator of the medium can oscillate in different modes and has therefore different resonance frequencies (ω_j) and damping coefficients (γ_{mj}). If N is the number of oscillators per unit volume and f_j is the number of possible oscillating states each with resonance frequency ω_j, the polarization of the medium is given by

$$\mathbf{P}(t) = \left[\frac{Nq^2}{m}\sum_{j}\frac{f_j}{(\omega_j^2-\omega^2)-i\omega\gamma_m}\right]\mathbf{E}(t) = \varepsilon_0\chi(\omega)\mathbf{E}(t) \tag{2.78}$$

where f_j is also known as the oscillator strength and χ is the linear electric susceptibility of the material. Solving for χ gives the result

$$\chi = \omega_p^2 \sum_{j} f_j \frac{(\omega_j^2-\omega^2)+i\omega\gamma_m}{(\omega_j^2-\omega^2)^2+\omega^2\gamma_m^2} \equiv \chi'+i\chi'' \tag{2.79}$$

where $\omega_p = \sqrt{Nq^2/m\varepsilon_0}$ is known as the plasma frequency. Figure 2.6 illustrates the real and imaginary parts of the susceptibility for a material having two resonances of different strengths.

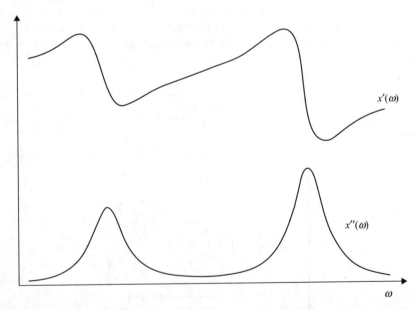

Figure 2.6   Real (χ') and imaginary (χ'') parts of the susceptibility for a material with two resonances.

2.7   THE REFRACTIVE INDEX

The complex refractive index of the material is given by

$$\hat{n} = (1 + \chi' + i\chi'')^{1/2} = n + in' \tag{2.80}$$

The imaginary part (n') of the complex refractive index is called the extinction coefficient. If $n' \ll n$, the real and imaginary parts of the complex refractive index can be approximated as

$$n \approx \sqrt{1 + \chi'}, \quad n' = \frac{\chi''}{2\sqrt{1 + \chi'}} \tag{2.81}$$

The real part of the complex refractive index leads to a phase shift of the wave in the material, which depends on the frequency difference between the external field ω and the resonance frequency of the oscillator ω_j. The imaginary part is responsible for an absorption of the wave in the material and it is particularly large near the resonance frequency, such that $\omega \approx \omega_j$. This explains why χ'' and χ' are sometimes called the absorptive and refractive index parts of the susceptibility, respectively.

Introducing the complex refractive index given in Eq. (2.80) into the solution of the wave equation corresponding to a plane wave that propagates in the positive z-direction gives

$$\mathbf{E} = \text{Re}\{\mathbf{E}_0 \, e^{i(kz - \omega t)}\} = \text{Re}\{\mathbf{E}_0 \, e^{i((n + in')k_0 z - \omega t)}\} = \text{Re}\{\mathbf{E}_0 \, e^{-n'k_0 z} e^{i(nk_0 z - \omega t)}\} \tag{2.82}$$

The exponential function $\mathbf{E}_0 \, e^{-n'k_0 z}$ describes the decrease in the wave amplitude with distance. Since the density of flux is proportional to the square of the wave amplitude, it becomes

$$I = \frac{1}{2}\varepsilon_0 cn |\mathbf{E}_0 \, e^{-n'k_0 z}|^2 = I_0 \, e^{-\alpha z} \tag{2.83}$$

where $I_0 = \varepsilon_0 cn |\mathbf{E}_0|^2 / 2$ and $\alpha = 2n'k_0$ is the so-called attenuation constant.

Equation (2.82) shows that the real part of the refractive index (n) is indeed responsible for an alteration of the phase of the wave. The frequency dependence of the refractive index leads to different velocities for the spectral components of the propagating signal. Far from resonance, we have from Eq. (2.79) that $\chi'' \approx 0$ and

$$\chi' \approx \omega_p^2 \sum_j \frac{f_j}{\omega_j^2 - \omega^2} \tag{2.84}$$

where the attenuation of the oscillators was neglected ($\gamma_{mj} \approx 0$). Writing Eq. (2.84) in terms of wavelength, it becomes

$$\chi' \approx \frac{\omega_p^2}{(2\pi c)^2} \sum_j \frac{f_j \lambda_j^2 \lambda^2}{\lambda^2 - \lambda_j^2} = \sum_j \frac{A_j \lambda^2}{\lambda^2 - \lambda_j^2} \tag{2.85}$$

where A_j is the magnitude of the jth resonance, whereas λ and λ_j are the free space wavelengths corresponding to ω and ω_j, respectively. From Eqs. (2.81) and (2.85), we obtain the following result, known as the *Sellmeier formula* for the refractive index:

$$n^2 = 1 + \sum_j \frac{A_j \lambda^2}{\lambda^2 - \lambda_j^2} \qquad (2.86)$$

Equation (2.86) together with Fig. 2.6 shows that away from any resonance, or between resonances, the refractive index decreases with increasing wavelength. This behavior is important to describe the material origins of group velocity dispersion.

2.8 THE LIMIT OF GEOMETRICAL OPTICS

Geometrical optics is an approximate description of the optical phenomena, which becomes acceptable in the limit as $\lambda \to 0$. In such a limit, Maxwell's equations provide an equation that describes the path taken by the normal to the wave surface of constant phase. This normal is called the *optical ray* and the corresponding equation of motion is known as the *eikonal equation*.

Let us consider a linear, isotropic, sourceless but nonhomogeneous medium, such that $\varepsilon = \varepsilon(r)$. Assuming that the electromagnetic fields vary with time with a frequency $f = \omega/2\pi$, Maxwell's equations can be written in the form

$$\nabla \cdot (\varepsilon_r \mathbf{E}_e) = 0 \qquad (2.87)$$

$$\nabla \cdot \mathbf{B}_e = 0 \qquad (2.88)$$

$$\nabla \times \mathbf{E}_e = i k_0 c \mathbf{B}_e \qquad (2.89)$$

$$\nabla \times \mathbf{B}_e = -i \frac{k_0}{c} \varepsilon_r \mathbf{E}_e \qquad (2.90)$$

where $\mathbf{E}_e$ and $\mathbf{B}_e$ are the complex amplitudes of the $\mathbf{E}$ and $\mathbf{B}$ fields, respectively, such that

$$\mathbf{E}(r, t) = \mathrm{Re}\{\mathbf{E}_e(r) e^{-i\omega t}\} \qquad (2.91)$$

$$\mathbf{B}(r, t) = \mathrm{Re}\{\mathbf{B}_e(r) e^{-i\omega t}\} \qquad (2.92)$$

On the other hand,

$$k_0 = \omega \sqrt{\varepsilon_0 \mu_0} = \frac{2\pi}{\lambda_0} \qquad (2.93)$$

denotes the free space propagation constant, whereas

$$\varepsilon_r = \frac{\varepsilon(r)}{\varepsilon_0} \qquad (2.94)$$

is the relative permittivity. Let us consider solutions of the form

$$\mathbf{E}_e(r) = \mathbf{E}_0(r)e^{ik_0 L(r)} \tag{2.95}$$

$$\mathbf{B}_e(r) = \mathbf{B}_0(r)e^{ik_0 L(r)} \tag{2.96}$$

where $L(r)$ is a function describing the phase front of the wave. Substituting Eqs. (2.95) and (2.96) in Eqs. (2.87)–(2.90), the following results are obtained:

$$\nabla L \cdot \mathbf{E}_0 = -\frac{1}{ik_0}\left(\varepsilon_r \nabla \cdot \mathbf{E}_0 + \nabla \varepsilon_r \cdot \mathbf{E}_0\right) \tag{2.97}$$

$$\nabla L \cdot \mathbf{B}_0 = -\frac{\nabla \cdot \mathbf{B}_0}{ik_0} \tag{2.98}$$

$$\nabla L \times \mathbf{E}_0 - c\mathbf{B}_0 = -\frac{\nabla \times \mathbf{E}_0}{ik_0} \tag{2.99}$$

$$\nabla L \times \mathbf{B}_0 + \frac{\varepsilon_r}{c}\mathbf{E}_0 = -\frac{\nabla \times \mathbf{B}_0}{ik_0} \tag{2.100}$$

Considering the limit $\lambda_0 \to 0$, or $k_0 \to \infty$, we have

$$\mathbf{E}_0 \cdot \nabla L = \mathbf{B}_0 \cdot \nabla L = 0 \tag{2.101}$$

$$\mathbf{E}_0 = \frac{c}{\varepsilon_r}\mathbf{B}_0 \times \nabla L \tag{2.102}$$

$$\mathbf{B}_0 = \frac{\nabla L \times \mathbf{E}_0}{c} \tag{2.103}$$

Eliminating $\mathbf{E}_0$ from (2.103) with the help of Eq. (2.102), or eliminating $\mathbf{B}_0$ from (2.102) using Eq. (2.103), we obtain

$$\varepsilon_r \mathbf{E}_0 = (\nabla L \times \mathbf{E}_0) \times \nabla L = |\nabla L|^2 \mathbf{E}_0 - \nabla L(\mathbf{E}_0 \cdot \nabla L) \tag{2.104}$$

$$\varepsilon_r \mathbf{B}_0 = \nabla L \times (\mathbf{B}_0 \times \nabla L) = |\nabla L|^2 \mathbf{B}_0 - \nabla L(\mathbf{B}_0 \cdot \nabla L) \tag{2.105}$$

Considering Eq. (2.101), the second terms on the right-hand side of Eqs. (2.104) and (2.105) vanish and we obtain the result

$$|\nabla L|^2 = \varepsilon_r = n^2 \tag{2.106}$$

The surfaces $L(r) = \text{constant}$ are surfaces of constant phase, that is, phase fronts. The normals to these surfaces have the direction of ∇L and represent the rays of geometrical optics. The equation

$$\nabla L = n\hat{\mathbf{s}} \tag{2.107}$$

is known as the *eikonal equation* and L is called the *eikonal*. The unit vector $\hat{s}$ is normal to the phase front and tangent to the optical ray, being given by

$$\hat{s} = \frac{\nabla L}{|\nabla L|} = \frac{\nabla L}{n} \tag{2.108}$$

Using Cartesian coordinates, Eq. (2.107) becomes

$$\left(\frac{\partial L}{\partial x}\right)^2 + \left(\frac{\partial L}{\partial y}\right)^2 + \left(\frac{\partial L}{\partial z}\right)^2 = n^2 \tag{2.109}$$

In the case of a homogeneous medium, the solution of Eq. (2.109) can be written as

$$L(r) = n(x \cos \theta_x + y \cos \theta_y + z \cos \theta_z) \tag{2.110}$$

where $\cos \theta_x$, $\cos \theta_y$, and $\cos \theta_z$ are the direction cosines defining the direction of the light ray straight line trajectory.

From Eqs. (2.28) and (2.101)–(2.103), the average Poynting vector can be written in the form

$$\langle S \rangle = \frac{1}{2c\mu_0}(\mathbf{E}_0 \times (\nabla L \times \mathbf{E}_0)) \tag{2.111}$$

Using Eq. (2.108), we have

$$\langle S \rangle = \frac{n|\mathbf{E}_0|^2}{2c\mu_0}\hat{s} \tag{2.112}$$

This result is similar to that shown in Eq. (2.32) and shows that the direction of the Poynting vector coincides with the optical ray direction, which is normal to the phase front.

PROBLEMS

2.1 The electric field of a plane electromagnetic wave has the following components:

$$E_x = 0, \quad E_y = E_0 \exp\{i(\beta x - \alpha t)\}, \quad E_z = 0$$

(a) Obtain the components of the magnetic flux density **B**.

(b) Derive the relation between α and β in order to satisfy the Maxwell's equations.

2.2 A 100 W monochromatic point source is radiating equally in all directions in vacuum. Calculate the flux density and the electric field amplitude 1 m away from the source.

2.3 Write an expression for a plane electromagnetic wave propagating in vacuum, with a wavelength of 600 nm and a flux density of $20\,\text{mW/cm}^2$. Assume that the wave propagates in the z-direction and that it is linearly polarized at an angle of $30°$ to the y-axis.

2.4 Obtain the Fresnel equations (2.59) and (2.60) for the reflectivity and transmissivity when the electric field is parallel to the plane of incidence (TM polarization).

2.5 Determine the sum of the incident and refracted angles for the Brewster angle of incidence. Show that no Brewster angle exists when the electric field vector is perpendicular to the plane of incidence (TE polarization).

2.6 Derive Eq. (2.112) for the average Poynting vector starting from Eqs. (2.101)–(2.103) and using Eq. (2.108).

REFERENCES

1. D. M. Cook, *The Theory of the Electromagnetic Field*, Prentice-Hall, Englewood Cliffs, NJ, 1975.
2. H. A. Haus, *Waves and Fields in Optoelectronics*, Prentice-Hall, Englewood Cliffs, NJ, 1984.
3. E. Hecht, *Optics*, 2nd ed., Addison-Wesley, Reading, MA, 1987.
4. M. Klein and T. Furtak, *Optics*, 2nd ed., Wiley, New York, 1986.
5. D. H. Staelin, A. W. Worgenthaler, and J. A. Kong, *Electromagnetic Waves*, Prentice-Hall, Englewood Cliffs, NJ, 1994.
6. S. Ramo, J. R. Whinnery, and T. Van Duzer, *Fields and Waves in Communication Electronics*, 3rd ed., Wiley, New York, 1994.

3

OPTICAL FIBERS

The phenomenon of total internal reflection, responsible for guiding of light in optical fibers, is known since the nineteenth century. In fact, the transmission of light along a curved dielectric cylinder was the subject of a lecture demonstration by John Tyndall in 1854 [1]. His light pipe was a stream of water emerging from a hole in the side of a tank that contained a bright light. Glass fibers were fabricated since the 1920s, but their use remained restricted to medical applications until the 1960s. The use of such glass fibers for optical communications was not practical, due to their high losses (~1000 dB/km).

In 1966, Kao and Hockham [2] suggested that optical fibers could be used in communication systems if their losses were reduced below 20 dB/km. Following an intense activity on the purification of fused silica, such main objective was achieved in 1970 [3]. Further technological progress allowed the reduction of fiber loss to 0.2 dB/km near the 1550 nm spectral region by 1979 [4]. Actually, optical fibers provide low-loss transmission over an enormous frequency range of about 25 THz, which is orders of magnitude more than the bandwidth available in copper cables or any other transmission medium. The superior properties of optical fibers led to a revolution in the field of lightwave communication systems.

The purpose of this chapter is to review the basic linear properties of optical fibers in the perspective of their use in communication systems. We start using a simple geometrical optics approach and continue with a more general wave theory model based on solving the Maxwell's equations. *Dispersion* refers to the phenomenon where different components of the signal travel at different velocities in the fiber and it became an important limiting factor as transmission systems evolved to longer distances and higher bit rates. Special attention will be paid in this chapter to both *chromatic dispersion* and *polarization mode dispersion*. Additional information concerning the topics discussed in this chapter can be found in several well-known books [5–12].

Nonlinear Effects in Optical Fibers. By Mário F. S. Ferreira.
Copyright © 2011 John Wiley & Sons, Inc. Published 2011 by John Wiley & Sons, Inc.

3.1 GEOMETRICAL OPTICS DESCRIPTION

In this section, some properties of dielectric waveguides are described using the geometrical optics approach. The discussion is applied to the cases of planar waveguides, step-index fibers, and graded-index fibers. It must be noticed, however, that such an approach is valid only if the thickness (radius) of the waveguide core is much greater than the light wavelength, λ. In the case of optical fibers, this situation corresponds to the so-called *multimode fibers*. When the magnitudes of the two quantities are similar, as happens with the so-called *monomode fibers*, it is necessary to use the electromagnetic theory to describe adequately the propagation of light along the waveguide.

3.1.1 Planar Waveguides

Let us consider a three-layer planar waveguide (Fig. 3.1), with refractive indices such that $n_2 > n_3 > n_1$, where n_2 is the refractive index of the guiding layer (core). The rays propagating in this layer will be guided only if the angle of incidence in the boundaries is greater than the angle for total internal reflection, that is,

$$\arcsin\left(\frac{n_3}{n_2}\right) < \theta_2 < \frac{\pi}{2} \tag{3.1}$$

Rays that satisfy Eq. (3.1) are associated with waves trapped in the region of index n_2, corresponding to the so-called *guided-wave modes*. On the other hand, rays with angles of incidence such that $\theta_2 < \arcsin(n_3/n_2)$ are transmitted to one or both of the adjacent layers, and correspond to the *radiation modes* of the waveguide.

Considering their polarization, waves undergoing total internal reflection can be classified in two categories: TE waves, presenting a longitudinal component of the magnetic field, and TM waves, which have a longitudinal component of the electric field. Since the results are qualitatively the same for the two polarizations, we will consider in the following only the TE waves.

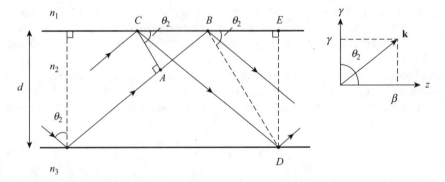

Figure 3.1 Geometry to analyze the propagation of the guided-wave modes in a planar waveguide.

As seen in Chapter 2, the geometric ray is parallel to the propagation vector **k**. The scalar components of this vector along the z-axis (direction of propagation) and along the y-axis (perpendicular to the dielectric layers) are given, respectively, by

$$\beta = n_2 k_0 \sin \theta_2 \tag{3.2}$$

and

$$\gamma = n_2 k_0 \cos \theta_2 = \sqrt{n_2^2 k_0^2 - \beta^2} \tag{3.3}$$

where k_0 is the vacuum wave number. The β component corresponds to the effective propagation constant of a mode of the waveguide. In the case of a guided mode, this propagation constant must satisfy the condition (corresponding to Eq. (3.1))

$$k_0 n_3 < \beta < k_0 n_2 \tag{3.4}$$

Using Fig. 3.1, let us consider a ray that travels from C to D and undergoes a reflection in each of these points. Meanwhile, a second ray goes from A to B and experiences no reflection. Since the wave fronts (represented by dashed lines) are surfaces of constant phase, the phase difference associated with the two propagation paths must be a multiple of 2π:

$$(n_2 k_0 \overline{CD} + \phi_1 + \phi_3) - n_2 k_0 \overline{AB} = 2p\pi \tag{3.5}$$

where p is an integer and ϕ_1 and ϕ_3 are phase shifts due to total reflections in the two boundaries. These phase shifts are given by

$$\phi_{1,3} = -2 \arctan\left(\frac{\sigma_{1,3}}{\gamma}\right) \tag{3.6}$$

where

$$\sigma_{1,3} = \sqrt{\beta^2 - (n_{1,3} k_0)^2} = \sqrt{(n_2^2 - n_{1,3}^2)k_0^2 - \gamma^2} \tag{3.7}$$

are the decay constants of the evanescent waves associated with total reflection at the guide boundaries. Considering the geometry of Fig. 3.1, we have

$$\overline{BE} = d/\tan \theta_2 \tag{3.8a}$$

$$\overline{CE} = d \tan \theta_2 \tag{3.8b}$$

$$\overline{CD} = d/\cos \theta_2 \tag{3.8c}$$

$$\overline{AB} = \overline{CB} \sin \theta_2 \tag{3.8d}$$

where $\overline{CB} = \overline{CE} - \overline{BE}$.

Introducing these results in Eq. (3.5), we obtain

$$2n_2 k_0 d \cos \theta_2 + \phi_1 + \phi_3 \equiv 2\gamma d + \phi_1 + \phi_3 = 2p\pi \qquad (3.9)$$

Equation (3.9) is the *dispersion relation* of the TE guided modes. This relation shows that the angle θ_2 can assume only certain discrete values, θ_p, determined by the integer values p. The rays associated with these angles θ_p correspond to the different modes of the waveguide.

From Eq. (3.1) we have

$$\cos \theta_p < \sqrt{1 - \left(\frac{n_3}{n_2}\right)^2} \qquad (3.10)$$

Equations (3.9) and (3.10) show that there is a maximum number of guided modes, p_{max}, given by

$$p_{max} \leq \frac{V}{\pi} + \frac{\phi_1 + \phi_3}{2\pi} \qquad (3.11)$$

The parameter V in Eq. (3.11) is given by

$$V = k_0 d \sqrt{n_2^2 - n_3^2} \qquad (3.12)$$

and it is known as the *normalized film thickness, normalized frequency*, or simply the *V number*. From Eqs. (3.11) and (3.12), we conclude that the number of modes of the waveguide increases when the thickness, d, or the refractive index, n_2, increases. The number of modes also increases when the refractive index of the substrate, n_3, or the light wavelength, λ, decreases.

Using Eqs. (3.6) and (3.7), Eq. (3.9) can be written in the form

$$\gamma d - \arctan\left(\frac{\sigma_1}{\gamma}\right) - \arctan\left(\frac{\sigma_3}{\gamma}\right) = p\pi \qquad (3.13)$$

or

$$\tan \gamma d = \frac{\gamma(\sigma_1 + \sigma_3)}{\gamma^2 - \sigma_1 \sigma_3} \qquad (3.14)$$

One can verify that there is a maximum value of γd that satisfies Eq. (3.14), above which one of the decay constants given by Eq. (3.7) becomes imaginary. This situation occurs first with the constant σ_3, since we have assumed that $n_3 > n_1$. When $\sigma_3 = 0$, the angle θ_2 does not satisfy the condition for total reflection in the boundary between layers 2 and 3. At this point, the guided modes disappear and substrate modes are produced. From Eq. (3.7) we obtain the maximum value of γd, which is given by

$$\gamma d = k_0 d \sqrt{n_2^2 - n_3^2} = V_c \qquad (3.15)$$

where V_c is the cutoff value of the normalized film thickness. Considering that $\sigma_3 = 0$ and using Eqs. (3.14) and (3.15), we obtain the following expression for cutoff condition:

$$\tan V_c = \tan \gamma d = \frac{\sigma_1}{\gamma} = \sqrt{\frac{n_3^2 - n_1^2}{n_2^2 - n_3^2}} \qquad (3.16)$$

For a given guide thickness d, there is a minimum wavelength that will propagate in the guide.

On the other hand, there is a minimum value of γd, $\gamma d = \pi/2$, below which there are no possible solutions of an asymmetric guide. This means that the asymmetric guide can only support modes in a band of wavelengths established by $\pi/2 \le \gamma d \le V_c$.

In the case of a symmetric guide, the refractive indices in media 1 and 3, as well as the decay constants, are equal ($\sigma_1 = \sigma_3 = \sigma$). The dispersion relation is given by Eq. (3.9) with $\phi_1 = \phi_2 = -2\arctan(\sigma/\gamma)$ and can be written in the form

$$\tan\left[\frac{1}{2}(\gamma d - p\pi)\right] = \frac{\sigma}{\gamma} \qquad (3.17)$$

In this case, Eq. (3.16) shows that there is not a minimum value for γd, since we have $n_3 = n_1$. The situation $\gamma d = 0$ corresponds to an angle $\theta_2 = \pi/2$, when the ray propagates along the z-axis. On the other hand, the maximum value of γd occurs when the decay constant σ becomes zero, corresponding to the absence of total internal reflection. As a consequence, a symmetric guide does not impose any maximum value, but only a minimum value to the wavelength of the guided modes.

3.1.2 Step-Index Fibers

In its simplest configuration, the optical fiber is constituted by a core with a uniform refractive index, surrounded by a cladding with a uniform refractive index, but slightly inferior to that of the core. Due to the profile of the radial dependence of the refractive index, these fibers are known as *step-index fibers*. Figure 3.2 shows a schematic of the refractive index variation in this case.

Figure 3.3 illustrates the ray propagation in a step-index fiber. Considering the geometry associated with ray (1), the refraction of the ray entering the fiber core is described by the equation

$$n_0 \sin \theta_i = n_1 \sin \theta_t \qquad (3.18)$$

where n_1 and n_0 are the refractive indices of the fiber core and the medium at the fiber entrance, respectively. The ray experiences a total reflection in the boundary core/cladding if the angle of incidence in this boundary is greater than the critical angle, ϕ_c, given by

$$\sin \phi_c = \frac{n_c}{n_1} \qquad (3.19)$$

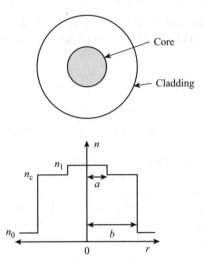

Figure 3.2 Schematic representation of the refractive index profile in the case of a step-index fiber.

where n_c is the cladding refractive index. All rays with $\phi > \phi_c$ remain confined to the core and correspond to the guided modes of the fiber.

We can use Eqs. (3.18) and (3.19) to determine the maximum angle of incidence at the fiber input to have a guided ray. Introducing $\theta_t = \pi/2 - \phi_c$ in Eq. (3.18) and using Eq. (3.19), we obtain

$$n_0 \sin \theta_{i,\max} = n_1 \cos \phi_c = (n_1^2 - n_c^2)^{1/2} \qquad (3.20)$$

The result given by Eq. (3.20) is used to define the numerical aperture (NA) of the optical fiber, given by

$$\mathrm{NA} = (n_1^2 - n_c^2)^{1/2} \qquad (3.21)$$

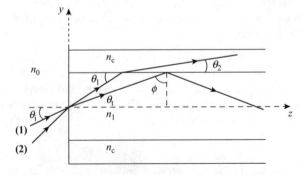

Figure 3.3 Ray propagation in a step-index fiber.

The numerical aperture is a measure of the light-gathering capacity of the fiber. Considering the typical values $n_1 = 1.48$ and $n_c = 1.46$, the numerical aperture is $NA = 0.242$ and the maximum angle of incidence at its entrance is $\theta_{i,max} \approx 14°$, assuming we are coupling from air ($n_0 = 1$).

In the case $n_1 \approx n_c$, the numerical aperture can be approximated as

$$NA = n_1 (2\Delta)^{1/2} \tag{3.22}$$

where

$$\Delta = \frac{n_1 - n_c}{n_1} \tag{3.23}$$

Equation (3.22) shows that the numerical aperture increases when Δ increases. However, an increase in Δ also determines an increase in the intermodal dispersion, which is undesirable from the viewpoint of the optical fiber communication systems.

To analyze this point, let us consider again the schematic representation in Fig. 3.3. Different rays in this figure have different trajectories and different travel times along the fiber. The shortest path occurs for $\theta_i = 0$ and is just equal to the fiber length L. The longest path occurs for an angle $\theta_i = \theta_{i,max}$, given by Eq. (3.20), and it has a length $L/\sin \phi_c$. Considering a velocity of propagation $v = c/n_1$, the time delay between the two rays is given by

$$\Delta t = \frac{n_1}{c} \left(\frac{L}{\sin \phi_c} - L \right) = \frac{L n_1^2}{c \, n_c} \Delta \tag{3.24}$$

This time delay is a measure of broadening by an impulse launched at the fiber input.

In an optical fiber communication system, the time delay Δt must be less than the bit slot $t_B = 1/R$, where R is the bit rate. This imposes a limit on the *bit rate–distance product*, RL, which is frequently used to measure the capacity of an optical communication system. Using Eq. (3.24), we obtain the following condition:

$$RL < \frac{n_c}{n_1^2} \frac{c}{\Delta} \tag{3.25}$$

As an example, $RL < 20$ (Mb/s) km for $\Delta = 0.01$ and $n_1 = 1.5$ ($\approx n_c$). Such fibers can communicate data at a bit rate of 2 Mb/s over distances up to 10 km. Equation (3.25) clearly shows the convenience of using a fiber with low numerical aperture, to achieve a sufficiently high value for the bit rate–distance product.

Considering the geometry associated with ray (2) in Fig. 3.3 and using Snell's law, we have

$$\sin \theta_2 = \sqrt{1 - \cos^2 \theta_2} = \sqrt{1 - \frac{n_1^2}{n_c^2} \cos^2 \theta_1} \tag{3.26}$$

When the angle of incidence in the boundary core/cladding exceeds the critical value for total reflection, the radicand in Eq. (3.26) becomes negative and we can write

$$\sin \theta_2 = \sqrt{1 - \frac{n_1^2}{n_c^2} \cos^2 \theta_1} = \pm i\delta \tag{3.27}$$

In this case, the amplitude of the transmitted wave in the cladding can be written in the form

$$E_t = A \exp[ik_2(z \cos \theta_2 + y \sin \theta_2)] = A \exp\left[ik_2\frac{n_1}{n_c}z \cos \theta_1\right] \exp(-\delta k_2 y) \tag{3.28}$$

Equation (3.28) shows that the amplitude of the transmitted wave decreases exponentially along the y-axis. The *penetration depth*, ξ, is given by the inverse of δk_2:

$$\xi = \left[k_2^2\left(\frac{n_1^2}{n_c^2}\cos^2 \theta_1 - 1\right)\right]^{-1/2} \tag{3.29}$$

Near the critical angle and considering a wavelength $\lambda = 1.3\,\mu m$, we have $\xi \approx 10\,\mu m$. This result shows that the cladding must be sufficiently thick in order that the amplitude of the evanescent wave be practically zero at its outside boundary. On the other hand, considering that a significant fraction of the transmitted energy is carried by this evanescent wave, the quality of the cladding glass must be as high as that of the core, in order to minimize the losses.

3.1.3 Graded-Index Fibers

Graded-index fibers were developed in order to minimize the effects of intermodal dispersion. In such fibers, the refractive index of the core is not uniform but decreases from a maximum value (n_1) at its axis to a minimum value (n_c) at the boundary core/cladding. Figure 3.4 illustrates the profile of the refractive index for this type of fibers.

The refractive index of a graded-index fiber can be described by the expression

$$n(r) = \begin{cases} n_1[1 - \Delta(r/a)^{\alpha}], & r < a \\ n_1(1 - \Delta) = n_c, & r \geq a \end{cases} \tag{3.30}$$

where a is the core radius, Δ is given by Eq. (3.23), and α is a constant. The case $\alpha = 2$ corresponds to a parabolic profile of the refractive index. A step-index fiber corresponds to the limit $\alpha \to \infty$.

To describe the propagation of a ray in a graded-index fiber, we consider the eikonal equation given by

$$\nabla L = n\hat{s} \tag{3.31}$$

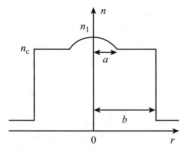

Figure 3.4 Schematic representation of the refractive index profile in the case of a graded-index fiber.

where

$$\hat{\mathbf{s}} = \frac{d\mathbf{R}}{|d\mathbf{R}|} \tag{3.32}$$

and $\mathbf{R}$ is the vector of position. Writing this vector in cylindrical coordinates and performing the differentiation, we obtain

$$d\mathbf{R} = dr\,\hat{\mathbf{u}}_r + r\,d\phi\,\hat{\mathbf{u}}_\phi + dz\,\hat{\mathbf{u}}_z \tag{3.33}$$

Using cylindrical coordinates, Eq. (3.31) gives

$$|\nabla L|^2 = \left(\frac{\partial L}{\partial r}\right)^2 + \left(\frac{1}{r}\frac{\partial L}{\partial \phi}\right)^2 + \left(\frac{\partial L}{\partial z}\right)^2 = n^2(r) \tag{3.34}$$

The eikonal of this problem can be written in the form

$$L(r,\phi,z) = \int \left(n^2 - \frac{c_2^2}{r^2} - c_1^2\right)^{1/2} dr + c_2\phi + c_1 z + c_0 \tag{3.35}$$

where c_0, c_1, and c_2 are constants.

From Eqs. (3.31)–(3.35), we obtain the following equations:

$$n\frac{dr}{dR} = \frac{\partial L}{\partial r} = \sqrt{n^2 - \frac{c_2^2}{r^2} - c_1^2} \tag{3.36}$$

$$nr^2\frac{d\phi}{dR} = \frac{\partial L}{\partial \phi} = c_2 \tag{3.37}$$

$$n\frac{dz}{dR} = \frac{\partial L}{\partial z} = c_1 \tag{3.38}$$

In the case of a parabolic profile for the refractive index, assuming that the point of incidence of the ray at the fiber entrance coincides with the origin of the reference system ($z=0$, $r=0$, $\phi=0$) and that the ray direction makes an angle θ_0 with the z-axis, the constants c_2 and c_1 are given by

$$c_2 = 0, \quad c_1 = n_1 \cos \theta_0 \tag{3.39}$$

It can be verified that, in this case, the ray remains in the meridian plane $\phi = 0$.

To obtain the ray trajectory, we divide each member of Eq. (3.36) by the corresponding member of Eq. (3.38), introduce Eq. (3.30) with $\alpha = 2$, and substitute the constants c_2 and c_1 using Eq. (3.39). Then, the following equation is obtained:

$$\frac{dr}{dz} = \frac{\sqrt{\sin^2 \theta_0 - 2\Delta (r/a)^2}}{\cos \theta_0} \tag{3.40}$$

The integration of Eq. (3.40) gives the result

$$r = h \sin \theta_0 \sin \left(\frac{z}{h \cos \theta_0} \right) \tag{3.41}$$

where $h = a/\sqrt{2\Delta}$. Equation (3.41) shows that the ray has a sinusoidal trajectory with an amplitude $A = h \sin \theta_0$ and a spatial period $z_p = 2\pi h \cos \theta_0$.

When the initial conditions of the incident ray are not those assumed above, the solution for the ray trajectory is more complex than that given by Eq. (3.41). In general, the guided ray is not restricted to the meridian plane and its trajectory is helical.

Figure 3.5 illustrates the evolution of three meridional guided rays, corresponding to three values of the angle θ_0. The ray propagating along the fiber axis has the shortest path, but its velocity is reduced due to the higher refractive index. On the other hand, a greater value of θ_0 determines a longer path of the ray. However, in such circumstances, the ray experiences a lower refractive index in regions far from the fiber axis, where it travels with a greater velocity. As a consequence, a significant reduction of

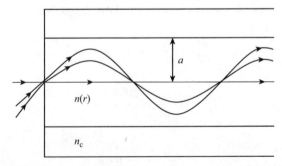

Figure 3.5 Trajectories of meridional guided rays in a graded-index fiber.

the intermodal dispersion can be achieved using graded-index fibers with an adequate refractive index profile.

3.2 WAVE PROPAGATION IN FIBERS

The geometrical optics provides only an approximate description of the guiding properties of optical fibers, which is reasonable when the core radius is much larger than the light wavelength. However, a more rigorous approach to describe propagation of light in fibers is given by the electromagnetic theory. This approach starts with the Maxwell's equations presented in Chapter 2.

3.2.1 Fiber Modes

The fiber modes can be classified as guided modes, leaky modes, and radiation modes [5]. Since the transmission of information in actual optical fiber communication systems makes use of the guided modes only, the following discussion will be restricted to them.

By introducing the Fourier transform of the electric field in the form

$$\tilde{\mathbf{E}}(\mathbf{r}, \omega) = \frac{1}{2\pi} \int\limits_{-\infty}^{\infty} \mathbf{E}(\mathbf{r}, t) \exp(i\omega t) \mathrm{d}t \tag{3.42}$$

the wave equation (2.13) becomes

$$\nabla^2 \tilde{\mathbf{E}} + n^2(\omega) k_0^2 \tilde{\mathbf{E}} = 0 \tag{3.43}$$

where

$$k_0 = \frac{\omega}{c} \tag{3.44}$$

is the free-space wave number.

Taking advantage of the cylindrical symmetry of fibers, Eq. (3.43) is written in the cylindrical coordinates r, ϕ, and z as

$$\frac{\partial^2 \tilde{\mathbf{E}}}{\partial r^2} + \frac{1}{r} \frac{\partial \tilde{\mathbf{E}}}{\partial r} + \frac{1}{r^2} \frac{\partial^2 \tilde{\mathbf{E}}}{\partial \phi^2} + \frac{\partial^2 \tilde{\mathbf{E}}}{\partial z^2} + n^2 k_0^2 \tilde{\mathbf{E}} = 0 \tag{3.45}$$

where $n = n_1$ for $r \leq a$ and $n = n_c$ for $r > a$. A similar equation can be written for the magnetic field $\tilde{\mathbf{H}}$. Since $\mathbf{E}$ and $\mathbf{H}$ satisfy Maxwell's equations (2.1)–(2.4), it results that of the six components only two are independent. $\tilde{E}_z$ and $\tilde{H}_z$ are usually chosen as the independent components and both satisfy Eq. (3.45). Writing $\tilde{E}_z$ as

$$\tilde{E}_z(r, \phi, z) = F(r)\Phi(\phi)Z(z) \tag{3.46}$$

and substituting in the wave equation (3.45), we obtain the solutions $Z(z) = \exp(i\beta z)$ and $\Phi(\phi) = \exp(im\phi)$, where β is the propagation constant and m is an integer. On the other hand, $F(r)$ satisfies the following ordinary differential equation:

$$\frac{\partial^2 F}{\partial r^2} + \frac{1}{r}\frac{\partial F}{\partial r} + \left(n^2 k_0^2 - \beta^2 - \frac{m^2}{r^2}\right)F = 0 \tag{3.47}$$

Equation (3.47) is a form of Bessel's equation. Its solution is given in terms of Bessel functions [13] and assumes the form

$$F(r) = \begin{cases} AJ_m(pr) + A'Y_m(pr), & r \leq a \\ CK_m(qr) + C'I_m(qr), & r > a \end{cases} \tag{3.48}$$

where A, A', C, and C' are constants, J_m and Y_m are ordinary Bessel functions of the first and second kind of order m, respectively, whereas K_m and I_m are modified Bessel functions. The parameters p and q are given by

$$p = \sqrt{n_1^2 k_0^2 - \beta^2} \tag{3.49}$$

$$q = \sqrt{\beta^2 - n_c^2 k_0^2} \tag{3.50}$$

and satisfy the important relation

$$p^2 + q^2 = (n_1^2 - n_c^2)k_0^2 \tag{3.51}$$

As $Y_m(pr)$ has a singularity at $r = 0$, $A' = 0$ for a physically meaningful solution. In the cladding region ($r > a$), the solution $F(r)$ must decay exponentially for large values of r, which happens only if $C' = 0$. As a result, the general solution of Eq. (3.47) becomes

$$E_z = \begin{cases} AJ_m(pr)\exp(im\phi)\exp(i\beta z), & r \leq a \\ CK_m(qr)\exp(im\phi)\exp(i\beta z), & r > a \end{cases} \tag{3.52}$$

Similarly, the solution for H_z can be written in the form

$$H_z = \begin{cases} BJ_m(pr)\exp(im\phi)\exp(i\beta z), & r \leq a \\ DK_m(qr)\exp(im\phi)\exp(i\beta z), & r > a \end{cases} \tag{3.53}$$

Maxwell's equations are used to find general expressions for all transverse components in terms of first derivatives of the z components. In cylindrical coordinates, we find

$$E_r = \frac{i}{p^2}\left(\beta\frac{\partial E_z}{\partial r} + \omega\mu\frac{1}{r}\frac{\partial H_z}{\partial \phi}\right) \tag{3.54}$$

$$E_\phi = \frac{i}{p^2}\left(\frac{\beta}{r}\frac{\partial E_z}{\partial \phi} - \omega\mu\frac{\partial H_z}{\partial r}\right) \tag{3.55}$$

$$H_r = \frac{i}{p^2}\left(\beta\frac{\partial H_z}{\partial r} - \omega\varepsilon\frac{1}{r}\frac{\partial E_z}{\partial \phi}\right) \tag{3.56}$$

$$H_\phi = \frac{i}{p^2}\left(\frac{\beta}{r}\frac{\partial H_z}{\partial \phi} + \omega\varepsilon\frac{\partial E_z}{\partial r}\right) \tag{3.57}$$

These equations are also valid in the cladding region if p^2 is replaced by $-q^2$.

The four constants $A, B, C,$ and D in Eqs. (3.52)–(3.57) are determined by using the requirement that all field components that are tangent to the core/cladding boundary at $r = a$ be continuous across it. All ϕ and z components apply, producing four equations in all. A nontrivial solution for $A, B, C,$ and D is obtained only if the determinant of the coefficient matrix vanishes. This condition provides the following eigenvalue equation [8,12]:

$$\left[\frac{J_m'(pa)}{pJ_m(pa)} + \frac{K_m'(qa)}{qK_m(qa)}\right]\left[\frac{n_1^2}{n_c^2}\frac{J_m'(pa)}{pJ_m(pa)} + \frac{K_m'(qa)}{qK_m(qa)}\right] = \frac{m^2}{a^2}\left(\frac{1}{p^2} + \frac{1}{q^2}\right)\left(\frac{n_1^2}{n_c^2}\frac{1}{p^2} + \frac{1}{q^2}\right) \tag{3.58}$$

where a prime indicates differentiation with respect to the argument.

The solutions of the eigenvalue equation determine the propagation constant β for the fiber modes. For each integer value of m, there are several solutions β_{mn} ($n = 1, 2, \ldots$). Each value β_{mn} corresponds to one specific mode of propagation whose field distribution is given by Eqs. (3.52)–(3.57). For $m = 0$, the fiber modes are denoted by TE0n and TM0n, since they correspond to transverse electric ($E_z = 0$) and transverse magnetic ($H_z = 0$) modes of propagation. However, for $m > 0$, fiber modes become hybrid, since all six components of the electromagnetic field are nonzero. These hybrid modes are denoted by HE$_{mn}$ or EH$_{mn}$, due to an early system in which contributions of H_z and E_z to a given transverse field component at some reference position are compared [14]. If, for example, E_z makes the larger contribution, then the mode is designated HE.

Using the weak-guidance approximation, stating that $n_1 \approx n_c$, the eigenvalue equation (3.58) can be simplified and written in the form

$$\frac{J_m'(pa)}{pJ_m(pa)} + \frac{K_m'(qa)}{qK_m(qa)} = \pm\frac{m}{a}\left(\frac{1}{p^2} + \frac{1}{q^2}\right) \tag{3.59}$$

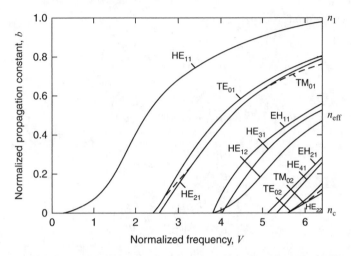

Figure 3.6 Normalized propagation constant b against normalized frequency V for some low-order fiber modes. The right axis shows the effective mode index n_{eff}. (After Ref. 15; © 1981 Academic Press; reprinted with permission)

Equation (3.59) with positive right-hand side is the eigenvalue equation for EH_{mn} modes, whereas with negative right-hand side it is the equation for HE_{mn} modes. The case in which the right-hand side is zero applies to TE and TM modes.

Each fiber mode has an effective refractive index $n_{\text{eff}} = \beta/k_0$, known as *effective mode index*, whose value is such that $n_1 > n_{\text{eff}} > n_c$. The mode is said to reach cutoff when $n_{\text{eff}} = n_c$, which from Eq. (3.50) corresponds to $q = 0$. It is useful to introduce a *normalized frequency* or V *number* parameter, similar to that given by Eq. (3.12) for planar waveguides, as

$$V = k_0 a (n_1^2 - n_c^2)^{1/2} \tag{3.60}$$

This parameter embodies both the fiber structural parameters and the optical wavelength. It is also useful to introduce a normalized propagation constant b, in the form

$$b = \frac{n_{\text{eff}} - n_c}{n_1 - n_c} \tag{3.61}$$

Figure 3.6 represents b against the V parameter for some low-order fiber modes obtained by solving Eq. (3.58). The V parameter determines the number of modes supported by a fiber, which is approximately given by $V^2/2$ for a multimode fiber with large value of V [11].

3.2.2 Single-Mode Fibers

The single-mode condition is given by the value of V at which the TE_{01} and TM_{01} modes reach cutoff. From the eigenvalue equations (3.58) or (3.59), it can be easily verified that the cutoff condition for both modes is given by $J_0(V) = 0$. The smallest value of V for which this condition is satisfied is 2.405. As seen from Fig. 3.6, for $V < 2.405$ the fiber supports only the HE_{11} mode, also referred to as the fundamental mode. This mode has no cutoff and is always supported by a fiber. A typical fiber with $n_1 = 1.45$, $\Delta = 3 \times 10^{-3}$, and $a = 4\,\mu m$ becomes single mode for $\lambda > 1.2\,\mu m$.

In the weak-guidance approximation, the axial components E_z and H_z of the fundamental mode are negligible and they are approximately linearly polarized. Assuming a linear polarization along the x-axis, the electric field for the HE_{11} mode is given by [16]

$$
E_x = \begin{cases} E_0 J_0(pr) \exp(i\beta z), & r \le a \\ E_0 [J_0(pa)/K_0(qa)]K_0(qr)\exp(i\beta z), & r > a \end{cases} \tag{3.62}
$$

The same fiber supports another mode linearly polarized along the y-axis. The two orthogonally polarized modes of a single-mode fiber are degenerate (i.e., they have the same propagation constant) under ideal conditions. However, in practice, a small deviation from the cylindrical geometry, or a slight anisotropy of the material, determines a breaking of the mode degeneration. In such a case, the propagation constant β turns out to be slightly different for the modes polarized along the two orthogonal directions. This is the origin of the polarization mode dispersion phenomenon, which will be discussed in Section 3.6.

Due to fluctuations in the fiber geometry or in the material anisotropy, the fiber birefringence does not remain constant along its length. As a consequence, when linearly polarized light is launched at the input, it generally acquires an arbitrary polarization during its propagation along the fiber. However, a proper design of the fiber can intentionally introduce a significant birefringence in order to make possible the propagation of light with a constant polarization. Fibers obtained in this way are called *polarization maintaining fibers*.

3.2.2.1 Gaussian Approximation for the Fundamental Mode Field   The field distribution for the fundamental mode given by Eq. (3.62) can be approximated by a Gaussian distribution having a width w, known as the *mode field radius*. The Gaussian field polarized along the x-axis is given by

$$
E_{gx} = E_{g0} \exp\left(-\frac{r^2}{w^2}\right) \exp(i\beta z) \tag{3.63}
$$

The mode field radius w is determined by fitting the exact distribution to the Gaussian function and it is given approximately by the following empirical formula [16]:

$$
\frac{w}{a} \approx 0.65 + \frac{1.619}{V^{3/2}} + \frac{2.879}{V^6} \tag{3.64}
$$

This formula approximates the exact result to better than 1% accuracy for $1.2 < V < 2.4$. The mode field radius w is used to define the effective core area, as

$$A_{eff} = \pi w^2 \qquad (3.65)$$

The parameter A_{eff} is important, since it indicates how tightly light is confined to the core.

3.3 FIBER ATTENUATION

Attenuation is one important characteristic of an optical fiber, since it determines the repeater spacing in a fiber transmission system. The lower the attenuation, the greater will be the required repeater spacing and lower will be the cost of that system. Representing by P_0 the power launched at the input of a fiber of length L, the output power is given by

$$P_t = P_0 \exp(-\alpha L) \qquad (3.66)$$

where α is the attenuation constant. Usually, the fiber attenuation is given in dB/km, using the relation

$$\alpha_{dB} = -\frac{10}{L} \log\left(\frac{P_t}{P_0}\right) = 4.343\alpha \qquad (3.67)$$

For example, if the output power is half the input power, then the loss is $10 \log(2) \approx 3$ dB.

A pure silica glass presents an attenuation below 1 dB/km for wavelengths in the range 0.8–1.8 μm. Outside this range, the attenuation increases rapidly. Figure 3.7 shows a plot of the spectra attenuation in a GeO_2–SiO_2 fiber. The attenuation presents minima values around 1.3 and 1.55 μm, achieving in the last case a value of 0.2 dB/km. The two bands around the above wavelengths correspond to the windows actually used in fiber communication systems.

The material absorption and Rayleigh scattering are the dominant contributions for the residual attenuation of fused silica. Pure silica absorbs both in the ultraviolet region and in the infrared region beyond 2 μm. However, even a relatively small quantity of impurity can determine a significant absorption in the range 0.5–2 μm. The most difficult extrinsic loss contributor to remove is the hydroxyl group, OH, which enters the glass through water vapor and whose main absorption peak occurs at 2.73 μm. The peak in Fig. 3.7 near 1.37 μm corresponds to the second harmonic of this value.

Rayleigh dispersion is a fundamental effect and results from random fluctuations of the material density. As a consequence, there are local fluctuations of the refractive index, which determine light scattering in all directions. This effect varies with λ^{-4}, becoming more significant for short wavelengths. The minimum of attenuation near 1.55 μm is mainly due to the Rayleigh scattering.

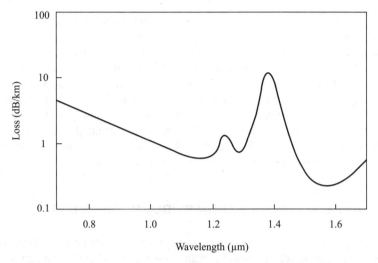

Figure 3.7 Typical spectral attenuation profile of a GeO_2–SiO_2 fiber.

3.4 MODULATION AND TRANSFER OF INFORMATION

The information transmitted in an optical communication system is carried by modulation of the light wave. The electric field of a modulated light wave can be represented in the form

$$E(z,t) = \frac{1}{2}[\bar{E}(z,t)\exp\{i(\beta_0 z - \omega_0 t)\} + \text{c.c.}] \qquad (3.68)$$

where $\bar{E}(z,t)$ represents the modulation, whereas β_0 and ω_0 are the propagation constant and the frequency of the unmodulated wave, respectively. The transmitted information may take an *analog format* or a *digital format*. In the analog format, the signal varies continuously with time, while in the digital format the signal takes only a few discrete values. For example, in the *binary representation*, the signal takes only two values. The two possibilities are called "bit 1" and "bit 0". The period of time t_B allocated to each bit is called *bit slot* and its inverse, $R = 1/t_B$, provides the *bit rate*, that is, the number of bits per second.

There are two possibilities concerning the modulation applied initially to the optical wave. When one pulse shorter than the bit slot is designed to represent bit 1, the format is called RZ (return to zero). On the other hand, if the optical pulse amplitude does not drop to zero between two or more successive 1 bits, the format is called NRZ (nonreturn to zero). In this case, the pulse width varies depending on the bit pattern, whereas it remains the same in the case of RZ format.

In practice, the NRZ format is often used because the bandwidth associated with the bit stream is smaller than that of the RZ format. However, the use of the RZ format began to attract a particular attention around 2000 when dispersion-managed optical solitons were found to constitute a good option to realize high-capacity lightwave communication systems [17,18].

The information transmitted along an optical fiber may be lost not only due to the fiber loss but also because the modulation $\bar{E}(z,t)$ is deformed due to various properties of the fiber. In particular, we must consider that $\bar{E}(z,t)$ propagates at the group velocity, $d\omega/d\beta$, discussed in Section 2.4. The amount of information that the light wave carries is approximately given by the spectral width associated with the modulation $\bar{E}(0,t)$ at the fiber input. To achieve an ultrahigh-speed information transfer in a fiber, we must consider the behavior of $\bar{E}(z,t)$ having a wide spectral width. The chromatic dispersion then becomes a main issue.

3.5 CHROMATIC DISPERSION IN SINGLE-MODE FIBERS

In digital optical fiber communication systems, information to be sent is first coded in the form of pulses, which are then transmitted through the fiber to the receiver, where the information is decoded. A pulse of light broadens in time as it propagates through the fiber, a phenomenon known as pulse dispersion.

A medium exhibits *chromatic dispersion* if the propagation constant β of a wave within it varies nonlinearly with frequency. Signal distortion arising from *group velocity dispersion* occurs as the frequency components of the signal propagate with different velocities. In an optical waveguide, this effect arises from two mechanisms: (i) refractive index variations with wavelength (material dispersion) and (ii) waveguide-related effects (waveguide dispersion).

Let us consider a pulse described by a function $\psi(0,t)$ at $z=0$. This pulse can be represented as a superposition of harmonic waves:

$$\psi(0,t) = \int_{-\infty}^{\infty} A(\omega)e^{-i\omega t}\, d\omega \tag{3.69}$$

After propagation along a distance z, the spectral component of frequency ω experiences a phase variation $\beta(\omega)z$, where $\beta(\omega)$ is the propagation constant at this frequency. The total field is then given by

$$\psi(z,t) = \int_{-\infty}^{\infty} A(\omega)e^{i(\beta(\omega)z - \omega t)}\, d\omega \tag{3.70}$$

The propagation constant $\beta(\omega)$ is related to the refractive index $n(\omega)$ in the form

$$\beta(\omega) = \frac{\omega n(\omega)}{c} \tag{3.71}$$

We can expand $\beta(\omega)$ in a Taylor series about the carrier frequency ω_0:

$$\beta(\omega) = \beta_0 + \frac{d\beta}{d\omega}(\omega - \omega_0) + \frac{1}{2}\frac{d^2\beta}{d\omega^2}(\omega - \omega_0)^2 + \cdots \tag{3.72}$$

where $\beta_0 = \beta(\omega_0)$,

$$\frac{d\beta}{d\omega} \equiv \beta_1 = \frac{1}{c}\left[n + \omega\frac{dn}{d\omega}\right] = \frac{n_g}{c} = \frac{1}{v_g} \tag{3.73}$$

and

$$\frac{d^2\beta}{d\omega^2} \equiv \beta_2 = -\frac{dv_g/d\omega}{v_g^2} \tag{3.74}$$

In Eq. (3.73), $n_g = n + \omega(dn/d\omega)$ and $v_g = c/n_g$ represent the group refractive index and the group velocity, respectively. The parameter β_2 characterizes the group velocity dispersion (GVD).

In the presence of GVD, the information carried by different frequency components of $\psi(0, t)$ propagating at different speeds thus arrives at different times. The relative delay Δt_D of arrival time of information at frequencies ω_1 and ω_2, after propagating a distance z, is given by

$$\Delta t_D = \frac{z}{v_g(\omega_1)} - \frac{z}{v_g(\omega_2)} = \frac{(dv_g/d\omega)(\omega_2 - \omega_1)z}{v_g^2} \tag{3.75}$$

Using Eq. (3.74), we have

$$\Delta t_D = -\beta_2(\omega_2 - \omega_1)z \tag{3.76}$$

Equation (3.76) shows that the difference of arrival time of information is proportional to the group velocity dispersion β_2, the difference of the frequency components $\omega_2 - \omega_1$, and the distance of propagation z. We note that if $\beta_2 < 0$ (anomalous dispersion regime), higher frequency components of information arrive first, whereas for $\beta_2 > 0$ (normal dispersion regime) the reverse is true. If the information at different frequency components arrives at different times, the information may be lost. This problem becomes more serious when $\omega_2 - \omega_1$ is large.

In practice, the group dispersion is usually given by a group delay parameter, D, defined by delay of arrival time in picosecond unit for two wavelength components separated by 1 nm over a distance of 1 km. From Eq. (3.76), the group delay may be described by

$$D = \beta_2 c\left(\frac{2\pi}{\lambda_1} - \frac{2\pi}{\lambda_2}\right)z = -\beta_2 c\frac{2\pi}{\lambda^2}\Delta\lambda z \tag{3.77}$$

Thus, by taking $\Delta\lambda = 1$ nm and $z = 1$ km, D in the unit of ps/(nm km) is related to β_2 through

$$D = -\frac{2\pi c\beta_2}{\lambda^2} \tag{3.78}$$

For a typical single-mode fiber, we have $D = 0$ (16 ps/(nm km)) at $\lambda = 1.31$ μm (1.55 μm). This indicates that two wavelength components of a pulse separated by

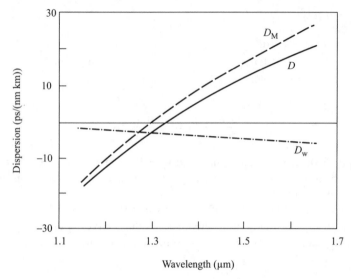

Figure 3.8 Relative contributions of material dispersion D_M and waveguide dispersion o, as well as total dispersion $D = D_M + D_w$ for a typical single-mode fiber.

1 nm arrive with a time delay on the order of a few picoseconds when they propagate over a distance of 1 km in a typical fiber.

The fiber dispersion is given by the sum of the material dispersion, D_M, and waveguide dispersion, D_w. The material dispersion D_M is related to the rate of variation of the group index with wavelength, $D_M = c^{-1}(dn_g/d\lambda)$. In the case of fused silica, $dn_g/d\lambda = 0$ and consequently $D_M = 0$ at $\lambda = \lambda_{ZD} = 1.276\,\mu m$, which is referred to as the material zero dispersion wavelength.

The waveguide dispersion D_w depends on the fiber parameters such as the core radius and the index difference between the core and the cladding, and is negative in the range 0–1.6 μm. Figure 3.8 shows D_M, D_w, and their sum $D = D_M + D_w$ for a typical single-mode fiber. The main effect of waveguide dispersion is to shift λ_{ZD} so that the total dispersion is zero near 1.31 μm.

It is possible to design the fiber such that the zero dispersion wavelength λ_{ZD} is shifted in the vicinity of 1.55 μm. Such fibers are referred to as dispersion-shifted fibers (DSFs). It is also possible to design the fiber such that the total dispersion D is relatively small over a wide wavelength range extending from 1.3 to 1.6 μm. Such fibers are called *dispersion-flattened fibers*. Figure 3.9 illustrates the fiber dispersion as a function of wavelength for several kinds of commercial fibers.

3.5.1 Unchirped Input Pulses

Let us consider the case of an input pulse with a Gaussian profile:

$$E(0,t) = E_0 \exp\left\{-\frac{1}{2}\left(\frac{t}{t_0}\right)^2\right\} \exp\left\{-i\omega_0 t\right\} \tag{3.79}$$

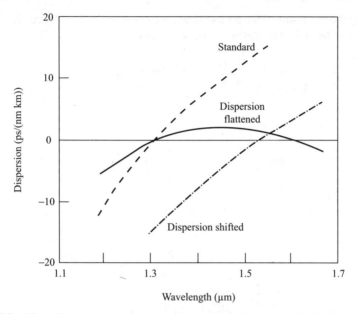

Figure 3.9 Fiber dispersion as a function of wavelength for several kinds of commercial fibers.

where E_0 is the peak amplitude, ω_0 is the carrier frequency, and the parameter t_0 represents the half-width at $1/e$ intensity point. It is related to the full width at half maximum (FWHM) by the relation

$$t_{\text{FWHM}} = 2(\ln 2)^{1/2}t_0 \approx 1.665t_0 \tag{3.80}$$

The amplitude spectrum of this pulse is obtained by calculating its Fourier transform:

$$E(0,\omega) = \frac{1}{2\pi} \int_{-\infty}^{\infty} E_0 \exp\left\{ -\frac{1}{2}\left(\frac{t}{t_0}\right)^2 \right\} \exp\left\{i(\omega - \omega_0)t\right\}dt$$

$$= \frac{E_0 t_0}{\sqrt{2\pi}} \exp\left\{ -\frac{t_0^2(\omega - \omega_0)^2}{2} \right\} \tag{3.81}$$

Each spectral component of (3.81) is propagated over distance z in the medium with propagation constant $\beta(\omega)$ and acquires a phase variation $\beta(\omega)z$. Thus, the pulse at z is found through the inverse Fourier transform of the electric field after propagation:

$$E(z,t) = \int_{-\infty}^{\infty} E(0,\omega) \exp\{i[\beta(\omega)z - \omega t]\}d\omega \tag{3.82}$$

Using the expansion of the propagation constant given by Eq. (3.72), and retaining terms up to second order in $(\omega - \omega_0)$, we obtain

$$E(z,t) = \frac{E_0}{(1+\sigma^2)^{1/4}} \exp\left\{ -\frac{(t-z/v_g)^2}{2t_p^2(z)} \right\} \exp\{i[\beta_0 z - \phi(z,t)]\} \qquad (3.83)$$

where

$$\phi(z,t) = \omega_0 t + \kappa\left(t - \frac{z}{v_g}\right)^2 - \frac{1}{2}\operatorname{tg}^{-1}(\sigma) \qquad (3.84)$$

$$\kappa = \frac{\sigma}{2(1+\sigma^2)t_0^2} \qquad (3.85)$$

$$\sigma = \frac{\beta_2 z}{t_0^2} \qquad (3.86)$$

$$t_p^2(z) = t_0^2(1+\sigma^2) \qquad (3.87)$$

Equation (3.87) shows that the half-width of the pulse $t_p(z)$ increases with the propagation distance by an amount that is governed by the GVD coefficient β_2. The *dispersion length*, L_D, is defined as the distance over which a Gaussian pulse broadens to $\sqrt{2}$ times its initial width t_0. Considering Eqs. (3.86) and (3.87), it can be seen that

$$L_D = \frac{t_0^2}{|\beta_2|} \qquad (3.88)$$

The broadening effect due to the fiber dispersion is illustrated in Fig. 3.10, which shows the Gaussian pulse profile at the input (dashed curve) and after the propagation over a distance $z = 3L_D$ (solid curve).

It is apparent from Eqs. (3.86) and (3.87) that an optimum initial pulse width exists for which the output width is minimized. By substituting Eq. (3.86) in Eq. (3.87) and differentiating with respect to t_0, we find that the optimum initial pulse width is given by $t_{opt} = \sqrt{\beta_2 z}$.

Equation (3.84) shows that the phase of the pulse varies with the square of time. As a consequence, the instantaneous frequency also varies with time and the pulse is said to be chirped. The instantaneous frequency is found by taking the time derivative of the phase:

$$\omega(t) = \omega_0 + 2\kappa\left(t - \frac{z}{v_g}\right) \qquad (3.89)$$

This result shows that in the normal dispersion regime ($\kappa > 0$) the instantaneous frequency at a fixed position z increases with time (positive linear chirp). As a

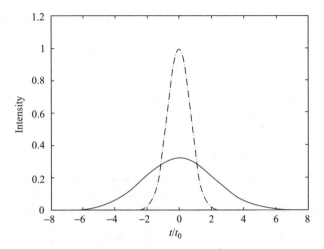

Figure 3.10 Broadening of a Gaussian pulse due to the fiber dispersion. The dashed curve corresponds to the initial pulse profile, whereas the solid curve gives the pulse profile after a propagation over a distance $z = 3L_D$.

consequence, the leading edge of the pulse is redshifted, whereas the trailing edge is blueshifted. On the other hand, in the anomalous dispersion regime ($\kappa < 0$), the instantaneous frequency decreases with time (negative chirp). In this case, the leading edge of the pulse is blueshifted and the trailing edge becomes redshifted. Figure 3.11 illustrates the pulse chirping in the normal and anomalous dispersion regimes.

The rate of frequency change with time over the pulse envelope is found by differentiating Eq. (3.89) with respect to time:

$$\frac{\partial \omega}{\partial t} = 2\kappa = \frac{\sigma}{t_p^2} \qquad (3.90)$$

where we identify the *dimensionless linear chirp parameter* σ.

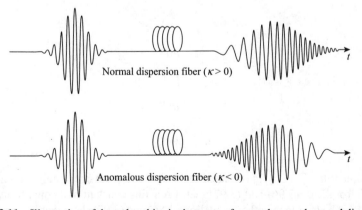

Figure 3.11 Illustration of the pulse chirp in the cases of anomalous and normal dispersion.

From Eq. (3.83), we find that the peak intensity of the broadened pulse is

$$|E_0'|^2 = \frac{|E_0|^2}{(1+\sigma^2)^{1/2}} \tag{3.91}$$

From Eqs. (3.87) and (3.91), we see that the pulse energy (proportional to $|E_0'|^2 t_p$) is independent of position, as expected.

3.5.2   Chirped Input Pulses

When a pulse having initial chirp is input to a dispersive medium, the pulse is either further broadened or compressed, depending on the sign of the dispersion. Let us consider a chirped Gaussian input pulse of width t_0 and chirp C_0, whose electric field is given by

$$E(0,t) = E_0 \exp\left\{ -\frac{1+iC_0}{2}\left(\frac{t}{t_0}\right)^2 \right\} \exp\left(-i\omega_0 t\right) \tag{3.92}$$

The pulse spectrum is given by the Fourier transform of Eq. (3.92), which is

$$E(0,\omega) = \frac{E_0 t_0}{\sqrt{2\pi(1+iC_0)}} \exp\left\{ -\frac{t_0^2(\omega-\omega_0)^2}{2(1+iC_0)} \right\} \tag{3.93}$$

Using Eq. (3.93) and considering that the propagation constant is given by the expansion (3.72) limited to second order in $(\omega - \omega_0)$, we obtain the output pulse field by calculating the inverse Fourier transform as indicated by Eq. (3.82). The result can be presented in the form

$$E(z,t) = \tilde{E}_0 \exp\left\{ -\frac{(1+iC)(t-z/v_g)^2}{2t_p^2(z)} \right\} \exp\left\{ i[\beta_0 z - \omega_0 t] \right\} \tag{3.94}$$

where

$$\tilde{E}_0 = \frac{E_0 t_0 [1 + i\sigma(1-iC_0)]^{1/2}}{t_p} \tag{3.95}$$

$$C = C_0 + \sigma(1+C_0^2) \tag{3.96}$$

$$t_p^2 = t_0^2[(1+C_0\sigma)^2 + \sigma^2] \tag{3.97}$$

The quantity $(1+C^2)/t_p^2$ is related to the spectral width of the pulse and it can be verified from Eqs. (3.96) and (3.97) that it remains constant and equal to its initial value $(1+C_0^2)/t_0^2$. On the other hand, Eqs. (3.86) and (3.97) show that the input pulse

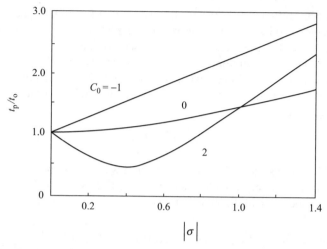

Figure 3.12 Evolution of the pulse width with the normalized distance $|\sigma|$, for three values of the chirp parameter C_0.

broadens monotonously during propagation if C_0 and β_2 are of the same sign. However, if these parameters are of opposite sign, the pulse compresses until a given location z_{min}, after which it starts to broaden. This behavior is illustrated in Fig. 3.12 for the case $\beta_2 < 0$. The location z_{min} of maximum compression is found from Eq. (3.96) by setting $C = 0$:

$$z_{min} = \left| \frac{C_0}{\beta_2} \right| \frac{t_0^2}{1 + C_0^2} \qquad (3.98)$$

Substituting Eq. (3.98) in Eq. (3.97), we obtain the following result for the minimum pulse width:

$$t_{min} = \frac{t_0}{\sqrt{1 + C_0^2}} \qquad (3.99)$$

3.5.3 Dispersion Compensation

As referred above, standard single-mode fibers have a zero dispersion wavelength of 1.31 μm. On the other hand, most of the installed communication systems using these fibers operate in the 1.55 μm, where the fiber loss is minimum, but the typical dispersion is $+16$ ps/(km nm). In the presence of this dispersion, the optical pulses will suffer from broadening and thus before they completely overlap, they have to be processed. Such task can be realized optically by using the concept of *dispersion compensation*.

In Section 3.5.1, it was demonstrated that when pulses at 1.55 μm propagate through a conventional fiber, experiencing anomalous dispersion, the leading edge of the pulse is blueshifted, while the trailing edge is redshifted. If such pulses are then propagated through an appropriate length of a fiber having normal dispersion ($D < 0$), they can be compressed to their original width. Such a fiber is called a dispersion-compensating fiber (DCF). If the transmission and compensating fibers have dispersions D_1 and D_2 and lengths L_1 and L_2, respectively, the requirement for a complete dispersion compensation at a given wavelength λ is

$$D_2 L_2 = -D_1 L_1 \tag{3.100}$$

In a multichannel system, there are a number of wavelengths propagating simultaneously through the fiber. In such a case, if Eq. (3.100) is satisfied for a channel of wavelength λ, the residual dispersion at a nearby wavelength $\lambda + \Delta\lambda$ will be

$$\text{Dispersion} = D_1(\lambda + \Delta\lambda)L_1 + D_2(\lambda + \Delta\lambda)L_2 \approx D_1' L_1 \Delta\lambda \left(1 - \frac{D_2'/D_2}{D_1'/D_1}\right) \tag{3.101}$$

where prime denotes differentiation with respect to wavelength. Thus, for the dispersion compensation over a band of wavelengths, the relative dispersion slopes D'/D of the two fibers must be equal:

$$\frac{D_2'}{D_2} = \frac{D_1'}{D_1} \tag{3.102}$$

If condition (3.102) is not satisfied, the dispersion may accumulate and reach unacceptable levels in multichannel systems.

One disadvantage of using DCF is the added loss associated with the increased fiber span. Concerning this aspect, a useful figure of merit is the ratio of the DCF dispersion magnitude to its loss in dB/km, expressed in units of ps/(nm dB). Clearly, a high value of this parameter is desired. Another disadvantage is related to nonlinear effects, which may degrade the signal over the long length of the fiber if its intensity is relatively high. This problem arises because the effective core area of a DCF is usually much smaller than that of a conventional single-mode fiber. However, this fact can be used with advantage to achieve Raman amplification along the DCF. In such a case, by pumping a DCF with a high-power Raman pump, simultaneous dispersion compensation and amplification is possible [19].

3.6   POLARIZATION MODE DISPERSION

Another pulse broadening mechanism in single-mode fibers besides chromatic dispersion is related to polarization mode dispersion (PMD) [20–24]. PMD has its origin in the optical fiber birefringence determined by imperfection in the manufacturing process and/or mechanical stress on the fiber after manufacture. PMD causes a difference in the delays corresponding to different polarizations and when this difference approaches a significant fraction of the bit period, pulse distortion and system penalties occur. Limitations due to PMD become particularly significant in high-speed long-haul optical fiber transmission systems [25–30].

3.6.1 Fiber Birefringence and the Intrinsic PMD

In a birefringent fiber, the effective mode index varies continuously with the field orientation angle in the transverse plane. The directions that correspond to the maximum and minimum mode indices are orthogonal and define the *principal axes* of the fiber. Let us assume that these principal axes coincide with the x- and y-axes. A linearly polarized field along x or y has a propagation constant $\beta_x = n_x \omega / c$ or $\beta_y = n_y \omega / c$, where n_x and n_y are the effective mode indices associated with x and y polarizations, respectively. The fiber *birefringence* is given by

$$\Delta\beta = \beta_x - \beta_y = \frac{\omega}{c}\Delta n_{\text{eff}} \tag{3.103}$$

where $\Delta n_{\text{eff}} = n_x - n_y$. Typically, Δn_{eff} range between 10^{-7} and 10^{-5}, which is much smaller than the index difference between core and cladding ($\sim 3 \times 10^{-3}$).

Let us consider an input field given by

$$\mathbf{E}(z) = \exp{(i\beta_x z)}[E_{0x}\hat{i} + E_{0y}\exp{(-i\Delta\beta z)}\hat{j}] \tag{3.104}$$

where E_{0x} and E_{0y} are the input field amplitudes along the principal axes. Equation (3.104) describes a wave whose polarization state transforms from linear to elliptical and then returns to the original linear state when $\Delta\beta z = 2\pi$. The spatial period of this evolution process, known as the *beat length*, is given by

$$L_b = \frac{2\pi}{\Delta\beta} = \frac{\lambda}{\Delta n_{\text{eff}}} \tag{3.105}$$

Equations (3.103) and (3.105) show that both the birefringence and the beat distance depend on the frequency. Therefore, the output polarization state also varies in a cyclic manner with frequency for a fixed fiber length. Changes in frequency or distance therefore produce equivalent effects on the polarization state.

A different behavior occurs when the input field is polarized along one of the two principal axes. In this case, the polarization remains along the given principal axis for the entire span, even if the frequency changes. This behavior provides the basis for the definition of what is called the *input principal polarization states* of the fiber. Such states correspond to the input polarizations that lead to output polarization states that are *invariant with frequency*.

The difference in the local propagation constants indicated by Eq. (3.103) is usually accompanied by a difference in the local group velocities for the two polarization modes. This differential group velocity is described by a differential

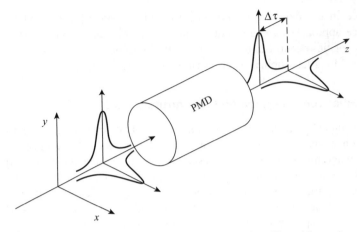

Figure 3.13   Effect of PMD when a signal is introduced in a short fiber with equal components along the two principal axes.

group delay (DGD), $\Delta\tau$, per unit length between the two modes, often referred to as intrinsic polarization mode dispersion (PMD$_i$), given by [20–24]

$$\text{PMD}_i = \frac{\Delta\tau}{L} = \frac{d}{d\omega}(\Delta\beta) = \frac{1}{c}\left(\Delta n_{\text{eff}} + \omega\frac{d\Delta n_{\text{eff}}}{d\omega}\right) \qquad (3.106)$$

The quantity PMD$_i$ is often expressed in units of picoseconds per kilometer of fiber length. The linear length dependence of DGD applies only when the birefringence can be considered uniform, as in a short fiber. Figure 3.13 illustrates the effect of PMD when a signal is introduced in such a short fiber with equal components along the two principal axes. DGD can be interpreted as the effective pulse spread corresponding to the group delay difference between the two orthogonally polarized components.

From Eq. (3.106) and ignoring the dispersion of Δn_{eff}, we can see that the DGD for a single beat length is given by

$$\Delta\tau_b = L_b\frac{\Delta n_{\text{eff}}}{c} = \frac{\lambda}{c} = \frac{1}{f} \qquad (3.107)$$

which corresponds to an optical cycle. At $\lambda = 1.55\,\mu\text{m}$, we have $\Delta\tau_b = 5.2\,\text{fs}$.

The equivalence between changes in frequency and distance in affecting polarization provides a method of measuring PMD. Over a distance L, the two principal polarizations establish a phase difference given by

$$\Delta\phi = \Delta\beta L = \frac{\omega}{c}\Delta n_{\text{eff}}L \qquad (3.108)$$

On the other hand, the change in frequency $\Delta\omega$ producing an additional phase shift of 2π is such that

$$\Delta\phi + 2\pi = \frac{\omega + \Delta\omega}{c} \left(\Delta n_{\text{eff}} + \Delta\omega \frac{dn_{\text{eff}}}{d\omega} \right) L \qquad (3.109)$$

Carrying out the product in Eq. (3.109), neglecting the term on the order of $(\Delta\omega)^2$, and subtracting Eq. (3.108), we obtain the result

$$\Delta\tau \approx \frac{2\pi}{\Delta\omega} \qquad (3.110)$$

where Eq. (3.106) was also used. Equation (3.110) provides a direct relation between the measurement of polarization evolution in frequency domain and the time-domain effects of PMD.

3.6.2 PMD in Long Fiber Spans

The assumption of constant birefringence and polarization axis orientation is no longer valid in the case of long fiber spans. In fact, a long span is likely to possess a very complicated progression of features along its length that will modify the polarization of light. As a consequence, even if the light is linearly polarized along a principal axis of the fiber at the input, it couples into the orthogonal polarization during its progress along the fiber. Long fibers are often modeled as a concatenation of birefringent sections whose birefringence axes (and magnitudes) change randomly along the fiber.

Due to mode coupling, the birefringence of each section may either add to or subtract from the total birefringence. As a result, PMD in long fiber spans does not accumulate linearly with fiber length, but accumulates in a random-walk-like process that leads to a square root of length dependence.

Since most of the perturbations that act on a fiber depend on the temperature, the transmission properties also vary with ambient temperature. This manifests, for example, as a random, time-dependent drifting of the state of polarization at the output of a fiber. Therefore, a statistical approach for PMD must be adopted.

To distinguish between the short- and long-length regimes, it is important to consider a parameter called the correlation length, L_c. One considers the evolution of the polarizations as a function of length in an ensemble of fibers with statistically equivalent perturbations. While the input polarization is fixed (e.g., along the x-axis), it is equally probable to observe any polarization state at large lengths. The correlation length is defined to be the length at which the average power in the orthogonal polarization mode, $\langle P_y \rangle$, is within $1/e^2$ of the power in the starting mode $\langle P_x \rangle$, that is,

$$\frac{\langle P_y \rangle - \langle P_x \rangle}{P_{\text{tot}}} = \frac{1}{e^2} \qquad (3.111)$$

where $P_{\text{tot}} = \langle P_x \rangle + \langle P_y \rangle$. When the fiber transmission distance L satisfies $L \ll L_c$, the fiber is in the short-length regime, whereas the case $L \gg L_c$ corresponds to the long-length regime.

The mean square DGD of the fiber for arbitrary values of the fiber length is given by [31,32]

$$\langle \Delta \tau^2 \rangle = 2 \left(\Delta \tau_b \frac{L_c}{L_b} \right)^2 \left(\frac{L}{L_c} + e^{-L/L_c} - 1 \right) \tag{3.112}$$

For $L \ll L_c$, Eq. (3.112) simplifies to

$$\sqrt{\langle \Delta \tau^2 \rangle} = \Delta \tau_{rms} = \frac{\Delta \tau_b L}{L_b} \tag{3.113}$$

On the other hand, for $L \gg L_c$, Eq. (3.112) gives

$$\Delta \tau_{rms} = \left(\frac{\Delta \tau_b}{L_b} \right) \sqrt{2LL_c} \tag{3.114}$$

reflecting the length dependence discussed earlier.

Equations (3.112)–(3.114) give the mean (and most likely) DGD values that would be observed. However, in practice, significant variations of DGD are observed about the mean, so the form of this distribution is important. When $L \gg L_c$, the PMD has a Maxwellian probability distribution given by [31,32]

$$p(\Delta \tau) = \frac{8}{\pi^2 \Delta \tau_{rms}} \left(\frac{2 \Delta \tau}{\Delta \tau_{rms}} \right)^2 \exp \left[\frac{1}{\pi} \left(\frac{2 \Delta \tau}{\Delta \tau_{rms}} \right)^2 \right], \quad \Delta \tau > 0 \tag{3.115}$$

The propagation of a pulse through a long span is very complicated due to the random sequence of perturbations along the length that produce local changes in the polarization. However, a surprising feature of PMD is that even for long fibers there are two *input principal states of polarization* (PSPs) [33]. These two principal states are orthogonal and are defined as those input polarizations for which the output states of polarization are independent of frequency to first order, that is, over a small frequency range. The later states, known as *output principal states of polarization*, are also orthogonal, and are not necessarily the same as the input states. Therefore, if a pulse having a sufficiently narrow frequency spectrum is coupled into the span in either one of the input PSPs, the output pulse (appearing in the associated output PSP) experiences no broadening or change of shape and presents a group delay τ_i ($i = 1, 2$). The differential group delay between the PSPs is $\Delta \tau = \tau_1 - \tau_2$.

The PSPs can be evaluated considering the expression of the fiber transfer matrix, given by [33]

$$\mathbf{T}(\omega) = e^{\gamma(\omega)} \begin{bmatrix} u_1(\omega) & u_2(\omega) \\ -u_2^*(\omega) & u_1^*(\omega) \end{bmatrix} \tag{3.116}$$

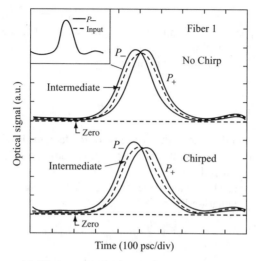

Figure 3.14   Measured DGD due to PMD in 10 km of dispersion-shifted fiber. (After Ref. 34; © 1988 OSA) The waveforms marked P_- and P_+ correspond to inputs in each input principal state.

where $\gamma(\omega)$, $u_1(\omega)$, and $u_2(\omega)$ are complex quantities, and the latter two satisfy the relation

$$|u_1|^2 + |u_2|^2 = 1 \tag{3.117}$$

The differential group delay $\Delta\tau$ between the PSPs is related to the matrix elements of $\mathbf{T}(\omega)$ as follows [33]:

$$\Delta\tau = 2\sqrt{\left|\frac{du_1}{d\omega}\right|^2 + \left|\frac{du_2}{d\omega}\right|^2} \tag{3.118}$$

Figure 3.14 shows experimental results that demonstrate the principal states model [34]. The three waveforms were observed at the output of 10 km of spooled single-mode fiber. The waveforms marked P_- and P_+ correspond to inputs in each input principal state. Each output pulse retains its shape and width, but the two are separated by the DGD of the fiber span (40 ps). The intermediate waveform was obtained when the pulse power was equally split between the two principal states. This pulse was slightly broadened, as indicted by an approximately 10% reduction in the pulse height.

Using the principal states model, PMD can be characterized by the PMD vector, τ, defined in the three-dimensional Stokes space and given by

$$\tau = \Delta\tau\hat{\mathbf{p}} \tag{3.119}$$

The magnitude of the PMD vector is the DGD, $\Delta\tau$, whereas $\hat{\mathbf{p}}$ is a unit vector pointing in the direction of the slower PSP.

Any output polarization, $\hat{t}$, is related to its input polarization, $\hat{s}$, by the 3×3 Müller rotation matrix, R, as $\hat{t} = R\hat{s}$. A similar relation is verified between the input PMD vector, τ_s, and the output PMD vector: $\tau = R\tau_s$. The frequency derivative of $\hat{t} = R\hat{s}$ leads to the law of infinitesimal rotation:

$$\frac{d\hat{t}}{d\omega} = \tau \times \hat{t} \tag{3.120}$$

where $\tau \times = R_\omega R^T$ and R^T denotes the transpose of R. Here, the PMD vector describes how, for a fixed input polarization, the output polarization $\hat{t}$ will precess around τ as the frequency is changed.

The PMD vector τ can be related to the microscopic birefringence in a fiber through the following vector equation [35]:

$$\frac{d\tau}{dz} = \frac{d\beta}{d\omega} + \beta \times \tau \tag{3.121}$$

where z represents the position along the fiber and

$$\beta = \Delta\beta\hat{n} \tag{3.122}$$

is a three-dimensional vector representing the local birefringence of the fiber [36]. This equation is the basis for the statistical theory of PMD [31].

PROBLEMS

3.1 The refractive index difference between the core and the cladding of an optical fiber is 1%. Assuming that the fiber numerical aperture is $NA = 0.18$, calculate the core refractive index.

3.2 Consider a step-index fiber, whose core has a radius $a = 20\,\mu m$ and a refractive index $n_1 = 1.475$, whereas the cladding refractive index is $n_c = 1.460$.
   (a) Obtain the maximum value of the angle between a guided ray and the fiber axis.
   (b) Calculate the number of reflections experienced by a guided ray propagating according the angle obtained in (a), along a fiber with a length of 5 km.
   (c) Assuming that there is a power attenuation of 0.01% in each reflection, calculate the total attenuation in dB/km for the fiber considered in (b).

3.3 Obtain the maximum transmission rate determined by the intermodal dispersion for a fiber with the characteristics specified in Problem 3.2 and a length $L = 20\,km$.

3.4 Using Maxwell's equations, obtain Eqs. (3.54)–(3.57) for the field components E_r, E_ϕ, H_r, and H_ϕ in terms of E_z and H_z.

3.5 The width of a Gaussian pulse propagating in a fiber increases with distance according to Eq. (3.87). Calculate the distance z after which the pulse width becomes $t_1(z) = 20t_0$, with $t_0 = 5$ps. Consider that the fiber GVD is $\beta_2 = 20\,\text{ps}^2/\text{km}$.

3.6 Explain the origin of PMD in optical fibers and give the reason why it leads to pulse broadening.

3.7 Estimate the intrinsic PMD at 1550 nm of a polarization maintaining fiber that has a beat length of 4 mm.

REFERENCES

1. J. Tyndall, *Proc. R. Inst.* **1**, 446 (1854).
2. K. C. Kao and G. A. Hockham, *Proc. IEE* **113**, 1151 (1966).
3. F. P. Kapron, D. B. Keck, and R. D. Maurer, *Appl. Phys. Lett.* **17**, 423 (1970).
4. T. Miya, Y. Terunuma, F. Hosaka, and T. Miyoshita, *Electron. Lett.* **15**, 106 (1979).
5. A. W. Snyder and J. D. Love, *Optical Waveguide Theory*, Chapman & Hall, London, 1983.
6. J. C. Palais, *Fiber Optic Communications*, Prentice-Hall, Englewood Cliffs, NJ, 1988.
7. L. Jeunhomme, *Single-Mode Fiber Optics*, 2nd ed., Marcel Dekker, New York, 1990.
8. D. Marcuse, *Theory of Dielectric Optical Waveguides*, 2nd ed., Academic Press, San Diego, CA, 1991.
9. A. Kaiser, *Optical Fibre Communications*, 2nd ed., McGraw-Hill, 1991.
10. J. M. Senior, *Optical Fibre Communications: Principles and Practice*, 2nd ed., Prentice-Hall, 1992.
11. J. Gower, *Optical Communication Systems*, 2nd ed., Prentice-Hall, London, 1993.
12. J. A. Buck, *Fundamentals of Optical Fibers*, Wiley, New York, 1995.
13. M. Abramowitz and I. A. Stegun (Eds.), *Handbook of Mathematical Functions*, Dover, New York, 1970.
14. E. Snitzer, *J. Opt. Soc. Am.* **51**, 491 (1961).
15. D. B. Keck, in M. K. Barnoski (Ed.), *Fundamentals of Optical Fiber Communications*, Academic Press, San Diego, CA, 1981.
16. D. Marcuse, *Bell Syst. Tech. J.* **56**, 703 (1977).
17. M. I. Hayee and A. E. Wilner, *IEEE Photon. Technol. Lett.* **11**, 991 (1999).
18. M. Nakazawa, H. Kubota, K. Suzuki, E. Yamada, and A. Sahara, *IEEE J. Sel. Top. Quantum Electron.* **6**, 363 (2000).
19. M. N. Islam, *IEEE J. Sel. Top. Quantum Electron.* **8**, 548 (2002).
20. C. D. Poole and J. Nagel, in *Optical Fiber Telecommunications*, Vol. IIIA, Academic Press, San Diego, CA, 1997.
21. I. P. Kaminow, *IEEE J. Quantum Electron.* **17**, 15 (1981).
22. H. Kogelnik and R. M. Jopson, and L. E. Nelson, *Optical Fiber Telecommunications*, Vol. IVB, Academic Press, San Diego, CA, 2002.

23. J. N. Damask, *Polarization Optics in Telecommunications*, Springer, New York, 2005.

24. A. Galtarrossa and C. R. Menyuk (Eds.), *Polarization Mode Dispersion*, Springer, New York, 2005.

25. F. Matera, M. Settembre, M. Tamburrini, F. Favre, D. Le Guen, T. Georges, M. Henry, G. Michaud, P. Franco, A. Shiffini, M. Romagnoli, M. Guglielmucci, and S. Casceli, *J. Lightwave Technol.* **17**, 2225 (1999).

26. E. Kolltveit, P. A. Andrekson, J. Brentel, B. E. Olsson, B. Bakhshi, J. Hansryd, P. O. Hedekvist, M. Karlsson, H. Sunnerud, and J. Li, *Electron. Lett.* **35**, 75 (1999).

27. H. Sunnerud, M. Karlsson, C. Xie, and P. A. Andrekson, *J. Lightwave Technol.* **20**, 2204 (2002).

28. A. Galtarrossa, P. Griggio, L. Palmieri, and A. Pizzinat, *J. Lightwave Technol.* **22**, 1127 (2004).

29. V. Chernyak, M. Chertkov, I. Gabitov, I. Kolokolov, and V. Lebedev, *J. Lightwave Technol.* **22**, 1155 (2004).

30. G. Ning, S. Aditya, P. Shum, Y. D. Gong, H. Dong, M. Tang, *Opt. Commun.* **260**, 560 (2006).

31. G. J. Foschini and C. D. Poole, *J. Lightwave Technol.* **9**, 1439 (1991).

32. P. K. A. Wai and C. R. Menyuk, *J. Lightwave Technol.* **14**, 148 (1996).

33. C. D. Poole and R. E. Wagner, *Electron. Lett.* **22**, 1029 (1986).

34. C. D. Poole and C. R. Giles, *Opt. Lett.* **13**, 155 (1988).

35. C. D. Poole, J. H. Winters, and J. A. Nagel, *Opt. Lett.* **16**, 372 (1991).

36. W. Eickoff, Y. Yen, and R. Ulrich, *Appl. Opt.* **20**, 3428 (1981).

THE NONLINEAR
SCHRÖDINGER EQUATION

In previous chapters it was assumed that the field intensities were relatively low, such that the discrete oscillators constituting the dielectric media would respond linearly to such fields. This is actually the case when using ordinary lamps or even light-emitting diodes. However, with the invention of the laser rather high field intensities became available and the limits for a linear response from the oscillator are easily achieved.

The nonlinearities of fused silica, from which optical fibers are made, are weak compared to those of many other materials. However, nonlinear effects can be readily observed in optical fibers due to two main reasons. On the one hand, the field cross section is rather small, which results in high field intensities, even if the total power carried by the fiber is relatively low. On the other hand, the optical fiber provides a long interaction length, which significantly enhances the efficiency of the nonlinear processes.

The first part of this chapter provides a brief introduction to nonlinear optics by presenting a basic model for the nonlinear susceptibility and showing how it is included to generalize the wave equation obtained in Section 2.1 to a nonlinear wave equation. Such an equation is used as the starting point to derive the central equation of this book: the celebrated nonlinear Schrödinger equation (NLSE). The last part of this chapter is dedicated to finding some soliton solutions of the NLSE using different approaches.

4.1 THE NONLINEAR POLARIZATION

The harmonic oscillator model was used in Section 2.6 to describe the wave propagation in a dielectric medium. The equation of motion (2.75) assumes that the

Nonlinear Effects in Optical Fibers. By Mário F. S. Ferreira.
Copyright © 2011 John Wiley & Sons, Inc. Published 2011 by John Wiley & Sons, Inc.

restoring force of this oscillator is a linear function of the electron displacement from the mean position. If the displacement is in the x-direction, the restoring force is given by

$$F_r = -kx \tag{4.1}$$

In physics, the linear dependence of one physical quantity on another is generally valid over only a limited range of values. In our case, the motion of the charged particles in a dielectric medium can be considered to be linear with the applied field only if this field is relatively weak and the displacement x is small. If these conditions are not satisfied, it becomes necessary to include additional anharmonic terms in the restoring force. The first correction term depends quadratically on x, with which the restoring force becomes

$$F_r = -kx + ax^2 \tag{4.2}$$

where the coefficient a describes the strength of the nonlinearity. The potential energy corresponding to this force is

$$U_p(x) = \frac{1}{2}kx^2 - \frac{1}{3}ax^3 \tag{4.3}$$

In materials with inversion symmetry, the carriers are isotropically distributed and the potential function must satisfy the symmetry condition $U_p(x) = U_p(-x)$. This is not the case for Eq. (4.3). In order to satisfy the above condition, the lowest order correction term in the restoring force must depend on x^3, with which it becomes

$$F_r = -kx + bx^3 \tag{4.4}$$

where the coefficient b describes the strength of the third-order nonlinearity.

   The above considerations can be applied to the propagation of light in a dielectric medium. In this case, the expansion of the restoring force in a power series given by Eq. (4.4) corresponds, on a macroscopic scale and in the frequency domain, to a power series expansion of the medium polarization **P** in the electric field vector **E** [1,2]:

$$\mathbf{P}(\mathbf{r}, \omega) = \varepsilon_0 [\chi^{(1)} \mathbf{E} + \chi^{(2)} \mathbf{EE} + \chi^{(3)} \mathbf{EEE} + \cdots] \tag{4.5}$$

where $\chi^{(j)}$ ($j = 1, 2, \ldots$) is the jth-order susceptibility. The first-order susceptibility $\chi^{(1)}$ determines the linear part of the polarization, $\mathbf{P}_L$, which is given by

$$\mathbf{P}_L(\mathbf{r}, \omega) = \varepsilon_0 \chi^{(1)} \mathbf{E} \tag{4.6}$$

On the other hand, terms of second and higher order in (4.5) determine the nonlinear polarization, $\mathbf{P}_{NL}$, given by

$$\mathbf{P}_{NL}(\mathbf{r}, \omega) = \varepsilon_0 \left[\chi^{(2)} \mathbf{EE} + \chi^{(3)} \mathbf{EEE} + \cdots \right] \tag{4.7}$$

In time domain, the form of the expansion of the polarization **P** is identical to (4.5) if the nonlinear response is assumed to be instantaneous. If not, then the successive terms in Eq. (4.5) become convolutions of one, two, and three dimensions.

To account for the light polarization effects, $\chi^{(j)}$ is a tensor of rank $j + 1$. For example, since the product **EEE** in general contains 27 terms, the susceptibility $\chi^{(3)}$ is a fourth-rank tensor that could contain 81 different terms. In practice, however, most materials of interest for technical applications exhibit symmetries that reduce drastically the number of independent tensor elements [3]. For an isotropic material, for example, the first-order susceptibility $\chi^{(1)}$ contains only one single element. This element leads to the refractive index and attenuation coefficient, as shown in Chapter 2.

A crystal having inversion symmetry is characterized by an inversion center, such that if the radial coordinate becomes symmetric, the crystal's atomic arrangement remains unchanged and so the crystal responds in the same way to a physical influence. In such a crystal, reversing the applied field direction should also reverse the direction of the second-order nonlinear polarization. However, because the polarization is proportional to the square of the field, such reversal would not be allowed mathematically. It therefore follows that in a material with inversion symmetry all 27 elements of the second-order susceptibility must be zero. Since SiO_2 is a symmetric molecule, the second-order susceptibility $\chi^{(2)}$ vanishes for silica glasses. As a consequence, virtually all nonlinear effects in optical fibers are determined by the third-order susceptibility.

In the case of third-order nonlinear polarization, the three fields can interact with each other. If the fields have different frequencies, they produce a polarization of the material at the sum and difference frequencies, or at multiples of the input frequencies. The third-order nonlinear polarization is given by

$$\mathbf{P}_{NL}^{(3)}(\omega = \omega_1 + \omega_2 + \omega_3) = \varepsilon_0 \chi^{(3)}(\omega_1, \omega_2, \omega_3) \mathbf{E}(\omega_1) \mathbf{E}(\omega_2) \mathbf{E}(\omega_3) \tag{4.8}$$

which can be written in the matrix form

$$
\begin{bmatrix} P_{NLx}^{(3)}(\omega) \\ P_{NLy}^{(3)}(\omega) \\ P_{NLz}^{(3)}(\omega) \end{bmatrix} = \varepsilon_0 \begin{bmatrix} \chi_{xxxx}^{(3)} & \chi_{xxxy}^{(3)} & \cdots & \chi_{xzzz}^{(3)} \\ \chi_{yxxx}^{(3)} & \chi_{yxxy}^{(3)} & \cdots & \chi_{yzzz}^{(3)} \\ \chi_{zxxx}^{(3)} & \chi_{zxxy}^{(3)} & \cdots & \chi_{zzzz}^{(3)} \end{bmatrix} \begin{bmatrix} E_x(\omega_1)E_x(\omega_2)E_x(\omega_3) \\ E_x(\omega_1)E_x(\omega_2)E_y(\omega_3) \\ \cdots \\ \cdots \\ E_z(\omega_1)E_z(\omega_2)E_y(\omega_3) \\ E_z(\omega_1)E_z(\omega_2)E_z(\omega_3) \end{bmatrix} \tag{4.9}
$$

A discussion of the form of $\chi^{(3)}$ in different types of media is provided in Ref. [4]. In particular, if the medium is isotropic, as happens with the silica glass, it is found that 60 of the 81 elements of $\chi^{(3)}$ are zero. Moreover, the remaining 21 elements are not independent of each other and the tensor $\chi^{(3)}$ can be reduced to only three elements. If, in addition, all frequencies are far from any resonance frequency, the Kleinman symmetry condition [5] holds and the third-order susceptibility possesses only one independent element.

4.1.1 The Nonlinear Wave Equation

By taking the curl of Eq. (2.3) and using Eqs. (2.4)–(2.6), the wave equation for the electric field can be obtained in the form

$$\nabla^2 \mathbf{E} - \frac{1}{c^2} \frac{\partial^2 \mathbf{E}}{\partial t^2} = \mu_0 \frac{\partial^2 \mathbf{P}}{\partial t^2} \tag{4.10}$$

where

$$c = \frac{1}{\sqrt{\varepsilon_0 \mu_0}} \tag{4.11}$$

is the velocity of light in vacuum. Equation (4.10) shows that the time-varying polarization serves as a driving term in the wave equation.

Considering Eqs. (4.5)–(4.7), the wave equation (4.10) can be written in the form

$$\nabla^2 \mathbf{E} - \frac{1}{c^2} \frac{\partial^2 \mathbf{E}}{\partial t^2} = \mu_0 \frac{\partial^2 \mathbf{P}_L}{\partial t^2} + \mu_0 \frac{\partial^2 \mathbf{P}_{NL}}{\partial t^2} \tag{4.12}$$

Using Eq. (4.6) and the relation $n = \sqrt{1 + \chi^{(1)}}$, Eq. (4.12) becomes

$$\nabla^2 \mathbf{E} - \frac{n^2}{c^2} \frac{\partial^2 \mathbf{E}}{\partial t^2} = \mu_0 \frac{\partial^2 \mathbf{P}_{NL}}{\partial t^2} \tag{4.13}$$

The electric field $\mathbf{E}$ in Eq. (4.13) represents the total field, that is, the sum of all incident fields and those generated by the nonlinear polarization.

4.2 THE NONLINEAR REFRACTIVE INDEX

Let us consider a plane optical wave propagating in a medium having inverse symmetry, with an electric field given by

$$E = \frac{1}{2} (\hat{E}(z) e^{-i\omega t} + \text{c.c.}) \tag{4.14}$$

The corresponding nonlinear third-order polarization is given by

$$P_{NL}^{(3)} = \varepsilon_0 \chi^{(3)} EEE = \frac{1}{8} \varepsilon_0 \chi^{(3)} (\hat{E} e^{-i\omega t} + \hat{E}^* e^{i\omega t})^3$$

$$= \frac{1}{8} \varepsilon_0 \chi^{(3)} ((\hat{E}^3 e^{-i3\omega t} + \text{c.c.}) + 3|\hat{E}|^2 (\hat{E} e^{-i\omega t} + \text{c.c.})) \tag{4.15}$$

where c.c. means the complex conjugate. If the necessary phase matching conditions for an efficient generation of the third harmonic generation are not provided, the first term on the right-hand side of Eq. (4.15) can be neglected and we have simply

$$P_{NL}^{(3)} \approx \frac{3}{4}\varepsilon_0 \chi^{(3)} |\hat{E}|^2 E \tag{4.16}$$

Taking into account the result given by Eq. (4.16), the polarization of the material becomes

$$P = \varepsilon_0 \left(\chi^{(1)} + \frac{3}{4}\chi^{(3)} |\hat{E}|^2 \right) E \tag{4.17}$$

Since the intensity of the plane wave is given by

$$I = \frac{1}{2} c\varepsilon_0 n_0 |\hat{E}|^2 \tag{4.18}$$

where n_0 is the linear refractive index, the polarization can be written as

$$P = \varepsilon_0 \left(\chi^{(1)} + \frac{3}{2} \frac{\chi^{(3)}}{c\varepsilon_0 n_0} I \right) E \tag{4.19}$$

From Eqs. (2.78) and (2.81) and neglecting the absorption by the medium, we have the following general relation between the polarization and refractive index:

$$P = \varepsilon_0 (n^2 - 1)E \tag{4.20}$$

Comparing Eqs. (4.19) and (4.20), the following result for the refractive index is obtained:

$$n = \sqrt{1 + \chi^{(1)} + \frac{3}{2} \frac{\chi^{(3)}}{c\varepsilon_0 n_0} I} \approx n_0 + n_2 I \tag{4.21}$$

where

$$n_0 = \sqrt{1 + \chi^{(1)}} \tag{4.22}$$

is the linear refractive index and

$$n_2 = \frac{3}{4} \frac{\chi^{(3)}}{c\varepsilon_0 n_0^2} \tag{4.23}$$

is the refractive index nonlinear coefficient, also known as the *Kerr coefficient*. The dependence of the refractive index on the field intensity is known as the *nonlinear Kerr effect*.

Since the linear refractive index n_0 and the third-order susceptibility $\chi^{(3)}$ depend on the frequency, the Kerr coefficient given by Eq. (4.23) is frequency dependent as well. In the case of fused silica, a value of $n_2 = 2.73 \times 10^{-20}$ m^2/W was measured at a wavelength of 1.06 μm [6]. Several measurements indicate that the nonlinear index coefficient decreases with increasing wavelengths [7]. Considering the telecommunications window, a value $n_2 = 2.35 \times 10^{-20}$ m^2/W was obtained for silica fibers [8]. Such value can be increased by doping the fiber core with germania [9].

Assuming a single-mode fiber with an effective mode area $A_{\text{eff}} = 50$ μm^2 carrying a power $P = 100$ mW, the nonlinear part of the refractive index is

$$n_2 I = n_2 \frac{P}{A_{\text{eff}}} \approx 4.7 \times 10^{-11} \qquad (4.24)$$

In spite of this very small value, the effects of the nonlinear component of the refractive index become significant due to very long interaction lengths provided by the optical fibers.

4.3 IMPORTANCE OF NONLINEAR EFFECTS IN FIBERS

The relevance of a nonlinear process depends not only on the intensity, I, of the propagating wave but also on the effective interaction length, L_{eff}, along which that intensity remains sufficiently high. In the case of a Gaussian beam propagating in bulk media, the effective interaction length is [10]

$$L_{\text{eff,bulk}} \sim Z_R = \frac{\pi w_0^2}{\lambda} \qquad (4.25)$$

where Z_R is the so-called *Rayleigh length* and w_0 is the spot size of the beam at the waist. The intensity of the focused beam can be increased by reducing the spot size w_0. However, a smaller spot size determines a faster divergence of the beam and a smaller Rayleigh length, as seen from Eq. (4.25). The product of the intensity and the effective length gives

$$(I L_{\text{eff}})_{\text{bulk}} = \frac{P}{\pi w_0^2} Z_R = \frac{P}{\lambda} \qquad (4.26)$$

In the case of an optical fiber, the field is guided and the power decreases only due to the attenuation, in the form

$$P(z) = P(0) e^{-\alpha z} \qquad (4.27)$$

The product of the intensity and effective length becomes

$$(IL_{\text{eff}})_{\text{fiber}} = \int_0^L \frac{P(0)}{A_{\text{eff}}} e^{-\alpha z} \, dz = \frac{P(0)}{A_{\text{eff}}} \frac{1 - e^{-\alpha L}}{\alpha} = \frac{P(0)}{A_{\text{eff}}} L_{\text{eff}} \tag{4.28}$$

where

$$L_{\text{eff}} = \frac{1 - e^{-\alpha L}}{\alpha} \tag{4.29}$$

is the effective length of interaction. If the fiber real length, L, is sufficiently small, such that $\alpha L \ll 1$, we have $L_{\text{eff}} \approx L$. However, the difference between the real and the effective length increases with L. When $\alpha L \gg 1$, the effective length approaches a limiting value, given by

$$L_{\text{eff}}^{\max} = \frac{1}{\alpha} \tag{4.30}$$

In the case of a fiber with an attenuation of 0.2 dB/km, the maximum effective length is $L_{\text{eff}}^{\max} = 21.72$ km. Figure 4.1 shows the effective length of interaction against

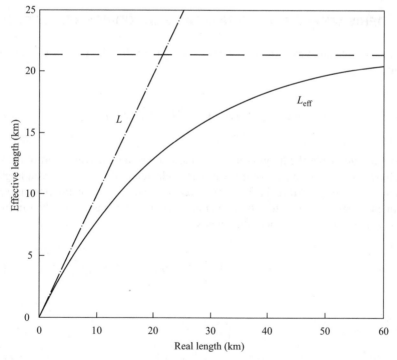

Figure 4.1 Effective length of interaction against the real fiber length for an attenuation of 0.2 dB/km.

the real fiber length for an attenuation of 0.2 dB/km. In this case and considering the results given by Eqs. (4.26) and (4.28), it can be verified that the efficiency of a nonlinear process in a typical single-mode fiber can be enhanced by a factor $\sim 10^9$ relatively to a bulk medium.

In actual transmission systems, the fiber attenuation is compensated by periodically located optical amplifiers. Assuming that the spacing between consecutive amplifiers is L_A and that the power after each amplifier is equal to that at the fiber input, $P(0)$, the effective length of the transmission system becomes

$$L_{\text{eff}} = N \frac{1 - e^{-\alpha L_A}}{\alpha} \tag{4.31}$$

where $N = L/L_A$ is the number of sections constituting the transmission link. Then, the limiting value of the effective length is N times greater than that in a nonamplified system:

$$L_{\text{eff}}^{\max} = \frac{N}{\alpha} \tag{4.32}$$

4.4   DERIVATION OF THE NONLINEAR SCHRÖDINGER EQUATION

The field propagating in a single-mode fiber along the z-direction can be written in the form

$$E = \frac{1}{2}(F(r, \phi)A(z)e^{i(\beta_0 z - \omega_0 t)} + \text{c.c.}) \tag{4.33}$$

where $A(z)$ describes the longitudinal part and $F(r, \phi)$ the transverse part of the field amplitude. The vector nature of the waves is neglected in this section, assuming that they are linearly polarized in the same direction. The power of the fiber mode is obtained integrating the intensity in the transverse plane. Considering the symmetry of the single-mode waveguide, this yields

$$P = \frac{1}{2}\varepsilon_0 c n |A(z)|^2 2\pi \int |F(r)|^2 r \, dr = B^2 |A(z)|^2 \tag{4.34}$$

where

$$B^2 = \pi \varepsilon_0 c n \int |F(r)|^2 r \, dr \tag{4.35}$$

Substituting Eq. (4.33) in the nonlinear wave equation (4.13), the first term on the left-hand side gives

$$\nabla^2 E = \nabla_T^2 E + \frac{\partial^2 E}{\partial z^2} = \nabla_T^2 E + \frac{1}{2}\left[\frac{\partial^2 A}{\partial z^2} + 2i\beta_0 \frac{\partial A}{\partial z} - \beta_0^2 A\right] F(r)e^{i(\beta_0 z - \omega_0 t)}$$

$$\approx \nabla_T^2 E + \frac{1}{2}\left[2i\beta_0 \frac{\partial A}{\partial z} - \beta_0^2 A\right] F(r)e^{i(\beta_0 z - \omega_0 t)} \tag{4.36}$$

In obtaining the last result, it was assumed that the second derivative of $A(z)$ is much smaller than the first one, which corresponds to the so-called *slowly varying envelope approximation* (SVEA). Moreover, assuming that the nonlinearity has no influence on the transversal component of the field, the first term in Eq. (4.36) will be also neglected in the following.

Concerning the second term on the left-hand side of Eq. (4.13), it gives

$$\frac{n^2}{c^2}\frac{\partial^2 E}{\partial t^2} = -\frac{1}{2}\beta^2 F(r)A(z)e^{i(\beta_0 z - \omega_0 t)} \tag{4.37}$$

where $\beta = n\omega/c$.

Since the fused silica used in optical fibers is a material with inversion symmetry, the first nonlinear polarization component to be considered is that of third order. Assuming that the phase matching condition for an efficient generation of the third harmonic is not provided, the third-order nonlinear polarization is given by Eq. (4.16). In such a case, the right-hand side of Eq. (4.13) gives

$$\mu_0 \frac{\partial^2 P_{NL}}{\partial t^2} = -\frac{3\omega^2}{8c^2}\chi^{(3)}|F(r)A(z)|^2 F(r)A(z)e^{i(\beta_0 z - \omega_0 t)} \tag{4.38}$$

Substituting Eqs. (4.36) (without the first term), (4.37), and (4.38) in Eq. (4.13) gives

$$\frac{1}{2}\left[2i\beta_0 \frac{\partial A}{\partial z} - \beta_0^2 A + \beta^2 A\right] = -\frac{3\omega^2}{8c^2}\chi^{(3)}|F(r)A|^2 A \tag{4.39}$$

or

$$\frac{\partial A}{\partial z} = \frac{i}{2\beta}\left[\frac{3\omega^2}{4c^2}\chi^{(3)}|F(r)|^2|A|^2 A + A(\beta^2 - \beta_0^2)\right]$$

$$\approx i\left[\frac{3\omega}{8cn}\chi^{(3)}|F(r)|^2|A|^2 A + A(\beta - \beta_0)\right] \tag{4.40}$$

where the approximations $\beta \approx \beta_0$ and $(\beta^2 - \beta_0^2) \approx 2\beta_0(\beta - \beta_0)$ were used.

Introducing a normalized amplitude $U = BA(z)$, where B is given by Eq. (4.35), and multiplying both sides of Eq. (4.40) by $\int |F(r)|^2 r \, dr$, we obtain

$$\frac{\partial U}{\partial z} = i\left[U(\beta - \beta_0) + \gamma |U|^2 U\right] \tag{4.41}$$

where γ is the nonlinearity coefficient, given by

$$\gamma = \frac{\omega n_2}{c A_{\text{eff}}} \tag{4.42}$$

In Eq. (4.42), n_2 is the Kerr coefficient, given by Eq. (4.23), whereas A_{eff} is the effective core area, given by

$$A_{\text{eff}} = \frac{2\pi \left(\int_0^\infty |F(r)|^2 r \, dr\right)^2}{\int_0^\infty |F(r)|^4 r \, dr} \tag{4.43}$$

Equation (4.41) can be written in the frequency domain as

$$\frac{\partial \widetilde{U}}{\partial z} = i\left[\widetilde{U}(\beta - \beta_0) + \gamma |\widetilde{U}|^2 \widetilde{U}\right] \tag{4.44}$$

where $\widetilde{U}$ is the Fourier transform of U. Using the expansion given by Eq. (3.72) for the propagation constant β and retaining only terms up to second order in $(\omega - \omega_0)$, Eq. (4.44) becomes

$$\frac{\partial \widetilde{U}}{\partial z} = i\widetilde{U}\left[(\omega - \omega_0)\beta_1 + \frac{1}{2}(\omega - \omega_0)^2 \beta_2\right] + i\gamma |\widetilde{U}|^2 \widetilde{U} \tag{4.45}$$

Equation (4.45) can be transformed into the time domain using the inverse Fourier transform and the relation

$$\Delta\omega \leftrightarrow i\frac{\partial}{\partial t} \tag{4.46}$$

Hence, we obtain

$$i\left(\frac{\partial}{\partial z} + \frac{1}{v_g}\frac{\partial}{\partial t}\right)U - \frac{1}{2}\beta_2 \frac{\partial^2}{\partial t^2}U + \gamma |U|^2 U = 0 \tag{4.47}$$

where $v_g = 1/\beta_1$ is the group velocity. Considering a moving frame propagating with the group velocity and using the new time variable

$$\tau = t - \frac{z}{v_g} \tag{4.48}$$

Eq. (4.47) can be written in the form

$$i\frac{\partial U}{\partial z} - \frac{1}{2}\beta_2\frac{\partial^2 U}{\partial \tau^2} + \gamma|U|^2 U = 0 \qquad (4.49)$$

Equation (4.49) is usually called the *nonlinear Schrödinger equation* due to its similarity with the Schrödinger equation of quantum mechanics. The NLSE describes the propagation of pulses in optical fibers taking into account both the group velocity dispersion and the fiber nonlinearity. In order to gain a better physical insight concerning both these effects, it will be useful to consider some particular cases of Eq. (4.49).

4.4.1 Propagation in the Absence of Dispersion and Nonlinearity

If there is no dispersion or fiber nonlinearity, Eq. (4.49) reduces to

$$\frac{\partial U(z,\tau)}{\partial z} = 0 \qquad (4.50)$$

or

$$\frac{\partial U}{\partial z} + \frac{1}{v_g}\frac{\partial U}{\partial t} = 0 \qquad (4.51)$$

Equation (4.50) has the solution

$$U = U_0(\tau) = U_0(t - z/v_g) \qquad (4.52)$$

It is clear from this result that, in the absence of dispersion and nonlinearity, the pulse propagates without any distortion in shape with the group velocity v_g. The pulse energy also propagates with the group velocity since from Eq. (4.51) we get that

$$\frac{\partial |U|^2}{\partial z} + \frac{1}{v_g}\frac{\partial |U|^2}{\partial t} = 0 \qquad (4.53)$$

4.4.2 Effect of Dispersion Only

Neglecting only the nonlinear term in Eq. (4.49), we have

$$i\frac{\partial U}{\partial z} - \frac{1}{2}\beta_2\frac{\partial^2 U}{\partial \tau^2} = 0 \qquad (4.54)$$

Using the method of separation of variables, one can obtain the following general solution:

$$U(z,\tau) = \int_0^\infty U(0,\omega)e^{i((1/2)\beta_2\omega^2 z - \omega\tau)}\, d\omega \qquad (4.55)$$

where $U(0, \omega)$ represents the frequency spectrum of the input pulse. The above equation is similar to Eq. (3.82), from which we have shown that the pulse becomes broadened and chirped due to dispersion.

4.4.3 Effect of Nonlinearity Only

Neglecting only the dispersion term in Eq. (4.49), we obtain

$$i\frac{\partial U}{\partial z} + \gamma |U|^2 U = 0 \tag{4.56}$$

Multiplying the above equation by U^* and its complex conjugate by U and subtracting the two equations, we obtain

$$\frac{\partial |U|^2}{\partial z} = 0 \tag{4.57}$$

which has the general solution

$$|U|^2 = f(\tau) = f(t - z/v_g) \tag{4.58}$$

The above result shows that, in the presence of nonlinearity and neglecting the dispersion, the absolute square of the wave envelope maintains its shape during propagation. Based on this fact, the solution of Eq. (4.57) can be written in the form

$$U(z, \tau) = U_0(\tau)e^{i\phi(z,\tau)} \tag{4.59}$$

where $U_0(\tau)$ and $\phi(z, \tau)$ are real functions. From Eqs. (4.56) and (4.59) we obtain

$$\phi(z, \tau) = \gamma |U_0(\tau)|^2 z \tag{4.60}$$

which indicates that the fiber nonlinearity leads to a phase modulation that is directly proportional to the pulse intensity and the distance of propagation. This phenomenon is called *self-phase modulation* (SPM) and will be discussed in Chapter 5.

4.4.4 Normalized Form of NLSE

Let us introduce a normalized amplitude Q given by

$$U(z, \tau) = \sqrt{P_0}Q(z, \tau) \tag{4.61}$$

where P_0 is the peak power of the incident pulse. Using Eqs. (4.49) and (4.61), $Q(z, \tau)$ is found to satisfy

$$i\frac{\partial Q}{\partial z} - \frac{1}{2}\beta_2 \frac{\partial^2 Q}{\partial \tau^2} + \gamma P_0 |Q|^2 Q = 0 \tag{4.62}$$

Let us define a *nonlinear length*, L_{NL}, as

$$L_{NL} = \frac{1}{\gamma P_0} \tag{4.63}$$

L_{NL} is the propagation distance required to produce a nonlinear phase change rotation of one radian at a power P_0. In Chapter 3, a *dispersion length*, L_D, was also defined as

$$L_D = \frac{t_0^2}{|\beta_2|} \tag{4.64}$$

These two characteristic distances provide the length scales over which nonlinear or dispersive effects become important for pulse evolution.

Using a distance Z normalized by the dispersion distance L_D and a timescale T normalized to the input pulse width t_0, Eq. (4.62) becomes

$$i\frac{\partial Q}{\partial Z} \pm \frac{1}{2}\frac{\partial^2 Q}{\partial T^2} + N^2|Q|^2 Q = 0 \tag{4.65}$$

where

$$N^2 = \frac{L_D}{L_{NL}} = \frac{\gamma P_0 t_0^2}{|\beta_2|} \tag{4.66}$$

In the second term of Eq. (4.65), the plus signal corresponds to the case of anomalous GVD $(\text{sgn}(\beta_2) = -1)$, whereas the minus signal corresponds to normal GVD $(\text{sgn}(\beta_2) = +1)$. The parameter N can be eliminated from Eq. (4.65) by introducing a new normalized amplitude $q = NQ$, with which Eq. (4.65) takes the standard form of the NLSE:

$$i\frac{\partial q}{\partial Z} + \frac{1}{2}\frac{\partial^2 q}{\partial T^2} + |q|^2 q = 0 \tag{4.67}$$

The case of anomalous GVD $(\beta_2 < 0)$ was assumed in writing Eq. (4.67).

4.5 SOLITON SOLUTIONS

The NLSE has pulse-like solutions called bright solitons, which have the potential of many technological applications. In the following, we find a simple analytical solution, corresponding to the so-called fundamental soliton. Other soliton solutions can be found by solving exactly the NLSE using the inverse scattering method [11–13].

4.5.1 The Fundamental Soliton

Let us consider a pulse solution of Eq. (4.62) in the form

$$Q(z, \tau) = S(\tau)e^{i\phi(z)} \tag{4.68}$$

where the envelope function $S(\tau)$ is assumed to be a real function of τ. Substituting Eq. (4.68) in Eq. (4.62) and rearranging, we obtain

$$\frac{d\phi}{dz} = \gamma P_0 S^2(\tau) - \frac{1}{2}\beta_2 \frac{d^2 S/d\tau^2}{S(\tau)} \tag{4.69}$$

Since the left-hand side of Eq. (4.69) depends only on z and the right-hand side depends only on τ, we can set each side equal to a constant, C_1. The equation resulting from the left-hand side has the solution

$$\phi(z) = C_1 z + \phi_0 \tag{4.70}$$

where ϕ_0 is a constant of integration, which we will assume to be equal to zero in the following. Concerning the right-hand side of Eq. (4.69), it gives the equation

$$\frac{d^2 S}{d\tau^2} - \frac{2\gamma}{\beta_2} P_0 S^3(\tau) + \frac{2C_1}{\beta_2} S(\tau) = 0 \tag{4.71}$$

We can integrate Eq. (4.71) to obtain

$$\left(\frac{dS}{d\tau}\right)^2 = \frac{\gamma}{\beta_2} P_0 S^4(\tau) - \frac{2C_1}{\beta_2} S^2(\tau) + K \tag{4.72}$$

where K is a constant of integration. However, this constant is zero, as for a localized solution we must have

$$\lim_{\tau \to \pm\infty} S(\tau) = \lim_{\tau \to \pm\infty} \frac{dS}{d\tau} = 0 \tag{4.73}$$

Moreover, considering that the envelope function presents a maximum such that $S(\tau) = 1$ and $dS/d\tau = 0$, we obtain

$$C_1 = \frac{\gamma P_0}{2} \tag{4.74}$$

Using the above result, we can rewrite Eq. (4.72) in the form

$$\frac{dS}{d\tau} = \alpha S \sqrt{1 - S^2} \tag{4.75}$$

where

$$\alpha = \sqrt{\gamma P_0 / |\beta_2|} \tag{4.76}$$

Equation (4.75) can be easily integrated, giving the result

$$S(\tau) = \text{sech}(\alpha\tau) \tag{4.77}$$

Thus, the soliton solution of Eq. (4.62) is given by

$$Q(z, \tau) = \text{sech}(\alpha\tau)\exp\left\{i\frac{\gamma P_0}{2}z\right\} \tag{4.78}$$

Considering the normalizations used to derive Eq. (4.67), the above result can be written in the form

$$q(Z, T) = \eta\,\text{sech}(\eta T)\exp\left\{i\frac{1}{2}\eta^2 Z\right\} \tag{4.79}$$

where

$$\eta = \left(\frac{\gamma P_0 t_0^2}{|\beta_2|}\right)^{1/2} \tag{4.80}$$

is simultaneously the soliton amplitude and the inverse of pulse width.

The canonical form of the fundamental soliton can be obtained from Eq. (4.79) considering $\eta = 1$ and is given by

$$q(Z, T) = \text{sech}(T)\exp(iZ/2) \tag{4.81}$$

4.5.2 Solutions of the Inverse Scattering Theory

The nonlinear Schrödinger equation given by Eq. (4.67) is a completely integrable equation that can be solved exactly using the inverse scattering method [11,12], as shown by Zakharov and Shabat [13]. The main goal of this method is to find an appropriate scattering problem whose potential is $q(Z, T)$. The initial scattering data are determined by the initial field $q(0, T)$. The field $q(Z, T)$ after propagation along a distance Z is obtained from the evolved scattering data by solving a linear integral equation.

For the NLSE given by Eq. (4.67), the suitable scattering problem becomes [14,15]

$$i\frac{\partial\psi_1}{\partial T} + q(Z, T)\psi_2 = \varsigma\psi_1 \tag{4.82}$$

$$i\frac{\partial\psi_2}{\partial T} + q^*(Z, T)\psi_1 = -\varsigma\psi_2 \tag{4.83}$$

where ψ_1 and ψ_2 are the amplitudes of the two waves scattered by the potential $q(Z, T)$, and ς is the Z-independent eigenvalue. In general, we can have N eigenvalues, which may be written as

$$\varsigma_n = \frac{1}{2}(\kappa_n + i\eta_n), \quad \text{for } n = 1, 2, \ldots, N \tag{4.84}$$

In the case $N = 1$, the fundamental soliton solution of Eq. (4.67) is obtained

$$q(T, Z) = \eta \text{sech}[\eta(T + \kappa Z - T_0)]\exp\left(-i\kappa T + \frac{i}{2}(\eta^2 - \kappa^2)Z + i\sigma\right) \tag{4.85}$$

which is characterized by four parameters: the amplitude η (also the pulse width), the frequency κ (also the pulse speed), the time position T_0 and the phase σ. We note that the amplitude and speed of the soliton are given by the imaginary and real parts, respectively, of the eigenvalue (4.84). The canonical form of the fundamental soliton, given by Eq. (4.81), can be obtained from Eq. (4.85) considering $\eta = 1$, $T_0 = 0$, $\kappa = 0$, and $\sigma = 0$.

In general, if the κ_n in Eq. (4.84) are all distinct, the N-soliton solutions that arise from the initial wave form are asymptotically given in the form of N separated solitons:

$$q(Z, T) = \sum_{j=1}^{N} \eta_j \text{sech}\left[\eta_j(T + \kappa_j Z - T_{0j})\right]\exp\left[-i\kappa_j T + i\frac{1}{2}(\eta_j^2 - \kappa_j^2)Z - i\sigma_j\right] \tag{4.86}$$

If the input pulse shape is symmetrical, it can be shown that the eigenvalues of Eqs. (4.82) and (4.83) are purely imaginary. For example, considering an input pulse shape given by

$$q(0, T) = N \text{sech}(T) \tag{4.87}$$

the corresponding eigenvalues are given by

$$\varsigma_n = \frac{i}{2}\eta_n = \left(N - n + \frac{1}{2}\right), \quad n = 1, 2, \ldots, N \tag{4.88}$$

Since $\kappa_n = 0$, all the output solitons propagate at exactly the same speed. This type of solution is called the *bound soliton solution* and its shape evolves periodically during the propagation due to the phase interference among the constituting solitons. This is illustrated in Fig. 4.2. Compared to the fundamental soliton, the shape of higher order solitons will change during propagation, returning to their initial shape after a certain distance.

For the second-order soliton ($N = 2$), the eigenvalues are $\eta_1 = 1/2$ and $\eta_2 = 3/2$ and the field distribution is given by [13]

$$q(Z, T) = \frac{4[\cosh(3T) + 3\exp(4iZ)\cosh(T)]\exp(iZ/2)}{[\cosh(4T) + 4\cosh(2T) + 3\cos(4Z)]} \tag{4.89}$$

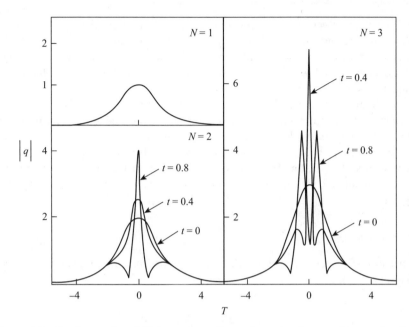

Figure 4.2 Fundamental and higher order solitons that arise from an input pulse shape $q(0, T) = N \text{sech}(T)$. (After Ref. [14]; reprinted with permission)

The period of oscillation of the higher order solitons resulting from the initial pulse given by Eq. (4.87) is

$$Z_0 = \frac{\pi}{2} \tag{4.90}$$

which can be written in terms of real units as

$$z_0 = \frac{\pi}{2} L_D = \frac{\pi}{2} \frac{t_0^2}{|\beta_2|} \tag{4.91}$$

The length z_0 is usually called the *soliton period*. Considering a pulse with a temporal width $t_0 = 10 \, \text{ps}$ propagating in a dispersion-shifted fiber with $\beta_2 = -1 \, \text{ps}^2/\text{km}$ in the C-band ($\lambda \approx 1550 \, \text{nm}$), we have a propagation period $z_0 \approx 150 \, \text{km}$.

According to Eq. (4.66), the peak power of a soliton in an optical fiber is

$$P_0 = \frac{N^2 |\beta_2|}{\gamma t_0^2} \tag{4.92}$$

Let us consider pulses with a temporal width $t_0 = 6 \, \text{ps}$. Using typical parameter values $\beta_2 = -1 \, \text{ps}^2/\text{km}$ and $\gamma = 3 \, \text{W}^{-1} \, \text{km}^{-1}$ for dispersion-shifted fibers, we obtain $P_0 \sim 10 \, \text{mW}$ for fundamental solitons ($N = 1$) in the C-band ($\lambda \approx 1550 \, \text{nm}$). The required power to launch the Nth-order soliton is N^2 times that for the fundamental one. Moreover, higher order solitons compress periodically (see Fig. 4.2), resulting in soliton chirping and spectral broadening. In contrast, fundamental solitons preserve their shape during propagation. This fact, together with the lower power required for their generation, makes fundamental solitons the preferred option in soliton communication systems.

The existence of fiber solitons was suggested for the first time in 1973 by Hasegawa and Tappert [16], whereas their first experimental observation was reported only in 1980 by Mollenauer et al. [17]. In this experiment, a mode-locked color-center laser was used to obtain short optical pulses ($t_0 \cong 4\,\text{ps}$) near 1550 nm, where the fiber loss is minimum. Due to its robust nature, the soliton pulse appeared soon as an ideal solution to the problem of pulse spreading caused by fiber dispersion.

4.5.3 Dark Solitons

In the normal dispersion regime ($\beta_2 > 0$), the normalized nonlinear Schrödinger equation becomes

$$i\frac{\partial q}{\partial Z} - \frac{1}{2}\frac{\partial^2 q}{\partial T^2} + |q|^2 q = 0 \tag{4.93}$$

Equation (4.93) differs from Eq. (4.67) only in the sign of its dispersion term. The soliton solution of Eq. (4.93) can be written in the form [18]

$$q(Z,T) = \sqrt{\rho}\,e^{i\sigma} \tag{4.94}$$

where

$$\rho = \rho_0[1 - a^2 \,\text{sech}^2(\sqrt{\rho_0}\,aT)], \quad a^2 \leq 1 \tag{4.95}$$

$$\sigma = [\rho_0(1-a^2)]^{1/2}T + \tan^{-1}\left[\frac{1}{\sqrt{1-a^2}}\tanh(\sqrt{\rho_0}\,aT)\right] - \frac{\rho_0(3-a^2)}{2}Z \tag{4.96}$$

The parameter a is known as depth of modulation. In fact, this kind of solitons appears in the section where light waves are absent, and thus they are called *dark solitons*. When $a = 1$, the solution becomes

$$q = \sqrt{\rho_0}\,\tanh(\sqrt{\rho_0}\,T) \tag{4.97}$$

The two types of dark solitons are illustrated in Fig. 4.3.

Dark solitons were first observed by Emplit et al. [19] and Kröbel et al. [20] independently by transmitting a light wave in the normal dispersion region of a fiber. Some numerical results have shown that dark solitons are more stable and are less affected than bright solitons in the presence of several perturbations, demonstrating their potential application for optical communication systems [21].

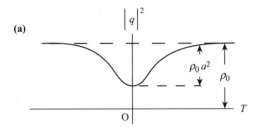

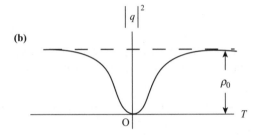

Figure 4.3   Schematic representation of dark solitons with $a < 1$ and $a = 1$.

4.6 NUMERICAL SOLUTION OF THE NLSE

It is generally difficult to obtain an analytical solution of the nonlinear Schrödinger equation in practical conditions. In such cases, the NLSE must be solved numerically, which is efficiently performed using the "split-step Fourier" analysis. In such an approach, the waveguide is partitioned into n equally long segments, whose lengths h are adjusted according to the desired accuracy (Fig. 4.4). For each segment, the nonlinear and dispersive effects are considered independently [22]. Since the dispersive effects are most naturally dealt in the frequency domain, whereas the nonlinear effects are easier to handle in the time domain, the propagation in each segment is treated in two consecutive steps. First, the pulse $q(Z, T)$ is transformed using the fast Fourier transform (FFT) into the frequency domain to $\tilde{q}(Z, \omega)$. Then, to take into account the dispersive effects, an intermediate state is calculated:

$$\tilde{q}_{\text{int}}(Z + h, \omega) = \tilde{q}(Z, \omega)e^{-i(1/2)\omega^2 h} \tag{4.98}$$

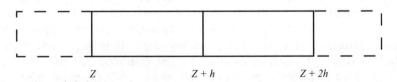

Figure 4.4   Illustration of the split-step Fourier method.

(a)   (b)

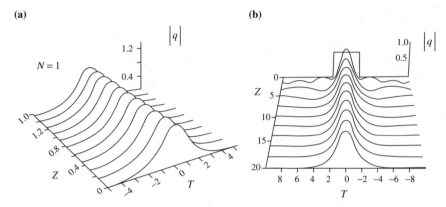

Figure 4.5   (a) Numerical solution of the NLSE for an initial condition $q(0,t) = \mathrm{sech}(T)$. (b) Numerical solution of the NLSE for an initial rectangular pulse of height 0.7626 and width 4.

In the second step, the inverse FFT is used to transform $\widetilde{q}_{int}(Z+h,\omega)$ into an intermediate state in the time domain, $q_{int}(Z+h,T)$. Then, the nonlinear effects of the segment h are taken into account by computing

$$q(Z+h,T) = q_{int}(Z+h,T)e^{i|q|^2 h} \qquad (4.99)$$

As long as the length of the segments is short enough, this method can provide very accurate results. Such accuracy can be improved using the symmetrized split-step Fourier method [23], in which the effect of nonlinearity is included in the middle of the segment rather than at the segment boundary.

Figure 4.5a shows the numerical solution of the NLSE obtained using the method described above for an initial condition $q(0,T) = \mathrm{sech}(T)$, which corresponds to the fundamental soliton. This figure illustrates the most important property of the fundamental soliton: as long as the fiber loss is negligible, it propagates undistorted without any change in shape for arbitrarily long distances. It is this feature of fundamental solitons that makes them attractive for optical communication systems. Figure 4.5b illustrates the evolution of an initially rectangular pulse toward the fundamental soliton, demonstrating the truly stable nature of this solution. In the first stage of propagation, some energy of the initial pulse is radiated until the pulses achieve the fundamental soliton profile.

Figure 4.6 shows the evolution of the second-order ($N=2$) and third-order ($N=3$) solitons, obtained numerically from Eq. (4.67). This figure illustrates the periodic evolution of higher order solitons. Such periodic evolution can be understood as follows. In the initial part of its evolution, Fig. 4.6 shows that the pulse contracts in the time domain and increases its peak power. As will be seen in Chapter 5, SPM generates a frequency chirp such that the leading edge of the pulse is redshifted, while its trailing edge is blueshifted from the central frequency, resulting in a broad spectrum in the frequency domain. In the absence of GVD, the pulse shape would have remained unchanged. However, in the presence of anomalous GVD the

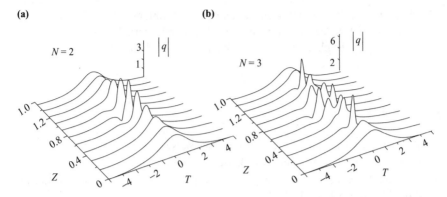

Figure 4.6 Numerical solution of the NLSE for an initial condition $q(0, t) = N \sech(T)$ for the cases $N = 2$ and $N = 3$.

redshifted components in the leading edge of the pulse move slower, whereas the blueshifted components in the trailing edge are faster. As a consequence, the pulse spectrum is compressed and the pulse reaches again its original width in the time domain. After that everything starts again. This mutual interaction between the SPM and GVD effects explains the periodic evolution for higher order solitons.

PROBLEMS

4.1 Determine the effective length of a 10 km fiber with an attenuation constant $\alpha = 0.3$ dB/km. Calculate the limiting value of the effective length for this type of fiber.

4.2 Show that Eq. (4.47) is converted into Eq. (4.49) using the transformation given by Eq. (4.48).

4.3 Verify by direct substitution that $q(Z, T)$ given by Eq. (4.79) or Eq. (4.81) is indeed a solution of the nonlinear Schrödinger equation (4.67).

4.4 Determine the required peak power of a 10 ps (FWHM) pulse of center wavelength 1550 nm to propagate as a fundamental soliton in a fiber having dispersion of $D = 2$ ps/(nm km) and a nonlinear parameter $\gamma = 5$ W^{-1} km^{-1}. What peak power is required to propagate a $N = 3$ soliton in such a fiber?

4.5 Verify by direct substitution that the dark soliton described by Eq. (4.97) is indeed a solution of the normalized NLSE given by Eq. (4.93).

4.6 Develop a computer program to solve Eq. (4.67) using the split-step Fourier method presented in Section 4.6. Check the accuracy of the numerical result by comparing it with the analytical solutions for the fundamental soliton, given by Eq. (4.81), and for the second-order soliton, given by Eq. (4.89).

4.7 Use the numerical program developed in Problem 4.6 to investigate the evolution of an input pulse given by Eq. (4.87) with $N = 0.6$, 1.3, and 1.8, respectively. Explain the observed behavior in each case.

REFERENCES

1. Y. R. Shen, *Principles of Nonlinear Optics*, Wiley, New York, 1984.

2. R. W. Boyd, *Nonlinear Optics*, 2nd ed., Academic Press, Boston, MA, 2003.

3. J. F. Nye, *Physical Properties of Crystals: Their Representation by Tensors and Matrices*, Oxford University Press, Oxford, 1985.

4. R. W. Hellwarth, in J. H. Sanders and S. Stenholm (Eds.), *Progress in Quantum Electronics*, Vol. **5**, Pergamon Press, Oxford, 1977.

5. D. A. Kleinman, *Phys. Rev.* **126**, 1977 (1962).

6. D. Milam and M. J. Weber, *J. Appl. Phys.* **47**, 2497 (1976).

7. D. Milam, *Appl. Opt.* **37**, 546 (1998).

8. N. G. R. Broderick, T. M. Monroe, P. J. Bennett, and D. J. Richardson, *Opt. Lett.* **24**, 1395 (1999).

9. K. Nakajima and M. Ohashi, *IEEE Photon. Technol. Lett.* **14**, 492 (2002).

10. A. Yariv, *Introduction to Optical Electronics*, Holt, Rinehart and Winston, New York, 1977.

11. C. S. Gardner, J. M. Greene, M. D. Kruskal, and R. M. Miura, *Phys. Rev. Lett.* **19**, 1095 (1967).

12. M. J. Ablowitz and P. A. Clarkson, *Solitons, Nonlinear Evolution Equations, and Inverse Scattering*, Cambridge University Press, New York, 1991.

13. V. E. Zakharov and A. Shabat, *Sov. Phys. JETP* **34**, 62 (1972).

14. J. Satsuma and N. Yajima, *Prog. Theor. Phys. Suppl.* **55**, 284 (1974).

15. H. Hasegawa and Y. Kodama, *Solitons in Optical Communications*, Oxford University Press, New York, 1995.

16. A. Hasegawa and F. D. Tappert, *Appl. Phys. Lett.* **23**, 142 (1973).

17. L. F. Mollenauer, R. H. Stolen, and J. P. Gordon, *Phys. Rev. Lett.* **45**, 1095 (1980).

18. A. Hasegawa and F. Tappert, *Appl. Phys. Lett.* **23**, 171 (1973).

19. P. Emplit, J. P. Hamaide, F. Reynaud, C. Froehly, and A. Barthelemy, *Opt. Commun.* **62**, 374 (1987).

20. D. Kröbel, N. J. Halas, G. Giuliani, and D. Grischkowsky, *Phys. Rev. Lett.* **60**, 29 (1988).

21. Y. S. Kivshar and G. P. Agrawal, *Optical Solitons: From Fibers to Photonic Crystals*, Academic Press, San Diego, CA, 2003.

22. R. A. Fisher and W. K. Bischel, *J. Appl. Phys.* **46**, 4921 (1975).

23. J. A. Fleck, J. R. Morris, and M. D. Feit, *Appl. Phys.* **10**, 129 (1976).

5

NONLINEAR PHASE MODULATION

As verified in Chapter 4, the refractive index of the fiber core increases with the intensity I of the transmitted field according to the relation

$$n = n_0 + n_2 I \tag{5.1}$$

where n_0 is the linear part of the refractive index, whereas n_2 is the nonlinear index coefficient. The numerical value of n_2 is about $n_2 = 2.35 \times 10^{-20} \, \text{m}^2/\text{W}$ [1] for silica fibers and varies somewhat with dopants used inside the core [2]. In spite of this relatively small value, the nonlinear part of the refractive index significantly affects modern lightwave systems due to the long fiber lengths. In particular, it leads to the phenomena of self-phase modulation (SPM) and cross-phase modulation (XPM).

Both SPM and XPM produce a phase alteration of the pulse, which leads to a change in its spectrum. SPM determines a broadening of the pulse spectrum due to its own intensity. In combination with the material dispersion, such spectral broadening can lead to an alteration of the temporal pulse width. The pulse width can be broadened or compressed, depending on the sign of the group velocity dispersion (GVD). The broadening of the pulse width can negatively impact the system performance, but it can be controlled using the dispersion management technique along the transmission link.

The phase of a pulse can also be affected by the other pulses propagating at the same time in the waveguide through the phenomenon of XPM. This is especially important in the case of WDM systems, where pulse trains of different wavelengths propagate simultaneously inside the same fiber. Since pulses belonging to different channels travel at different speeds, they overlap from time to time and mutually influence each other. The phase of each pulse in a given channel is affected by both the average power and the bit pattern of all other channels, which is completely random. This fact makes the cancellation of XPM virtually impossible in practice.

Nonlinear Effects in Optical Fibers. By Mário F. S. Ferreira.
Copyright © 2011 John Wiley & Sons, Inc. Published 2011 by John Wiley & Sons, Inc.

5.1 SELF-PHASE MODULATION

SPM is a nonlinear phenomenon affecting the phase of a propagating wave and does not involve other waves. It arises because the refractive index of the fiber has an intensity-dependent component, as shown in Eq. (5.1). The nonlinear Schrödinger equation (NLSE) derived in Section 4.5 provides an adequate base for the theoretical description of SPM.

5.1.1 SPM-Induced Phase Shift

Assuming that the pulse temporal width is sufficiently large, we can neglect the dispersive term in the NLSE in Eq. (4.49), which reduces to

$$\frac{\partial U}{\partial z} = i\gamma P U \tag{5.2}$$

where $P = |U|^2$ is the pulse power. Neglecting the fiber loss, this corresponds to the input power of the pulse. Equation (5.2) has the following general solution for the field amplitude at the output of a fiber of length L:

$$U(L,\tau) = U(0,\tau)\exp[i\phi_{NL}(L,\tau)] \tag{5.3}$$

where $U(0,\tau)$ is the input pulse envelope and

$$\phi_{NL}(L,\tau) = \gamma P(\tau)L \tag{5.4}$$

is the nonlinearity-induced phase change. The form of this general solution clearly shows that the fiber nonlinearity modifies the phase shift across the pulse, but not the intensity envelope.

Using Eq. (5.3), the electric field transmitted in the fiber can be written in the form

$$E(L,t) = \frac{1}{2}\left[\hat{E}(0,\tau)\exp\{i[(\beta_0 + \gamma P)L - \omega_0 t]\} + \text{c.c.}\right] \tag{5.5}$$

where ω_0 is the pulse central frequency. The phase of the wave described by Eq. (5.5) is

$$\phi = (\beta_0 + \gamma P)L - \omega_0 t \tag{5.6}$$

According to Eq. (5.6), the phase alteration due to the nonlinearity is proportional to the power P of the wave. If the incident wave is a pulse with a power temporal profile given by $P(t)$, the power variation within the pulse leads to its own phase modulation. Hence, this phenomenon is appropriately called *self-phase modulation*.

The instantaneous frequency within the pulse described by Eq. (5.5) is given by

$$\omega(t) = -\frac{\partial \phi}{\partial t} = \omega_0 - \gamma L \frac{\partial P}{\partial t} \tag{5.7}$$

According to Eq. (5.7), the instantaneous frequency decreases in the increasing pulse wing, reaches the central frequency at the pulse maximum, and increases when the pulse power decreases. The time dependence of the frequency shift $\Delta\omega_{SPM} = \omega(t) - \omega_0$ is referred to as frequency chirping.

Let us consider the particular case of Gaussian pulse, with an electric field given by

$$E(0,t) = \frac{1}{2}E_0\exp\left\{ -\frac{1}{2}\left(\frac{t}{t_0}\right)^2 \right\}\exp\{-i\omega_0 t\} + \text{c.c.} \tag{5.8}$$

where E_0 is its peak amplitude. After propagating a distance z, the pulse becomes

$$E(z,t) = \frac{1}{2}E_0\exp\left\{ -\frac{1}{2}\left(\frac{\tau}{t_0}\right)^2 \right\}\exp\{i[\beta_0 + \gamma P(\tau)]z - i\omega_0 t\} + \text{c.c.} \tag{5.9}$$

where

$$P(\tau) = P_0\exp\left[-\left(\frac{\tau}{t_0}\right)^2 \right] \tag{5.10}$$

and

$$\tau = t - \frac{z}{v_g} \tag{5.11}$$

The instantaneous frequency is given by

$$\omega(\tau) = \omega_0 + \gamma z\frac{2\tau}{t_0^2}P_0 e^{-\tau^2/t_0^2} \tag{5.12}$$

Figure 5.1 shows the temporal variation of $P(\tau)$ and $dP/d\tau$ for a Gaussian pulse. In the leading edge of the pulse, where $dP/d\tau > 0$, the instantaneous frequency is

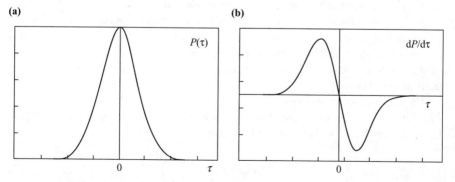

(a)   **(b)**

Figure 5.1   Temporal variation of $P(\tau)$ (a) and $dP/d\tau$ (b) for o Gaussian pulse.

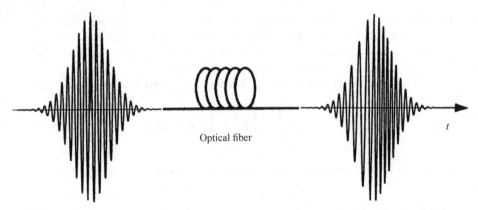

Figure 5.2 Input unchirped and output chirped pulses generated due to self-phase modulation.

downshifted from ω_0, whereas in the tailing edge, where $dP/d\tau < 0$, the instantaneous frequency is upshifted from ω_0. The frequency at the center of the pulse remains unchanged from ω_0. Figure 5.2 illustrates the effect of SPM on an input unchirped pulse.

In the presence of fiber losses and periodic lumped amplification, the distance L in Eqs. (5.4) and (5.7) must be replaced by the effective distance L_{eff}, as seen in Section 4.3:

$$L_{\text{eff}} = N\frac{1 - \exp(-\alpha L_{\text{A}})}{\alpha} \tag{5.13}$$

where α is the fiber loss, N is the number of sections constituting the fiber link, and L_{A} is the amplifier spacing. Taking into account the effective link length, the SPM-induced frequency shift $\Delta\omega_{\text{SPM}}$ becomes

$$\Delta\omega_{\text{SPM}} = -\gamma L_{\text{eff}}\frac{\partial P}{\partial\tau} \tag{5.14}$$

As a consequence of the SPM-induced frequency chirping, new frequency components are generated as the optical pulse propagates down the fiber, broadening the spectrum of the bit stream. The spectral broadening depends not only on the effective length but also on the temporal variation of the power of the input pulse. As a consequence, short pulses are much more affected by SPM than long pulses. The use of short pulses arises as a consequence of the increasing bit rates in modern optical transmission systems.

The dependence of the SPM-induced frequency chirp on the pulse shape is well illustrated considering the case of super-Gaussian pulses. The amplitude of a super-Gaussian pulse can be described in the form

$$U(0, \tau) = U_0\exp\left\{-\frac{1}{2}\left(\frac{\tau}{t_0}\right)^{2m}\right\} \tag{5.15}$$

(a) 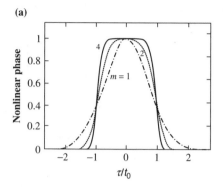 **(b)**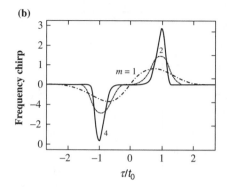

Figure 5.3 Temporal variations of the nonlinear phase shift (a) and the SPM-induced frequency chirp (b) at a distance $L_{eff} = L_{NL}$ for Gaussian ($m = 1$) and super-Gaussian ($m = 2$, 4) pulses.

where t_0 is the input pulse width and the parameter m controls the pulse shape. The Gaussian pulse corresponds to the case $m = 1$. For larger values of m, the pulses become nearly rectangular and develop steeper leading and trailing edges, which considerably affects the SPM-induced frequency chirp. Figure 5.3 shows the numerical results obtained for the nonlinear phase shift and the SPM-induced frequency chirp at a distance $L_{eff} = L_{NL}$, where $L_{NL} = (\gamma P_0)^{-1}$ is the nonlinear length . The results shown in Fig. 5.3 correspond to Gaussian ($m = 1$) and super-Gaussian ($m = 2$, 4) pulses. As the nonlinear phase shift, given by Eq. (5.4), is proportional to the pulse intensity, Fig. 5.3a also provides the intensity profile for the considered Gaussian and super-Gaussian pulses. Figure 5.3b shows that the frequency chirp is negative (downshift) in the leading edge of the pulse, whereas in the trailing edge it is positive (upshift). In contrast with the Gaussian pulses, the frequency chirp is larger and occurs only near the leading and trailing edges in the case of super-Gaussian pulses.

The chirping due to nonlinearity without any increase in pulse width leads to increased spectral broadening. Figure 5.4 illustrates the SPM-induced spectral broadening experienced by a Gaussian pulse, at a distance $L_{eff} = 11L_{NL}$. The spectrum develops multiple peaks such that the two outermost peaks are the most intense, in accordance with the experimental observations [3].

5.1.2 The Variational Approach

The variational technique is well known in several areas [4] and it is particularly useful in nonlinear optics for describing the propagation of chirped pulses in nonlinear media. The variational formulation was first used in 1977 to describe pulse propagation in nonlinear diffractive media [5]. Its use to describe pulse propagation in nonlinear dispersive media, such as optical fibers, was first realized by Anderson in 1983 [6]. This approach can be quite accurate even when the effect of nonlinearity is large [7,8]. The accuracy depends essentially on the ansatz chosen to describe the evolution of the field experiencing nonlinearity. In this

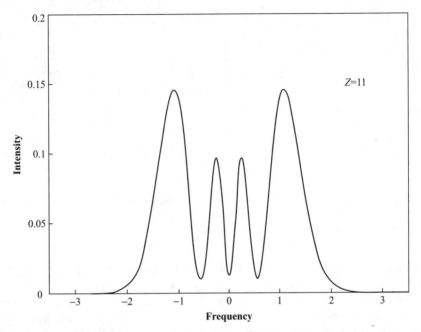

Figure 5.4   SPM-induced spectral broadening of an initially unchirped Gaussian pulse at a distance $L_{eff} = 11L_{NL}$.

section we apply this approach to discuss some combined effects of GVD, initial chirp, and SPM. The same approach is especially useful to describe the behavior of nonlinear pulses propagating in dispersion-managed links [9–14], as will be seen in Chapter 8.

The "action" functional is given by

$$S = \int L\,dz \tag{5.16}$$

where L the Lagrangian, which is related to the Lagrangian density L_d as

$$L = \int_{-\infty}^{\infty} L_d(v, v^*)d\tau \tag{5.17}$$

In Eq. (5.17), $v(z)$ and $v^*(z)$ represent the generalized coordinates. Minimization of the action S requires that L_d satisfy the Euler–Lagrange equation

$$\frac{\partial L_d}{\partial v} - \frac{\partial}{\partial \tau}\left(\frac{\partial L_d}{\partial v_\tau}\right) - \frac{\partial}{\partial z}\left(\frac{\partial L_d}{\partial v_z}\right) = 0 \tag{5.18}$$

where v_τ and v_z represent the derivative of v relative to τ and z, respectively. The Lagrangian density corresponding to Eq. (4.49) is

$$L_d = \frac{i}{2}\left(U\frac{\partial U^*}{\partial z} - U^*\frac{\partial U}{\partial z}\right) - \frac{\beta_2}{2}\left|\frac{\partial U}{\partial \tau}\right|^2 - \frac{\gamma}{2}|U|^4 \qquad (5.19)$$

In fact, Eq. (4.49) can be obtained using Eq. (5.19) in Eq. (5.18) and considering U^* as the generalized coordinate v. Minimization of the action S with respect to the pulse parameters provides the reduced Euler–Lagrange equations

$$\frac{\partial L}{\partial v} - \frac{d}{dz}\left(\frac{\partial L}{\partial v_z}\right) = 0 \qquad (5.20)$$

In the case of a chirped Gaussian pulse, we have

$$U(z,\tau) = U_p(z)\exp\left[-\frac{1}{2}(1+iC(z))\left(\frac{\tau}{t_p(z)}\right)^2 + i\psi(z)\right] \qquad (5.21)$$

where the parameters $U_p(z)$, $C(z)$, $t_p(z)$, and $\psi(z)$ represent the amplitude, the chirp parameter, the width, and the phase of the pulse, respectively. Substituting Eq. (5.21) into Eq. (5.19) and performing the time integration in Eq. (5.17) gives the following result for the Lagrangian:

$$L = \frac{\beta_2 E_p}{4t_p^2}(1+C^2) + \frac{\gamma E_p^2}{\sqrt{8\pi}t_p} + \frac{E_p}{4}\left(\frac{dC}{dz} - \frac{2C}{t_p}\frac{dt_p}{dz}\right) - E_p\frac{d\psi}{dz} \qquad (5.22)$$

where $E_p = \sqrt{\pi}U_p^2 t_p$ is the pulse energy. Using $v = \psi$ in Eq. (5.20) gives $E_p = E_0 = $ constant. On the other hand, using $v = C$, t_p, and E_p provides the following equations for t_p, C, and ψ, respectively:

$$\frac{dt_p}{dz} = \frac{\beta_2 C}{t_p} \qquad (5.23)$$

$$\frac{dC}{dz} = \frac{\gamma E_0}{\sqrt{2\pi}t_p} + (1+C^2)\frac{\beta_2}{t_p^2} \qquad (5.24)$$

$$\frac{d\psi}{dz} = \frac{5\gamma E_0}{4\sqrt{2\pi}t_p} + \frac{\beta_2}{2t_p^2} \qquad (5.25)$$

Equations (5.23) and (5.24) show that the SPM affects the pulse width via the chirp parameter C. The two terms in Eq. (5.24) are due to the SPM and GVD effects and they have the same or the opposite sign, depending on whether the dispersion is normal or anomalous, respectively. As a consequence, the combination of the SPM and GVD effects results in a reduced pulse broadening in the case of anomalous dispersion compared to the case of normal dispersion, because of smaller values of the chirp parameter in the first case.

Neglecting the fiber nonlinearity, it can be verified from Eqs. (5.23) and (5.24) that the quantity $(1 + C^2)/t_p^2$, which is related to the spectral width of the pulse, remains constant, as already observed in Chapter 3. Replacing this quantity with its initial value $(1 + C_0^2)/t_0^2$ in Eq. (5.24) and integrating, one obtains the solution

$$C = C_0 + (1 + C_0^2)\frac{\beta_2 z}{t_0^2} \tag{5.26}$$

which is similar to Eq. (3.96). Using this solution in Eq. (5.23), we find the result given by Eq. (3.97) for the pulse width.

Keeping the fiber nonlinearity but neglecting the GVD, Eq. (5.23) shows that the pulse width t_p remains constant, equal to its initial value t_0. Concerning the chirp parameter, it varies with distance as

$$C = C_0 + \frac{1}{\sqrt{2}}\gamma P_0 z \tag{5.27}$$

Equation (5.27) shows that the SPM-induced chirp is always positive. It adds to the initial pulse chirp when $C_0 > 0$, resulting in an enhanced oscillatory structure of the pulse spectrum. However, if $C_0 < 0$, the two contributions have opposite signs, which produces a spectral narrowing.

Let us consider, instead of the Gaussian ansatz given by Eq. (5.21), a sech ansatz of the form

$$U(z, \tau) = U_p(z)\mathrm{sech}\left(\frac{\tau}{t_p(z)}\right)\exp\left[-iC(z)\left(\frac{\tau}{t_p(z)}\right)^2 + i\psi(z)\right] \tag{5.28}$$

Substituting Eq. (5.28) into Eq. (5.19) and performing the time integration in Eq. (5.17) provides the following result for the Lagrangian:

$$L = \frac{\beta_2 E_p}{6t_p^2}\left(1 + \frac{\pi^2}{4}C^2\right) + \frac{\gamma E_p^2}{6t_p} + \frac{\pi^2 E_p}{24}\left(\frac{dC}{dz} - \frac{2C}{t_p}\frac{dt_p}{dz}\right) - E_p\frac{d\psi}{dz} \tag{5.29}$$

Using this expression in Eq. (5.20) and considering $v = C, t_p$, we reobtain Eq. (5.23) for the pulse width, whereas the equation for the chirp parameter becomes

$$\frac{dC}{dz} = \left(C^2 + \frac{4}{\pi^2}\right)\frac{\beta_2}{t_p^2} + \gamma P_0\frac{4}{\pi^2}\frac{t_0}{t_p} \tag{5.30}$$

Considering an unchirped initial pulse, such that $C(0) = C_0 = 0$ and $t_p(0) = t_0$, we can verify from Eq. (5.30) that $dC/dz = 0$ in the case of anomalous dispersion ($\beta_2 < 0$) if the peak power of the pulse satisfies the condition

$$\frac{L_D}{L_{NL}} = \frac{\gamma P_0 t_0^2}{|\beta_2|} = 1 \qquad (5.31)$$

In this case, the SPM- and GVD-induced chirps balance each other and the chirp parameter C remains zero. On the other hand, the pulse width given by Eq. (5.23) also remains constant at $t_p = t_0$ during propagation. Such a pulse that propagates with an undistorted sech shape was discussed in Section 4.5 and is known as *fundamental soliton*.

5.1.3 Impact on Communication Systems

The maximum phase shift due to SPM is given from Eqs. (5.4) and (5.13) as

$$\phi_{NL} = \gamma P_0 L_{eff} \qquad (5.32)$$

This phase shift becomes significant ($\sim \pi/2$) when the power times the net effective length of the system reaches 1 W km or 1 mW Mm. The first set of units is appropriate for repeaterless systems and the second for long amplified systems. In the first case, the effects of SPM are of little concern, since other nonlinear effects, namely, stimulated Brillouin scattering, limit them to power levels below 10 mW [15,16]. In the second case, however, SPM can be a major limiting factor, since its effects accumulate over the entire link and the maximum phase shift increases linearly with the number of amplifiers, N. Considering $L_{eff} \approx N/\alpha$ and using typical values, we find that the peak power is limited to below 3 mW for links with only 10 amplifiers.

The impact of the SPM effects on the transmission system depends on the modulation format of the carrier. For example, in the case of phase binary shift keying (PSK) systems the information lies in the carrier phase, which changes between $+\pi/2$ and $-\pi/2$. Phase noise leads to a reduction of the signal-to-noise ratio (SNR), which can be significant if semiconductor lasers are directly phase modulated, due to their strong intensity fluctuations.

In the presence of dispersion, the spectral broadening due to SPM determines two situations qualitatively different. As discussed in Section 3.5, in the normal dispersion region (wavelength shorter than the zero dispersion wavelength) the chirping due to dispersion corresponds to a downshift of the leading edge and to an upshift of the trailing edge of the pulse, which is a similar effect to that due to SPM. Thus, in this regime the chirping due to dispersion and SPM act in the same direction and lead to a stronger temporal broadening of the pulse than the dispersion alone, thus determining a more significant reduction of the system capacity.

If the pulse, spectrally broadened by SPM, is transmitted in the anomalous dispersion regime, the redshifted leading edge travels more slowly, and moves toward the pulse center. Similarly, the trailing edge of the pulse, which has been blueshifted, travels more quickly, and also moves toward the center of the pulse. Therefore, GVD and SPM act in different directions, resulting in a compression of the pulse.

In the range of anomalous dispersion, nonlinearity- and dispersion-induced chirpings can partially or even completely cancel each other. When this cancellation is total, the pulse neither broadens in time nor in its spectrum and such a pulse is called a fundamental soliton, as mentioned above. However, the pulse narrowing in the anomalous dispersion regime can be observed even with conventional nonreturn to zero (NRZ) pulses [17].

Dispersion management is a common technique to combat the fiber dispersion effects. However, it is important to note that an optimum dispersion compensation scheme designed according to the linear case will not work perfectly when SPM effects are significant. In fact, the SPM leads to an additional compression in the fiber spans with negative GVD, while it determines an additional broadening in the fiber spans with positive GVD. Because the power variations in the two types of fibers are different, the contribution of SPM does not cancel perfectly. It has been observed experimentally that an improved system performance is achieved when GVD is undercompensated [18,19]. In general, the optimization of the system design requires extensive computer simulations, including the nonlinearity of the fiber and the fine-tuning of the system parameters.

5.1.4 Modulation Instability

A continuous wave (CW) in the anomalous dispersion regime is known to produce modulation instability (MI), a phenomenon in which the sideband components of the amplitude-modulated light grows exponentially [20]. To investigate analytically the modulational instability, we add a small perturbation u to the CW solution, such that

$$U = (\sqrt{P_0} + u)\exp(i\phi_{\text{NL}}) \tag{5.33}$$

where P_0 is the incident power and $\phi_{\text{NL}} = \gamma P_0 z$ is the nonlinear phase shift induced by SPM. Substituting Eq. (5.33) into Eq. (4.49) under the assumption that the perturbation is sufficiently small, we obtain the following differential equation for u:

$$i\frac{\partial u}{\partial z} = \frac{\beta_2}{2}\frac{\partial^2 u}{\partial \tau^2} - \gamma P_0(u + u^*) \tag{5.34}$$

We assume the general solution for $u(z, \tau)$ of the form

$$u(z, \tau) = a\exp[i(Kz - \Omega\tau)] + b\exp[-i(Kz - \Omega\tau)] \tag{5.35}$$

where Ω and K are the angular frequency and the wave number of the perturbations, respectively. The following dispersion relation between K and Ω is obtained by substituting Eq. (5.35) into Eq. (5.34):

$$K^2 = \left(\frac{\beta_2 \Omega}{2}\right)^2 \left[\Omega^2 + \frac{4\gamma P_0}{\beta_2}\right]$$ (5.36)

Equation (5.36) shows that K becomes a pure imaginary number when $\beta_2 < 0$ (anomalous GVD) and

$$\Omega < \Omega_c \equiv \left[\frac{4\gamma P_0}{|\beta_2|}\right]^{1/2}$$ (5.37)

In these circumstances, the CW solution is unstable, since the perturbation $u(z, \tau)$ grows exponentially with z. Such a phenomenon is called the *modulation instability*.

The power gain of modulation instability is given from Eq. (5.36) by

$$g(\Omega) = |\beta_2 \Omega|[\Omega_c^2 - \Omega^2]^{1/2}$$ (5.38)

The gain vanishes at $\Omega = 0$ and becomes maximum at $\Omega = \pm\Omega_c/\sqrt{2}$, with a peak value

$$g_{max} = \frac{1}{2}|\beta_2|\Omega_c^2 = 2\gamma P_0$$ (5.39)

Even in the absence of other perturbations, the modulation instability can be seeded by the broadband noise added by optical amplifiers. The growth of this noise can degrade the signal-to-noise ratio considerably at the receiver end, both in the normal and in the anomalous dispersion regimes of the transmission fiber link [21–25].

Figure 5.5 shows the experimental observation of the growth of sideband components by the modulational instability in optical fiber [26]. The experiment was performed by injecting the pulses of a mode-locked Nd:YAG laser with a wavelength of 1319 nm, a pulse width of 100 ps, and a repetition rate of 100 MHz into a 1 km long fiber. The 100 ps pulses were used instead of a constant amplitude wave in order to suppress the stimulated Brillouin scattering. The fiber used in this experiment presented an effective core area $A_{eff} = 60\,\mu m^2$, a GVD of 2.4 ps/(nm km), and a loss rate of 0.67 dB/km at the operating wavelength. Figure 5.5a shows that, for a low input power, the output spectrum coincides with that of the input spectrum. When the input power is increased to 5.5 W (b), 6.1 W (c), and 7.1 W (d), sidebands are found to be generated at angular $\Omega_c/\sqrt{2}$ (b), then to grow (c), and higher order sidebands are also found to be generated (d). The spacing between each sideband and the carrier frequency becomes wider as the input power increases, since the sideband frequency $\Omega_c/\sqrt{2}$ is proportional to $\sqrt{P_0}$, as given by Eq. (5.37).

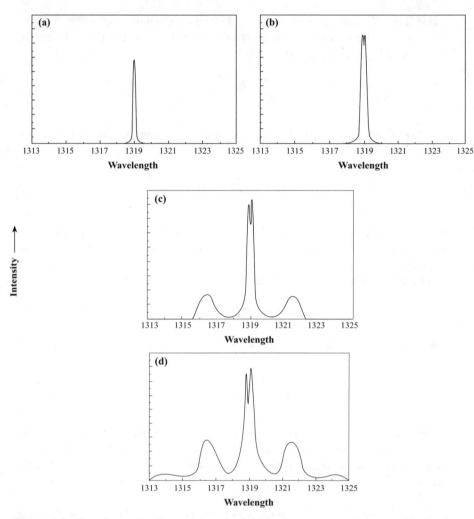

Figure 5.5 Experimentally observed growth of sideband components by the modulational instability in optical fibers. (After Ref. [21]; © 1986 APS.)

When a small modulation having the angular frequency in the vicinity of $\Omega_m = \Omega_c / \sqrt{2}$ is applied to an input signal, modulational instability can be induced. We call this an *induced modulational instability*. An ultrahigh bit rate pulse train can be generated by using such induced modulational instability [27]. The repetition period of this pulse train is equal to the inverse of the given modulation frequency $\Omega_m / 2\pi$. Figure 5.6 shows the numerical evaluation of equation (4.67) for which the input signal is modulated with a period of 12. One can see that a soliton-like pulse train can be generated from this initial condition. A simple technique of generating induced modulational instability consists in injecting into a fiber two continuous waves with different wavelengths [28].

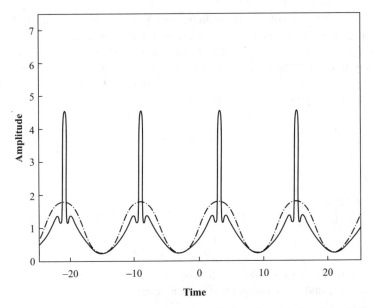

Figure 5.6 Result of computer simulation of induced modulational instability.

5.2 CROSS-PHASE MODULATION

Waves with different wavelengths propagating in the same fiber can interact with each other since the refractive index that a wave experiences depends on the intensities of all other waves. Hence, a pulse at one wavelength has an influence on the phase of a pulse at another wavelength. This nonlinear phenomenon is known as cross-phase modulation, and it can significantly limit the performance of WDM systems [29–37]. In fact, the phase of each optical channel is affected by both the average power and the bit patterns of all other channels, which are naturally stochastic and cannot be compensated by system design. Phase variations are converted in amplitude fluctuations due to the fiber dispersion, which considerably affects the SNR of multichannel systems.

5.2.1 XPM-Induced Phase Shift

To discuss the origin of XPM as clearly as possible, we will consider in the following only two channels propagating at the same time in the fiber. As assumed in Section 4.4, the two optical fields can be written in the form

$$E_j(r, z, t) = \frac{1}{2} F(r) \frac{U_j(z)}{B_j} \exp\left[i(\beta_j z - \omega_j t)\right] + \text{c.c.}, \quad j = 1, 2 \tag{5.40}$$

where $F(r)$ gives the transversal distribution of the single mode supported by the fiber, $U_j(z)$ is the normalized slowly varying amplitude, such that the square of its absolute value is equal the power of the wave with frequency ω_j, B_j is given by Eq. (4.35), and β_j is the propagation constant. The dispersive effects are taken into account by expanding the frequency-dependent propagation constant, given by Eq. (3.72). Retaining terms only up to the quadratic term in such an expansion and using the procedure of Section 4.4, the following coupled differential equations are obtained for the slowly varying amplitudes U_1 and U_2:

$$\frac{\partial U_1}{\partial z} + \frac{1}{v_{g1}}\frac{\partial U_1}{\partial t} + i\frac{\beta_{21}}{2}\frac{\partial^2 U_1}{\partial t^2} = i\gamma_1\left[(|U_1|^2 + 2|U_2|^2)U_1 + U_1^2 U_2^*\exp\{i(\Delta\beta z - \Delta\omega t)\}\right]$$

(5.41)

$$\frac{\partial U_2}{\partial z} + \frac{1}{v_{g2}}\frac{\partial U_2}{\partial t} + i\frac{\beta_{22}}{2}\frac{\partial^2 U_2}{\partial t^2} = i\gamma_2\left[(|U_2|^2 + 2|U_1|^2)U_2 + U_2^2 U_1^*\exp\{i(-\Delta\beta z + \Delta\omega t)\}\right]$$

(5.42)

where v_{gj} is the group velocity, β_{2j} is the GVD coefficient, $\Delta\omega = \omega_1 - \omega_2$, $\Delta\beta = \beta_1 - \beta_2$, and γ_j is the nonlinear parameter at frequency ω_j, defined as in Eq. (4.42):

$$\gamma_j = \frac{n_2\omega_j}{cA_{\text{eff}}}$$

(5.43)

The last terms in Eqs. (5.41) and (5.42) result from the phenomenon of four-wave mixing. However, since the efficiency of this process depends on the matching of the phases, these terms are neglected in this chapter after assuming that such phase matching does not occur. Moreover, assuming that the pulse duration is long or the dispersion in the fiber is low, we can also neglect the linear terms in Eqs. (5.41) and (5.42), leading to the following simplified system of equations:

$$\frac{\partial U_1}{\partial z} = i\gamma_1(P_1 + 2P_2)U_1$$

(5.44)

$$\frac{\partial U_2}{\partial z} = i\gamma_2(P_2 + 2P_1)U_2$$

(5.45)

where $P_j = |U_j|^2$ is the power of the wave with frequency ω_j, $j = 1, 2$. If, for simplicity, the powers are assumed to be constant, the solution of Eq. (5.44) is given by

$$U_1(L) = U_1(0)\exp\{i\gamma_1(P_1 + 2P_2)L\}$$

(5.46)

A similar solution can be obtained from Eq. (5.45) for U_2. From Eq. (5.46) it is apparent that the phase of the signal at frequency ω_1 is modified not only due to its own power—this is the SPM effect—but also due to the power of the signal at frequency

ω_2. This phenomenon is referred to as XPM and Eq. (5.46) shows that it is twice as effective as SPM.

Using (5.46), the electric field of the wave with frequency ω_1 transmitted in the fiber can be written in the form

$$E_1(z,t) = \frac{1}{2}\left[\hat{E}_1(0,t)\exp\{i[(\beta_1 + \gamma_1 P_1 + 2\gamma_1 P_2)z - \omega_1 t]\} + \text{c.c.}\right] \qquad (5.47a)$$

or

$$E_1(z,t) = \frac{1}{2}\left[\hat{E}_1(0,t)\exp\left\{i\left[\left(n_0 + n_2\frac{P_1}{A_{\text{eff}}} + 2n_2\frac{P_2}{A_{\text{eff}}}\right)\beta_{10}z - \omega_1 t\right]\right\} + \text{c.c.}\right]$$

$$(5.47b)$$

where $\hat{E}_1(0,t)$ is the wave amplitude of the first wave at the fiber input and Eq. (5.43) was used in writing Eq. (5.47b). Changing the indices, a similar equation can be written for the wave $E_2(z,t)$. From Eq. (5.47b), the total refractive index of the fiber seen by the wave with frequency ω_1 can be written as

$$n_{\text{tot}} = n_0 + n_2 I_1 + 2n_2 I_2 \qquad (5.48)$$

where $I_j = P_j/A_{\text{eff}}$ is the intensity of the wave with frequency ω_j. Equation (5.48) shows that the second wave affects the refractive index experienced by the first wave via its intensity. The phase of the wave in Eq. (5.47a) is

$$\phi_1 = [\beta_1 + \gamma_1(P_1 + 2P_2)]z - \omega_1 t \qquad (5.49)$$

The nonlinear phase shift resulting from the combination of SPM and XPM at the output of a fiber with length L is given by

$$\phi_{1\text{NL}} = \gamma_1(P_1 + 2P_2)L_{\text{eff}} \qquad (5.50)$$

where the attenuation was included through L_{eff}.

The XPM-induced frequency shift $\Delta\omega_{1\text{XPM}}$ of the signal at frequency ω_1 is given by

$$\Delta\omega_{1\text{XPM}} = -2\gamma_1 L_{\text{eff}}\frac{\partial P_2}{\partial t} \qquad (5.51)$$

Equation (5.51) shows that the part of the signal at ω_1 that is affected by the leading edge of the signal at ω_2 will be downshifted in frequency, whereas the part overlapping with the trailing edge will be upshifted in frequency. This determines a spectral broadening of the signal at ω_1, which is twice the spectral broadening caused by SPM.

5.2.2 Impact on Optical Communication Systems

The interaction among M WDM channels can be described by a set of M differential coupled equations similar to Eqs. (5.41) and (5.42). If only the effects of SPM and XPM are considered, these equations are given by

$$\frac{\partial U_j}{\partial z} = -i\gamma_j \left(|U_j|^2 + 2\sum_{m \neq j}^{M} |U_m|^2 \right) U_j \tag{5.52}$$

where $j = 1$ to M. In the CW case, the nonlinear phase shift experienced by the wave j resulting from the combination of SPM and XPM is given by

$$\phi_j = \gamma_j L_{\text{eff}} \left[P_j + 2\sum_{m \neq j}^{M} P_m \right] \tag{5.53}$$

where P_j is the input power of wave and L_{eff} is given by Eq. (5.13). Equation (5.53) shows that the impact of XPM is much more important than that of SPM in multichannel communication systems. However, the dispersion of the fiber generally attenuates such impact.

The effect of XPM is different in amplitude- and phase-modulated systems. In the last case, since the power in each channel is the same for all bits, the main limitation results from arbitrary phase fluctuations, which lead directly to a deterioration of the signal-to-noise ratio. Such phase fluctuations can be induced via the XPM by intensity variations, as happen if semiconductor lasers are directly phase modulated.

In the case of amplitude-modulated direct detection systems, the XPM has no effect on the system performance if the dispersion is neglected. Actually, since the phase alteration due to XPM is associated with a frequency alteration, the dispersion determines an additional temporal broadening or compression of the spectrally broadened pulses, which affects the system performance. Figure 5.7 shows XPM-induced fluctuations for a CW probe launched with a 10 Gb/s pump channel modulated using the NRZ format [34]. The intensity-modulated pump signal induces a phase modulation on the probe signal and the dispersion of the medium converts the phase modulation to power fluctuation of the probe. Figure 5.7 shows that after a propagation distance of 320 km, the probe power experiences a fluctuation by as much as 6%.

The impact of XPM is particularly significant in the case of amplitude-modulated coherent communication system, employing a phase-sensitive detection scheme. In fact, the phase in a given channel depends on the bit pattern of neighboring channels. In the worst case, in which all channels have "1" bits in their time slots, the XPM-induced phase shift is maximum. Assuming a repeaterless system such that the power P in each channel is the same, this phase shift is given by

$$\phi_{\max} = \frac{\gamma}{\alpha}(2M - 1)P \tag{5.54}$$

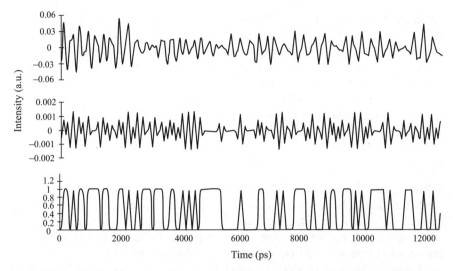

Figure 5.7 XPM-induced power fluctuations on a CW probe for a 130 km link (middle) and a 320 km link (top). At the bottom is represented the NRZ bit stream in the pump channel. (After Ref. [34]; © 1999 IEEE.)

where it was assumed that $\alpha L \gg 1$. Considering a maximum tolerable phase shift $\phi_{max} = 0.1$, the power in each channel is limited to

$$P < \frac{\alpha}{10\gamma(2M - 1)} \qquad (5.55)$$

For typical values of α and γ, P should be below 1 mW even for five channels.

The impact of XPM would be negligible in frequency- or phase-modulated coherent systems if the channel powers were really constant in time. However, this is not the case in practice, since the intensity noise of the transmitters or the amplified spontaneous emission (ASE) noise added by the optical amplifiers causes fluctuations of the channel powers. XPM converts such fluctuations into phase fluctuations, which degrade the performance of the coherent receiver.

The XPM effect determines a mutual influence between two pulses only if they overlap to some extent. However, in the presence of finite dispersion, the two pulses with different wavelengths will move with different velocities and thus will walk off from each other. If the pulses enter the fiber separately, walk through each other, and again become separated, it is said that they experience a complete collision. In a lossless fiber, such a collision is perfectly symmetric and no residual phase shift remains, since the pulses would have interacted equally with both the leading and the trailing edge of the other pulse. However, in case the pulses enter the fiber together the result is a partial collision, since each pulse will see only the trailing or the leading edge of the other pulse, which will lead to chirping. Moreover, in the case of a periodically amplified system, power variations also make complete collisions asymmetric, resulting in a net frequency shift that depends on the wavelength difference between the interacting pulses. Such frequency shifts lead to timing jitter in multichannel systems, since their magnitude depends on the bit pattern as well as on channel wavelengths. The combination of amplitude and timing jitter significantly degrades the system performance [31].

We can define a parameter called the interaction length, L_I, which gives the distance between the beginning and the end of the overlap at half the maximum power of the two pulses, given by

$$L_I = \frac{2t_{FWHM}}{D\Delta\lambda} \tag{5.56}$$

where t_{FWHM} is the pulse full width at half maximum (FWHM), D is the dispersion coefficient, and $\Delta\lambda$ is the wavelength spacing between the interfering channels. Using dispersion-shifted fibers, which have a reduced dispersion, will increase the interaction length and hence enhance the XPM effects. Equation (5.56) also shows that closely spaced channels will interact over longer fiber lengths and hence suffer greater XPM effects. This is actually the case when $L_I \ll L_{eff}$, where L_{eff} is the effective fiber length. In this case, the pulse power does not change significantly over the interaction length. However, for $L_I \gg L_{eff}$ the interaction lengths are determined by the fiber losses and the XPM effects become almost independent of the channel spacing, as observed experimentally [38].

Figure 5.8 shows the experimental results for the variation of the rms value of probe intensity modulation with the wavelength separation between the intensity-

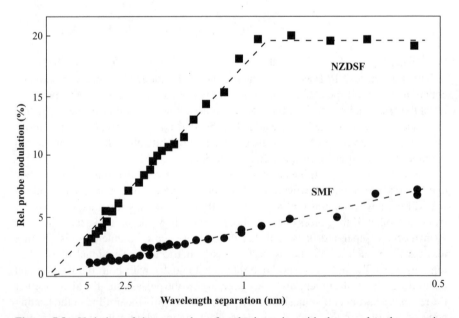

Figure 5.8 Variation of the rms value of probe intensity with the wavelength separation between the intensity-modulated pump signal and the probe. (After Ref. [38]; © 2000 IEEE.)

modulated pump and the probe [38]. Both standard single-mode fiber (SMF) and nonzero dispersion-shifted fiber (NZDSF) were used in this experiment. In the case of SMF, it was found that the probe modulation decreases approximately linearly with $1/\Delta\lambda$ for all values of $\Delta\lambda$. However, in the case of NZDSF, the modulation becomes independent of $\Delta\lambda$ for values of $\Delta\lambda$ smaller than 1 nm, decreasing linearly with $1/\Delta\lambda$ for larger channel spacing values. For $\Delta\lambda = 0.8$ nm, the relative probe modulation is approximately 5% in SMF and 20% in NZDSF.

The XPM effects in WDM systems can be reduced launching alternate channels with orthogonal state of polarizations (SOPs) [39]. Such a scheme is usually referred to as the polarization channel interleaving technique. The propagation of two orthogonally polarized waves can still be described by Eqs. (5.41) and (5.42) provided that the factor 2 appearing in the XPM term is replaced with 2/3. As a consequence, the magnitude of the XPM-induced phase shift is reduced, which improves the system performance.

5.2.3 Modulation Instability

As realized in Section 5.1.4, to investigate the XPM-induced modulational instability we add a small perturbation u_n to the CW solution, such that

$$U_n = (\sqrt{P_n} + u_n)\exp(i\phi_{\mathrm{NL},n}) \tag{5.57}$$

Substituting this into Eqs. (5.41) and (5.42) under the assumption that the perturbation is sufficiently small, the following differential equations for u_n are obtained:

$$\frac{\partial u_1}{\partial z} + \frac{1}{v_{g1}}\frac{\partial u_1}{\partial t} + i\frac{\beta_{21}}{2}\frac{\partial^2 u_1}{\partial t^2} = i\gamma_1(u_1 + u_1^*)P_1 + i2\gamma_1(u_2 + u_2^*)\sqrt{P_1 P_2} \tag{5.58}$$

$$\frac{\partial u_2}{\partial z} + \frac{1}{v_{g2}}\frac{\partial u_2}{\partial t} + i\frac{\beta_{22}}{2}\frac{\partial^2 u_2}{\partial t^2} = i\gamma_2(u_2 + u_2^*)P_2 + i2\gamma_2(u_1 + u_1^*)\sqrt{P_1 P_2} \tag{5.59}$$

Let us assume a general solution for $u_j(z,t)$ of the form

$$u_j(z,t) = a_j\exp[i(Kz - \Omega t)] + ib_j\exp[-i(Kz - \Omega t)] \tag{5.60}$$

where $j = 1, 2$, and Ω and K are the angular frequency and the wave number of the perturbations, respectively. Equations (5.58) and (5.59) provide four homogeneous equations for a_j and b_j, which have a nontrivial solution only if the determinant of the coefficient matrix vanishes. Such condition provides the following dispersion relation:

$$[\hat{K}_1^2 - f_1][\hat{K}_2^2 - f_2] = F_{\mathrm{XPM}} \tag{5.61}$$

where

$$\hat{K}_j = K - \Omega/v_{gj} \tag{5.62}$$

$$f_j = \gamma_j P_j c_j |\beta_{2j}| \Omega^2 \left(1 + \frac{c_j \Omega^2}{\Omega_{cj}^2}\right) \tag{5.63}$$

$$\Omega_{cj}^2 = \frac{4\gamma_j P_j}{|\beta_{2j}|} \tag{5.64}$$

for $c_j = \operatorname{sgn}(\beta_{2j})$ and $j = 1, 2$. The parameter F_{XPM} is given by

$$F_{\mathrm{XPM}} = 4\beta_{21}\beta_{22}\gamma_1\gamma_2 P_1 P_2 \Omega^4 \tag{5.65}$$

If the group velocity mismatch is neglected, we have $\hat{K}_1 \approx \hat{K}_2 \equiv \hat{K}$ and Eq. (5.61) becomes a quadratic equation of $\hat{K}^2$, whose solution is [40,41]

$$\hat{K}^2 = \frac{1}{2}(f_1 + f_2) \pm [(f_1 - f_2)^2/4 + F_{\mathrm{XPM}}]^{1/2} \tag{5.66}$$

If $\hat{K}$ has a nonzero imaginary part, the perturbation $u_j(z, t)$ grows exponentially with z, as seen from Eq. (5.60). As a consequence, the CW solution becomes unstable and transforms into a pulse train.

The modulational instability governed by Eq. (5.66) can appear in two cases. One case corresponds to a purely imaginary value of $\hat{K}$, which occurs when the condition

$$F_{\mathrm{XPM}} > f_1 f_2 \tag{5.67}$$

is satisfied. The other case corresponds to a complex $\hat{K}$ ($\hat{K} = a + ib$, with $a \neq 0$), which occurs when

$$(f_1 - f_2)^2 + 4F_{\mathrm{XPM}} < 0 \tag{5.68}$$

An analysis of Eq. (5.66) reveals that, when one wave propagates in the anomalous GVD regime and the other in the normal GVD regime ($c_1 c_2 = -1$), the modulation instability condition is given by Eq. (5.68). However, when both waves propagate in the anomalous ($c_1 = c_2 = -1$) or normal ($c_1 = c_2 = 1$) GVD regime, the modulation instability condition is given by Eq. (5.67) [41]. The possibility of occurrence of modulation instability for normal GVD in the presence of two waves with different frequencies is in contrast with the single wave case, in which the modulation instability requires anomalous GVD.

The gain of modulation instability is obtained from Eq. (5.66) as

$$g(\Omega) = 2\text{Im}(\hat{K}) = \sqrt{2}\{[(f_1 - f_2)^2 + 4F_{\text{XPM}}]^{1/2} - (f_1 + f_2)\}^{1/2} \qquad (5.69)$$

From Eq. (5.69) and considering the case in which both waves experience normal GVD ($c_1 = c_2 = 1$), the bandwidth of the gain spectrum is given by [41]

$$\Omega_{\text{BW}} = \sqrt{2}\{\sqrt{(\Omega_{C1}^2 + \Omega_{C2}^2)^2 + 12\Omega_{C1}^2\Omega_{C2}^2} + (\Omega_{C1}^2 + \Omega_{C2}^2)\}^{1/2} \qquad (5.70)$$

In practice, it is not easy to observe experimentally the modulation instability using two waves with different wavelength in the normal GVD. This is due to the influence of four-wave mixing phenomenon, which was neglected in the above analysis. However, several experiments have been realized demonstrating the XPM-induced modulation instability when one of the waves propagates in the normal GVD region, whereas the other wave propagates in the anomalous GVD region [42–44]. Using a pump wave in the form of intense pulses, a weak CW probe wave can be easily converted into a train of ultrashort pulses [43,44]. The new probe pulses reproduce the sequence of the pump pulses, since the XPM interaction requires the presence of a pump pulse. This principle of operation can be used for wavelength conversion of signals in optical communication systems, as discussed in Chapter 14 [45–49].

5.2.4 XPM-Paired Solitons

Several studies have shown that the coupled NLSEs given by Eqs. (5.41) and (5.42) have solitary wave solutions that maintain their shape during propagation through the XPM interaction [50–52]. Since such solutions always occur in pairs, they are referred to as XPM-paired solitons. Such solitons can have the same group velocity ($v_{g1} = v_{g2} = v_g$) if their wavelengths are chosen appropriately on opposite sides of the fiber zero dispersion wavelength.

Assuming that the phase matching condition for the FWM process does not occur, the last terms in Eqs. (5.41) and (5.42) can be neglected. Under these conditions, let us look for soliton solutions of these equations in the form

$$U_j(z,t) = V_j(t - z/v_e)\exp[i(K_j z - \Omega_j t + \phi_j)], \quad j = 1, 2 \qquad (5.71)$$

where v_e is the effective group velocity of the soliton pair, V_j describes the soliton shape, whereas K_j, Ω_j, and ϕ_j represent the propagation constant, frequency, and phase of the two solitons, respectively. The solution for the soliton shape has the form [52]

$$V_1(t - z/v_e) = V_{10}[1 - a^2 \text{sech}^2(B(t - z/v_e))]^{1/2} \qquad (5.72)$$

$$V_2(t - z/v_e) = V_{20}\text{sech}(B(t - z/v_e)) \qquad (5.73)$$

where

$$B^2 = \frac{3\gamma_1\gamma_2}{\gamma_1\beta_{22} - 2\gamma_2\beta_{21}} V_{20}^2 \tag{5.74}$$

$$a^2 = \frac{2\gamma_1\beta_{22} - \gamma_2\beta_{21}}{\gamma_1\beta_{22} - 2\gamma_2\beta_{21}} \frac{V_{20}^2}{V_{10}^2} \tag{5.75}$$

and the effective velocity of the soliton pair, v_e, is given by

$$v_e^{-1} = v_g^{-1} + \beta_{21}\Omega_1 = v_g^{-1} + \beta_{22}\Omega_2 \tag{5.76}$$

Equations (5.72) and (5.73) correspond to an XPM-coupled bright gray soliton pair, which have the same width but different amplitudes. The parameter a indicates the depth of the intensity dip of the gray soliton. If $a = 1$, $\beta_{21} < 0$, and $\beta_{22} > 0$, a bright dark soliton pair is obtained [50]. An interesting feature of such a solution is that the bright soliton propagates in the normal dispersion regime, whereas the dark soliton propagates in the anomalous dispersion regime. This behavior is opposite to what would normally be expected and is due solely to the effect of XPM. In fact, XPM is twice as strong as SPM and induces a mutual frequency shift between the two pulses. The leading edge of the bright pulse is upshifted in frequency, while its trailing edge is downshifted in frequency by XPM. The reverse occurs for the dark pulse. In the normal (anomalous) dispersion regime, the group velocity decreases (increases) when the frequency increases; therefore, the leading edges of the pulses are retarded and the trailing edges are advanced. As a consequence, XPM counteracts the temporal spreading of the pulses induced by SPM and GVD, and the final result is that the coupled pulses propagate with their shape undistorted.

Besides the above-mentioned soliton pair solutions, Eqs. (5.41) and (5.42) can also support pairs with two bright or two dark solitons [53], as well as periodic solutions representing two pulse trains [54,55]. XPM-coupled soliton pairs can be generalized to multicomponent solitons, constituted by multiple pulses at different carrier frequencies. Such multicomponent solitons can occur in multichannel communication systems [56].

PROBLEMS

5.1 Calculate the maximum value of the SPM-induced nonlinear phase shift experienced by 10 ps pulses with 300 mW peak power and 1 ps pulses with 1 W peak power at the output of a 3 km long fiber having a loss of 1 dB/km and an effective mode area of 40 μm^2.

5.2 Consider a super-Gaussian pulse with amplitude given by Eq. (5.15). Derive expressions for the maximum frequency shifts from the carrier induced by self-phase modulation. Determine the positions on the pulse envelope at which the maximum shifts occur.

5.3 Show that the NLSE (4.49) is obtained from the Euler–Lagrange equation (5.18) considering the Lagrangian density given by Eq. (5.19).

5.4 Obtain the Lagrangian given by Eq. (5.22) from Eq. (5.17) considering the Lagrangian density in Eq. (5.19) and $U(z, \tau)$ in Eq. (5.21).

5.5 Considering the Lagrangian given by Eq. (5.22), reproduce Eqs. (5.23–5.25) from the reduced Euler–Lagrangian equation (5.20).

5.6 Consider two pulses with a FWHM width of 30 ps belonging to two adjacent WDM channels with spacing of 50 GHz. Assuming that the two pulses are introduced simultaneously at the fiber input, calculate the distance after which they are walked off for the cases of a standard transmission fiber ($D = 17\,\text{ps/(nm km)}$) and a nonzero dispersion-shifted fiber with $D = 3\,\text{ps/(nm km)}$.

5.7 Derive the dispersion relation given by Eq. (5.61) starting from Eqs. (5.58–5.60).

5.8 Confirm by direct substitution that bright gray soliton pair given by Eqs. (5.72) and (5.73) indeed satisfies the coupled NLSEs given by Eqs. (5.41) and (5.42) when the FWM terms are neglected.

REFERENCES

1. N. G. R. Broderick, T. M. Monroe, P. J. Bennett, and D. J. Richardson, *Opt. Lett.* **24**, 1395 (1999).
2. K. Nakajima and M. Ohashi, *IEEE Photon. Technol. Lett.* **14**, 492 (2002).
3. R. H. Stolen and C. Lin, *Phys. Rev. A* **17**, 1448 (1978).
4. R. K. Nesbet, *Variational Principles and Methods in Theoretical Physics and Chemistry*, Cambridge University Press, New York, 2003.
5. W. J. Firth, *Opt. Commun.* **22**, 226 (1977).
6. D. Anderson, *Phys. Rev. A* **27**, 3135 (1983).
7. D. Anderson, M. Lisak, and T. Reichel, *Phys. Rev. A* **38**, 1618 (1988).
8. M. Desaix, D. Anderson, and M. Lisak, *Phys. Rev. A* **40**, 2441 (1989).
9. S. K. Turitsyn and E. G. Shapiro, *Opt. Fiber Technol.* **4**, 151 (1998).
10. S. K. Turitsyn, I. Gabitov, E. W. Laedke, V. K. Mezentsev, S. L. Musher, E. G. Shapiro, T. Shafer, and K. H. Spatschek, *Opt. Commun.* **151**, 117 (1998).
11. M. H. Sousa, M. F. Ferreira, and E. M. Panameño, *SPIE Proc.* **5622**, 944 (2004).
12. R. Jackson, C. Jones, and V. Zharnitsky, *Physica D* **190**, 63 (2004).
13. M. F. Ferreira, M. V. Facão, S. V. Latas, and M. H. Sousa, *Fiber Integr. Opt.* **24**, 287 (2005).
14. S. Konar, M. Mishra, and S. Jana, *Chaos Solitons Fractals* **29**, 823 (2006).

15. R. G. Smith, *Appl. Opt.* **11**, 2489 (1972).

16. X. P. Mao, R. W. Tkach, A. R. Chraplyvy, R. M. Jopson, and R. M. Derosier, *IEEE Photon. Technol. Lett.* **4**, 66 (1992).

17. M. J. Potasek and G. P. Agrawal, *Electron. Lett.* **22**, 759 (1986).

18. M. Suzuki, I. Morita, N. Edagawa, S. Yamamoto, H. Taga, and S. Akiba, *Electron. Lett.* **31**, 2027 (1995).

19. S. Wabnitz, I. Uzunov, and F. Lederer, *IEEE Photon. Technol. Lett.* **8**, 1091 (1996).

20. A. Hasegawa, *Opt. Lett.* **9**, 288 (1984).

21. K. Kikuchi, *IEEE Photon. Technol. Lett.* **5**, 221 (1993).

22. A. Mecozzi, *J. Opt. Soc. Am. B* **11**, 462 (1994).

23. N. J. Smith and N. J. Doran, *Opt. Lett.* **21**, 570 (1996).

24. C. Lorattanasane and K. Kikuchi, *IEEE J. Quantum Electron.* **33**, 1084 (1997).

25. E. Ciaramella and M. Tamburrini, *IEEE Photon. Technol. Lett.* **11**, 1608 (1999).

26. K. A. Tai, A. Hasegawa, and A. Tomita, *Phys. Rev. Lett.* **56**, 135 (1986).

27. A. Hasegawa, *Opt. Lett.* **9**, 288 (1984).

28. K. Tai, A. Tomita, J. L. Jewel, and A. Hasegawa, *Appl. Phys. Lett.* **49**, 236 (1986).

29. A. R. Chraplyvy and J. Stone, *Electron. Lett.* **20**, 996 (1984).

30. J. Wang and K. Petermann, *J. Lightwave Technol.* **10**, 96 (1992).

31. D. Marcuse, A. R. Chraplyvy, and R. W. Tkach, *J. Lightwave Technol.* **12**, 885 (1994).

32. T. K. Chiang, N. Kagi, M. E. Marhic, and L. G. Kazovsky, *J. Lightwave Technol.* **14**, 249 (1996).

33. G. Belloti, M. Varani, C. Francia, and A. Bononi, *IEEE Photon. Technol. Lett.* **10**, 1745 (1998).

34. R. Hui, K. R. Demarest, and C. T. Allen, *J. Lightwave Technol.* **17**, 1018 (1999).

35. L. E. Nelson, R. M. Jopson, A. H. Gnauck, and A. R. Chraplyvy, *IEEE Photon. Technol. Lett.* **11**, 907 (1999).

36. S. Betti and M. Giaconi, *IEEE Photon. Technol. Lett.* **13**, 1304 (2001).

37. H. J. Thiele, R. I. Killey, and P. Bayvel, *Opt. Fiber Technol.* **8**, 71 (2002).

38. M. Shtaif, M. Eiselt, and L. D. Garrett, *IEEE Photon. Technol. Lett.* **12**, 88 (2000).

39. D. I. Kovsh, L. Liu, B. Bakhashi, A. N. Pilipetskii, E. A. Golovchenko, and N. S. Gergano, *IEEE J. Sel. Top. Quantum Electron.* **8**, 597 (2002).

40. G. P. Agrawal, *Phys. Rev. Lett.* **59**, 880 (1987).

41. S. Zhang, F. Lu, W. Xu, and J. Wang, *Opt. Fiber Technol.* **11**, 193 (2005).

42. A. S. Gouveia-Neto, M. E. Faldon, A. S. Sombra, P. G. Wigley, and J. R. Taylor, *Opt. Lett.* **13**, 901 (1988).

43. E. J. Greer, D. M. Patrick, P. G. Wigley, and J. R. Taylor, *Opt. Lett.* **15**, 851 (1990).

44. D. M. Patrick and A. D. Ellis, *Electron. Lett.* **29**, 227 (1993).

45. J. Yu and P. Jeppese, *IEEE Photon. Technol. Lett.* **13**, 833 (2001).

46. J. Yu, X. Zheng, C. Peucheret, A. Clausen, H. N. Poulsen, and P. Jeppesen, *J. Lightwave Technol.* **18**, 1007 (2000).

47. W. Wang, H. N. Poulsen, L. Rau, H.-F. Chou, J. E. Bowers, and D. J. Blumenthal, *J. Lightwave Technol.* **23**, 1105 (2005).

48. M. Galili, L. K. Oxenlowe, H. C. H. Hansen, A. T. Clausen, and P. Jeppesen, *IEEE J. Sel. Top. Quantum Electron.* **14**, 573 (2008).

49. K. Igarashi and K. Kikuchi, *IEEE J. Sel. Top. Quantum Electron.* **14**, 551 (2008).

50. S. Trillo, S. Wabnitz, E. M. Wright, and G. I. Stegeman, *Opt. Lett.* **13**, 871 (1988).

51. V. V. Afanasjev, Y. S. Kivshar, V. V. Konotop, and V. N. Serkin, *Opt. Lett.* **14**, 805 (1989).

52. M. Lisak, A. Höök, and D. Anderson, *J. Opt. Soc. B* **7**, 810 (1990).

53. V. V. Afanasjev, Y. S. Kivshar, V. V. Konotop, and V. N. Serkin, *Opt. Lett.* **14**, 805 (1989).

54. M. Florjanczyk and R. Tremblay, *Phys. Lett.* **141**, 34 (1989).

55. F. T. Hioe, *Phys. Rev.* **56**, 2373 (1997).

56. Y. S. Kivshar and G. P. Agrawal, *Optical Solitons: From Fibers to Photonic Crystals*, Academic Press, San Diego, CA, 2003, Chapter 9.

6

FOUR-WAVE MIXING

Four-wave mixing (FWM) or four-photon mixing (FPM) is a parametric process in which four waves or photons interact with each other due to the third-order nonlinearity of the material. As a result, when several channels with frequencies $\omega_1, \ldots, \omega_n$ are transmitted simultaneously over the same fiber, the intensity dependence of the refractive index not only leads to phase shifts within a channel, as discussed in Chapter 5, but also gives rise to signals at new frequencies such as $2\omega_i - \omega_j$ and $\omega_i + \omega_j - \omega_k$. For example, three copropagating waves give rise, by FWM, to nine new optical waves. In a wavelength division multiplexed (WDM) system with only 10 channels, hundreds of new components are generated in this way. If the WDM channels are equally spaced, all the new components generated within the bandwidth of the system fall at the original channel frequencies, giving rise to crosstalk [1–6]. In contrast to self-phase modulation (SPM) and cross-phase modulation (XPM), which become more significant for high bit rate systems, the FWM does not depend on the bit rate. The efficiency of this phenomenon depends strongly on phase matching conditions, as well as on the channel spacing, chromatic dispersion, and fiber length. The occurrence of the FWM phenomenon in optical fibers was observed for the first time by Stolen et al. [7] using a 9 cm long multimode fiber pumped by a double-pulsed YAG laser at 532 nm.

In this chapter, the FWM process is described mathematically starting from the nonlinear Schrödinger equation (NLSE). A special attention is paid to the important problem of phase matching and the impact of FWM in multichannel systems. Moreover, some important applications of FWM such as parametric amplification, parametric oscillation, optical phase conjugation, and the generation of squeezed states of light will also be discussed. Other important applications of FWM such as wavelength conversion and optical switching will be discussed in Chapter 14.

Nonlinear Effects in Optical Fibers. By Mário F. S. Ferreira.
Copyright © 2011 John Wiley & Sons, Inc. Published 2011 by John Wiley & Sons, Inc.

6.1 WAVE MIXING

Three waves with frequencies ω_i, ω_j, and ω_k propagating in a fiber can interact nonlinearly and give rise to new waves with the frequencies

$$\omega_{i,j,k} = \omega_i + \omega_j - \omega_k \qquad (6.1)$$

In principle, three waves with different frequencies would produce 27 possible mixing frequencies. However, considering Eq. (6.1) we can see that changing the places of the first two frequencies ($\omega_i \leftrightarrow \omega_j$), or if the third frequency (ω_k) is equal to the first or the second frequency (ω_i, ω_j), no new frequencies are generated. As a consequence, only nine new waves will be produced from the 27 possible variations. If the original frequencies are ω_1, ω_2, and ω_3, the frequencies of the nine new waves generated due to the FWM process will be, according to Eq. (6.1), ω_{112}, ω_{113}, ω_{123}, ω_{132}, ω_{221}, ω_{223}, ω_{231}, ω_{331}, and ω_{332}. In general, considering M original channels of a WDM system, the number of possible mixing products is given by [8]

$$N = \frac{1}{2}M^2(M-1) \qquad (6.2)$$

Figure 6.1 illustrates the variation of N with M.

If the channels of a WDM system have an equal frequency spacing, the new generated waves will have the same spacing and a large number of them will coincide

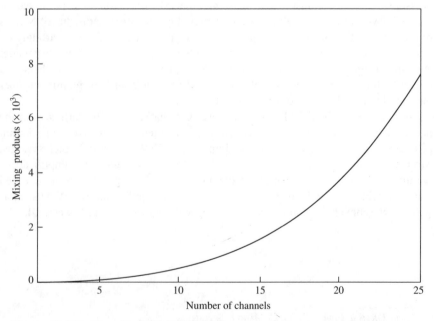

Figure 6.1 Number of possible FWM mixing terms against the number of channels.

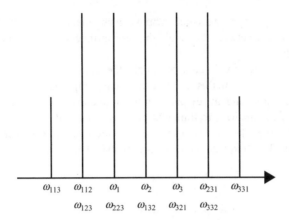

Figure 6.2 Four-wave mixing terms for three channels equally spaced in frequency.

exactly with the original channels or with each other. This possibility is illustrated in Fig. 6.2, where the mixing frequencies due to FWM are represented considering the existence of three original channels. In this case, from the expected nine new waves, only four new frequencies are really generated.

A different situation occurs if the original channels have an equal wavelength spacing. In this case, as the frequency spacing depends on the wavelength, the frequency differences between adjacent channels are not exactly equal. As a consequence, the new generated frequencies do not coincide exactly with those of the original channels.

The efficiency of the four-wave mixing process depends on the relative phase among the interacting optical waves. In the quantum mechanical description, FWM occurs when photons from one or more waves are annihilated and new photons are created at different frequencies. In this process, the rules of conservation of energy and momentum must be fulfilled. The conservation of momentum leads to the phase matching condition.

Considering the case in which two photons at frequencies ω_1 and ω_2 are annihilated with simultaneous creation of two photons at frequencies ω_3 and ω_4, the conservation of energy imposes the condition

$$\omega_1 + \omega_2 = \omega_3 + \omega_4 \tag{6.3}$$

On the other hand, the phase matching is given by the condition $\Delta k = 0$, where

$$\begin{aligned} \Delta k &= \beta_3 + \beta_4 - \beta_1 - \beta_2 \\ &= (n_3\omega_3 + n_4\omega_4 - n_1\omega_1 - n_2\omega_2)/c \end{aligned} \tag{6.4}$$

The FWM efficiency is significantly higher in dispersion-shifted fibers than in standard fibers due to the low value of the dispersion in the first case. That efficiency is reduced in WDM systems when the channel spacing is increased, since this corresponds to a higher phase mismatch between channels. As a result, the intensity of the mixing products between channels that are farther away in dispersive fibers is relatively small.

In practice, it is relatively easy to satisfy the phase matching condition in the degenerate case $\omega_1 = \omega_2$. In this situation, a strong pump at $\omega_1 = \omega_2 \equiv \omega_p$ generates a low-frequency sideband at ω_3 and a high-frequency sideband at ω_4, when we assume $\omega_4 > \omega_3$. In analogy to Raman scattering, these sidebands are referred to as the Stokes and anti-Stokes bands, respectively, which are often also called the *signal* and *idler* bands. The frequency shift of the two sidebands is given by

$$\Omega_s = \omega_p - \omega_3 = \omega_4 - \omega_p \tag{6.5}$$

6.2 MATHEMATICAL DESCRIPTION

The nonlinear Schrödinger equation given by Eq. (4.49) can be used as the starting point to describe the interaction among the waves involved in the FWM process. Assuming that the pulse durations are sufficiently long, we can neglect the linear term corresponding to the second-order dispersion in Eq. (4.49), which becomes

$$\frac{\partial U}{\partial z} = i\gamma |U|^2 U \tag{6.6}$$

In order to consider the FWM mixing process and considering the nondegenerate case ($\omega_1 \neq \omega_2$), the amplitude U in Eq. (6.6) is assumed to be the result of the superposition of four waves:

$$U = U_1\, e^{i(\beta_1 z - \omega_1 t)} + U_2\, e^{i(\beta_2 z - \omega_2 t)} + U_3\, e^{i(\beta_3 z - \omega_3 t)} + U_4\, e^{i(\beta_4 z - \omega_4 t)} \tag{6.7}$$

In the following, we will consider that the four frequencies in Eq. (6.7) satisfy the relation given by Eq. (6.3).

Substituting Eq. (6.7) into Eq. (6.6) and considering Eq. (6.3), the following set of four coupled equations for the normalized amplitudes U_j is obtained:

$$\frac{\partial U_1}{\partial z} = i\gamma \left[\left(|U_1|^2 + 2\sum_{j\neq 1} |U_j|^2 \right) U_1 + 2U_3 U_4 U_2^* \, e^{i\Delta kz} \right] \tag{6.8}$$

$$\frac{\partial U_2}{\partial z} = i\gamma \left[\left(|U_2|^2 + 2\sum_{j\neq 2} |U_j|^2 \right) U_2 + 2U_3 U_4 U_1^* \, e^{i\Delta kz} \right] \tag{6.9}$$

$$\frac{\partial U_3}{\partial z} = i\gamma \left[\left(|U_3|^2 + 2\sum_{j\neq 3} |U_j|^2 \right) U_3 + 2U_1 U_2 U_4^* \, e^{-i\Delta kz} \right] \tag{6.10}$$

$$\frac{\partial U_4}{\partial z} = i\gamma\left[\left(|U_4|^2 + 2\sum_{j\neq 4}|U_j|^2\right)U_4 + 2U_1U_2U_3^* \, e^{-i\Delta kz}\right] \qquad (6.11)$$

where Δk is the wave vector mismatch, given by Eq. (6.4), and

$$\gamma = \frac{n_2\omega}{cA_{\text{eff}}} \qquad (6.12)$$

is an averaged nonlinear parameter, which ignores the small variation due to the slightly different frequencies of the involved waves. In deriving Eqs. (6.8)–(6.11), only nearly phase-matched terms were kept. Fiber loss may be included by adding the term $-(\alpha/2)U_j, j = 1, 2, 3, 4$, to the right-hand side of each equation, respectively. The first term inside the brackets in Eqs. (6.8)–(6.11) describes the effect of SPM, whereas the second term is responsible for XPM. Since these terms can lead to a phase alteration only, the generation of new frequency components is provided by the remaining FWM terms. If only two pump waves (with frequencies ω_1 and ω_2) are present at the fiber input, the signal and idler waves are generated from the wideband noise that is always present in communication systems. However, in some cases the signal wave can already be present at the fiber input. If the phase matching condition is satisfied, both the signal and the idler waves grow during propagation due to the optical power transferred from the two pump waves.

A numerical approach is necessary to solve Eqs. (6.8)–(6.11) exactly. However, some useful approximate results can be obtained under certain conditions, as discussed in the following sections.

6.3 PHASE MATCHING

Let us assume that the pump waves are much more intense than the signal and idler waves and that their phases are not matched. In such a case, the power transfer between these waves is very ineffective and we can assume that the pump waves remain undepleted during the FWM process. The solutions of Eqs. (6.8) and (6.9) are then given by

$$U_j(z) = \sqrt{P_j}\exp[i\gamma(P_j + 2P_{3-j})z], \quad j = 1, 2 \qquad (6.13)$$

where $P_j = |U_j(0)|^2$ are the input pump powers. In this approximation, the pump waves experience only a phase shift due to SPM and XPM. We will also consider the case in which the signal wave has already a finite value at the fiber input. However, the input signal power is assumed to be much less than the input pump powers. In these circumstances, due to the phase mismatch between the involved waves, Eq. (6.10) has the following approximate solution:

$$U_3(z) = \sqrt{P_3}\exp[i\gamma 2(P_1 + P_2)z] \qquad (6.14)$$

Substituting Eqs. (6.13) and (6.14) in Eq. (6.11), we obtain the following equation for the idler field:

$$\frac{dU_4^*}{dz} = -2i\gamma\left[(P_1+P_2)U_4^* + \sqrt{P_1P_2P_3}\exp\{i[\Delta k - \gamma(P_1+P_2)]z\}\right] \quad (6.15)$$

Using the transformation

$$V_4 = U_4 \exp\{-2i\gamma(P_1+P_2)z\} \quad (6.16)$$

Eq. (6.15) becomes

$$\frac{dV_4^*}{dz} = -2i\gamma\sqrt{P_1P_2P_3}\exp(i\kappa z) \quad (6.17)$$

where the parameter

$$\kappa = \Delta k + \gamma(P_1+P_2) \quad (6.18)$$

corresponds to the effective phase mismatch and $P_1 + P_2 = P_p$ is the total pump power.

If Eq. (6.18) is integrated from $z = 0$ to $z = L$, and considering that the amplitude of the idler is zero at the fiber input, we obtain the result

$$V_4^*(L) = \frac{2\gamma}{\kappa}\sqrt{P_1P_2P_3}[1 - \exp(i\kappa L)] \quad (6.19)$$

Thus, the power of the idler wave is given by

$$P_4(L) = |V_4(L)|^2 = 4\gamma^2 P_1 P_2 P_3 L^2 \left[\frac{\sin(\kappa L/2)}{\kappa L/2}\right]^2 \quad (6.20)$$

If the phase matching condition $\kappa = 0$ is fulfilled, Eq. (6.20) shows that the power of the idler increases quadratically with the fiber length, as represented in Fig. 6.3 (solid curve). However, in such a case it is not reasonable to consider that the amplitudes of the pump and signal waves do remain constant along the fiber, as assumed in deriving Eq. (6.20). If the phases are not matched, the intensity of the idler wave shows a periodic evolution along the fiber, as described by Eq. (6.20) (dashed curve in Fig. 6.3).

Let us consider the degenerate case $\omega_1 = \omega_2 \equiv \omega_p$, where ω_p denotes the pump frequency. In this case, the effective phase mismatch is given by

$$\kappa = \Delta k + 2\gamma P_p \quad (6.21)$$

where P_p is the pump power and Δk is the linear phase mismatch. The following discussion is limited to the case of single-mode fibers, for which the phase mismatch contribution due to the waveguide dispersion is negligible. As a consequence, the

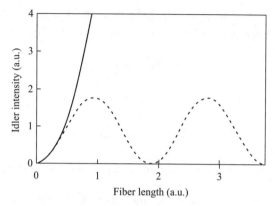

Figure 6.3 Idler against the fiber length for phase matching (solid curve) and a phase mismatch between the waves (dashed curve).

linear phase mismatch depends only on the material dispersion, as given by Eq. (6.4). Such a phase mismatch can be expressed in terms of the frequency shift Ω_s given by Eq. (6.5) if we use an expansion of Δk about the pump frequency ω_p. By retaining up to terms quadratic in Ω_s in this expansion, one obtains

$$\Delta k = \beta_2 \Omega_s^2 \tag{6.22}$$

where $\beta_2 = \mathrm{d}^2\beta/\mathrm{d}\omega^2$ is the group velocity dispersion (GVD) coefficient at the pump frequency.

One can define a coherence length, given by

$$L_{\text{coh}} = 2\pi/\kappa_{\text{max}} \tag{6.23}$$

where κ_{max} is the maximum tolerable phase mismatch. Significant four-wave mixing occurs only if $L < L_{\text{coh}}$. For λ_p below the zero of dispersion wavelength (normal dispersion regime), we have $\beta_2 > 0$. In such a case, in order to maximize L_{coh}, one has to keep both Δk and γP_0 small, by using small frequency shifts Ω_s and low pump powers. When Δk dominates, we have

$$L_{\text{coh}} \approx \frac{2\pi}{|\Delta k|} = \frac{2\pi}{|\beta_2|\Omega_s^2} \tag{6.24}$$

Considering standard single-mode fibers in the visible range, one has $\beta_2 \approx 50\text{--}60$ ps^2/km, and $L_{\text{coh}} \approx 1$ km for $f_s = \Omega_s/2\pi \approx 50$ GHz.

Phase matching with the pump positioned in the normal dispersion regime can be effectively achieved by using a birefringent fiber [9]. In this case, the nonlinear and the material phase mismatch contributions will have the same sign. However, by placing the pump in the slow propagation axis of the fiber, while positioning the idler and the signal in the fast axis, the sign of the waveguide mismatch contribution will cancel the material and nonlinear mismatch contributions.

For a pump wavelength in the anomalous dispersion region ($\beta_2 < 0$), the negative value of Δk can be compensated by the fiber nonlinearity. For phase matching, one has

$$\kappa \approx \beta_2 \Omega_s^2 + 2\gamma P_p = 0 \tag{6.25}$$

From this, we obtain that the frequency shift is given by

$$\Omega_s = \sqrt{2\gamma P_p / |\beta_2|} \tag{6.26}$$

which coincides with the gain peak frequency of the modulation instability (see Section 5.1.4).

When $|\beta_2|$ is small, we must take into account the fourth-order term in the expansion of Δk, which becomes

$$\Delta k \approx \beta_2 \Omega_s^2 + \frac{\beta_4}{12} \Omega_s^4 \tag{6.27}$$

where $\beta_4 = \mathrm{d}^4 \beta / \mathrm{d}\omega^4$. In this case, the phase matching condition is given by

$$\kappa \approx \beta_2 \Omega_s^2 + \frac{\beta_4}{12} \Omega_s^4 + 2\gamma P_0 = 0 \tag{6.28}$$

We observe from Eq. (6.28) that phase matching can be achieved even for a pump wavelength in the normal GVD regime ($\beta_2 > 0$) if $\beta_4 < 0$. This condition can be easily realized in tapered and microstructured fibers (MFs) [10,11].

6.4 IMPACT AND CONTROL OF FWM

In the case of WDM systems with equal channel spacing, the degradation due to FWM is particularly severe, since in this case most new frequencies coincide with the original channel frequencies. The interference between the original and the new generated waves depends on the bit pattern and leads to significant fluctuations in the detected signal at the receiver, thus increasing the bit error rate (BER) in the system. Note that in systems with channels equally spaced in wavelength, the frequency spacing will not be uniform. However, the unequal frequency spacing in this case is not sufficient to prevent interference. The difference in frequency spacing, and hence the offset of mixing product from the channel, must be at least twice the bit rate to avoid interference [12]. To prevent the coincidence of the mixing products with any channel, the difference between any two channel frequencies must be unique [12]. Such an objective can be achieved with a computer search.

In the case of WDM systems with unequal channel spacing, crosstalk due to FWM is suppressed, since the new frequencies fall in between the existing channel frequencies and only add to overall noise. The use of unequal channel spacings to reduce the FWM-induced degradation was shown to be effective in a 1999

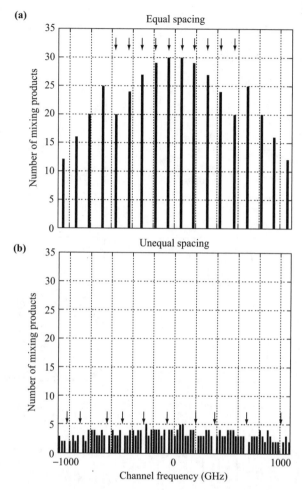

Figure 6.4   Generation of new mixing products due to the FWM process in a system with 10 channels. (a) Equally spaced channels. (b) Unequally spaced channels. Arrows indicate the channel frequencies. (After Ref. [13]; © 1995 IEEE.)

experiment, in which 22 channels, each operating at 10 Gb/s, were transmitted over 320 km of dispersion-shifted fiber with 80 km amplifier spacing [5].

Figure 6.4 shows the number of mixing products at various frequencies for equal and unequal spacing for a 10-channel system, where the mixing products are offset from the channels by at least 25 GHz and the channels are separated by at least 1 nm. Figure 6.5 shows simulation results for the channel placements of Fig. 6.4 [13]; the improvement achieved using unequally spaced channels is clearly demonstrated.

If three waves with the frequencies ω_i, ω_j, and ω_k copropagate inside the fiber and assuming that they are not depleted by the generation of the mixing products, the time-

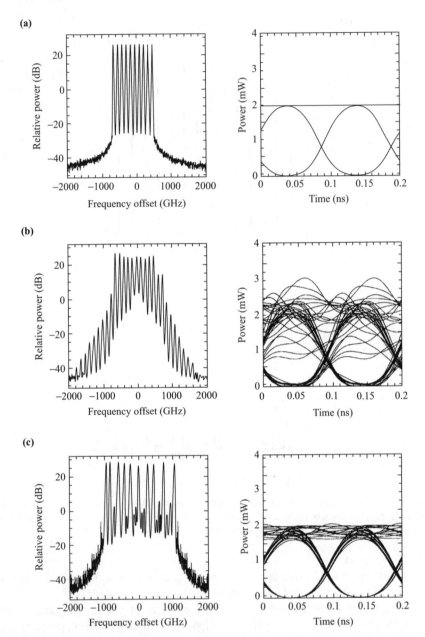

Figure 6.5   Simulation results for the channel placements of Fig. 6.4. The system consists of 500 km of fiber with the dispersion zero located at zero frequency offset. Optical amplifiers are spaced every 50 km. The channels are modulated at 10 Gb/s and are launched into each span of fiber with 1 mW average power. (a) System input power spectrum and eye diagram. (b) System output, equally spaced channels. (c) System output, unequally spaced channels. (After Ref. [13]; © 1995 IEEE.)

averaged optical power for the new generated frequency component ω_F can be written as [8]

$$P_F(L) = (D_F \gamma L_{\text{eff}})^2 P_i P_j P_k \eta \, e^{-\alpha L} \tag{6.29}$$

where P_i, P_j, and P_k are the input powers of the channels, α is the fiber attenuation coefficient, L_{eff} is the effective length of the fiber, γ is the nonlinear parameter, and D_F is the degeneracy factor, defined such that its value is 1 when $i = j$ but doubles when $i \neq j$. The parameter η is a measure of the FWM efficiency and is given by [1]

$$\eta = \frac{\alpha^2}{\alpha^2 + \Delta k^2} \left[1 + \frac{4e^{-\alpha L} \sin^2(\Delta k L/2)}{(1 - e^{-\alpha L})^2} \right] \tag{6.30}$$

The FWM efficiency depends on the channel spacing through the phase mismatch $\Delta k = \beta_F + \beta_k - \beta_i - \beta_j$. For a two-tone product with channel spacing Δf, we have [1]

$$\Delta k = \frac{2\pi \lambda^2}{c} \Delta f^2 \left[D + \Delta f \frac{\lambda^2}{c} \frac{dD}{d\lambda} \right] \tag{6.31}$$

where the dispersion D and its slope are computed at λ_k. Considering a value $D \approx 22 \, \text{ps/(nm km)}$, which corresponds to the dispersion of pure silica at a wavelength of 1550 nm, η nearly vanishes for typical spacings of 50 GHz or more. In contrast, $\eta \approx 1$ when the channels lie around the zero dispersion wavelength (ZDW) of the fiber, resulting in a considerable fraction of power transferred to the FWM component.

Assuming that all waves have the same input power P_{in} and a perfect phase matching ($\eta = 1$), the ratio of the generated power P_F due to the FWM process to the exiting power of one channel, $P_{\text{out}} = P_{\text{in}} e^{-\alpha L}$, is

$$\frac{P_F}{P_{\text{out}}} = (D_F \gamma L_{\text{eff}})^2 P_{\text{in}}^2 \tag{6.32}$$

Considering a nonlinear parameter $\gamma = 1.6 \times 10^{-3} \, \text{mW}^{-1} \, \text{km}^{-1}$ and an effective length $L_{\text{eff}} = 22 \, \text{km}$, we obtain

$$\frac{P_F}{P_{\text{out}}} \approx 5 \times 10^{-3} P_{\text{in}}^2 \quad (\text{mW}) \tag{6.33}$$

if all the waves have different frequencies ($D_F = 2$). Assuming an input power of 1 mW in each channel, the FWM-generated output power will be about 0.5% of the power of one channel at the fiber output. This indicates the level of crosstalk among channels due to FWM. The efficiency of the FWM process can be reduced by decreasing the channel power in the fiber. However, using such an approach the signal-to-noise ratio will also decrease, leading to a deterioration of the system performance.

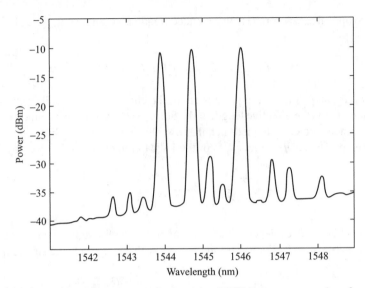

Figure 6.6 Generation of new frequencies because of FWM when waves at three frequencies are incident in the fiber. (After Ref. [13]; © 1995 IEEE.)

Figure 6.6 shows the optical spectrum measured at the output of a 25 km long dispersion-shifted fiber ($D = -0.2$ ps/(nm km) at the central channel) when three 3 mW channels are launched into it. The nine new frequencies due to FWM are clearly seen. Due to the unequal channel spacing used in this experiment, none of the newly generated waves coincides with the original channels.

Figure 6.7 shows the ratio of the FWM-generated power to the output power as a function of channel spacing for different dispersion coefficients. We observe that FWM efficiency can be significantly reduced by choosing a nonzero value of dispersion. For a given level of crosstalk, a larger value of the dispersion allows a smaller channel spacing.

Even for a nonzero value of dispersion, the FWM process can be resonantly enhanced for certain values of channel spacing due to the contribution of SPM and XPM [14]. In fact, as shown by Eq. (6.21), both these effects can produce phase matching when the GVD is in the anomalous regime. The resonance enhancement of FWM occurs if the frequency of the gain peak of modulation instability nearly coincides with the channel spacing in a WDM system. Using Eq. (6.26), such channel spacing is approximately given by

$$\Delta f_{ch} = \frac{1}{2\pi}\left(\frac{2\gamma P_{ch}}{|\beta_2|}\right)^{1/2} \tag{6.34}$$

Considering the values $P_{ch} = 5$ mW, $\beta_2 = -0.1$ ps^2/km, and $\gamma = 2$ W^{-1} km^{-1}, we obtain a channel spacing $\Delta f_{ch} \approx 70$ GHz, which is within the range usually considered in modern WDM systems.

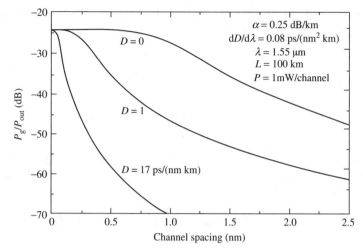

Figure 6.7   Ratio of FWM-generated power to the output power as a function of channel spacing for different dispersion coefficients. (After Ref. [13]; © 1995 IEEE.)

In spite of the advantages of using fibers with high local dispersion to reduce the FWM efficiency, it is also very important to have a small dispersion of the fiber span in the case of high bit rate communication systems. For example, in the case of a WDM system operating at 10 Gb/s over 1000 km, an average dispersion below 1 ps/(nm km) is required [13]. A solution for the above dilemma is provided by the technique of dispersion management. In this case, fibers with normal and anomalous dispersion are combined to form a periodic dispersion map, such that the local GVD is high but its average value is kept low. Due to its simplicity of implementation, the dispersion management technique became quite common since 1996 to control the FWM-induced limitations in WDM systems [15].

6.5   FIBER PARAMETRIC AMPLIFIERS

Four-wave mixing can be used with advantage to realize fiber-optic parametric amplifiers (FOPAs). In fact, the main difference between the four-wave mixing experiments and the parametric amplification experiments is whether a signal at the phase-matched frequency is copropagated with the pump or not. A FOPA offers a wide gain bandwidth and may be tailored to operate at any wavelength [16–20]. Besides the discrete FOPAs, distributed parametric amplification, that is, the use of a transmission fiber itself for parametric amplification of communication signals, has also been proposed and demonstrated [21].

6.5.1   FOPA Gain and Bandwidth

Using the result given by Eq. (6.13) for the pump fields and the transformation

$$V_j = U_j \exp\{-2i\gamma(P_1 + P_2)z\}, \quad j = 3, 4 \tag{6.35}$$

we obtain from Eqs. (6.10) and (6.11) the following coupled equations governing the evolution of the signal and idler waves:

$$\frac{dV_3}{dz} = 2i\gamma\sqrt{P_1 P_2}\exp(-i\kappa z)V_4^* \tag{6.36}$$

$$\frac{dV_4^*}{dz} = -2i\gamma\sqrt{P_1 P_2}\exp(i\kappa z)V_3 \tag{6.37}$$

where the effective phase mismatch κ is given by Eq. (6.18). If only the signal and the pumps are launched at $z = 0$, that is, assuming that $V_4^*(0) = 0$, Eqs. (6.36) and (6.37) have the following solutions:

$$V_3(z) = V_3(0)(\cosh(gz) + (i\kappa/2g)\sinh(gz))\exp(-j\kappa z/2) \tag{6.38}$$

$$V_4^*(z) = -i(\gamma/g)\sqrt{P_1 P_2}V_3(0)\sinh(gz)\exp(j\kappa z/2) \tag{6.39}$$

where g is the parametric gain, given by

$$g = \sqrt{4\gamma^2 P_1 P_2 - \left(\frac{\kappa}{2}\right)^2} \tag{6.40}$$

The maximum gain occurs for perfect phase matching ($\kappa = 0$) and is given by

$$g_{max} = 2\gamma\sqrt{P_1 P_2} \tag{6.41}$$

From Eq. (6.38) we can write the following result for the signal power, $P_3 = |V_3|^2$:

$$P_3(z) = P_3(0)\left[1 + \left(1 + \frac{\kappa^2}{4g^2}\right)\sinh^2(gz)\right] = P_3(0)\left[1 + \frac{4\gamma^2 P_1 P_2}{g^2}\sinh^2(gz)\right] \tag{6.42}$$

The idler power $P_4 = |V_4|^2$ can be obtained by noting from Eqs. (6.38) and (6.39) that $P_3(z) - P_4(z) = \text{constant} = P_3(0)$. The result is

$$P_4(z) = P_3(0)\frac{4\gamma^2 P_1 P_2}{g^2}\sinh^2(gz) \tag{6.43}$$

The unsaturated single-pass gain of a FOPA of length L becomes

$$G_p = \frac{P_3(L)}{P_3(0)} = 1 + \frac{4\gamma^2 P_1 P_2}{g^2}\sinh^2(gL) \tag{6.44}$$

According to Eq. (6.40), amplification ($g > 0$) occurs only in the case

$$|\kappa| < 4\gamma\sqrt{P_1 P_2} \qquad (6.45)$$

Since $|\kappa|$ must be small, this also means good phase matching. For $|\kappa| > 4\gamma\sqrt{P_1 P_2}$, the parametric gain becomes imaginary and there is no longer amplification, but rather a periodical power variation of the signal and idler waves.

In the case of a single pump, the effective phase mismatch is given by Eq. (6.21) and the parametric gain coefficient becomes

$$g^2 = \left[(\gamma P_p)^2 - (\kappa/2)^2\right] = -\Delta k\left(\frac{\Delta k}{4} + \gamma P_p\right) \qquad (6.46)$$

where P_p is the pump power. The maximum value of the parametric gain occurs when $\Delta k = -2\gamma P_p$ and is given by

$$g_{max} = \gamma P_p \qquad (6.47)$$

The unsaturated single-pass gain may be written as

$$G_p = \frac{P_3(L)}{P_3(0)} = 1 + \left[\frac{\gamma P_p}{g}\sinh(gL)\right]^2$$

$$= 1 + (\gamma P_p L)^2\left[1 + \frac{gL^2}{6} + \frac{gL^4}{120} + \cdots\right]^2 \qquad (6.48)$$

For signal wavelengths close to the pump wavelength, $\Delta k \approx 0$ and

$$G_p \approx 1 + (\gamma P_p L)^2 \qquad (6.49)$$

In this case of perfect phase matching ($\kappa = 0$) and assuming that $\gamma P_p L \gg 1$, the unsaturated single-pass gain becomes

$$G_p \approx \sinh^2(gL) \approx \frac{1}{4}\exp(2\gamma P_p L) \qquad (6.50)$$

A very simple expression for the FOPA peak gain may be obtained if Eq. (6.50) is rewritten in decibel units:

$$G_p(dB) = 10\log_{10}\left[\frac{1}{4}\exp(2\gamma P_p L)\right] = P_p L S_p - 6 \qquad (6.51)$$

where

$$S_p = 10\log_{10}(e^2)\gamma \approx 8.7\gamma \qquad (6.52)$$

is the parametric gain slope in dB/(W km).

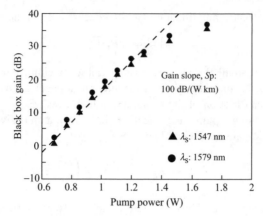

Figure 6.8 Measured parametric gain slope S_p using 500 m HNLF with $\gamma = 11\,\mathrm{W}^{-1}\,\mathrm{km}^{-1}$. (After Ref. [20]; © 2002 IEEE.)

Figure 6.8 shows the small signal gain against the pump power provided by a parametric amplifier constituted by a 500 m long highly nonlinear fiber (HNLF) with $\gamma = 11\,\mathrm{W}^{-1}\,\mathrm{km}^{-1}$ [18,20]. The gain slope in the undepleted pump region is 100 dB/(W km). The input signal power was -19.5 dBm and the maximum achieved output signal power was 21 dBm, limited by signal-induced stimulated Brillouin scattering (SBS). Figure 6.9 shows the amplifier gain versus the wavelength difference between pump and signal for the same fiber of Fig. 6.8, considering a pump power of 1.4 W. The region with exponential gain (corresponding to perfect phase matching) and the region with quadratic gain (corresponding to $\lambda_3 \approx \lambda_1$) are marked in the figure.

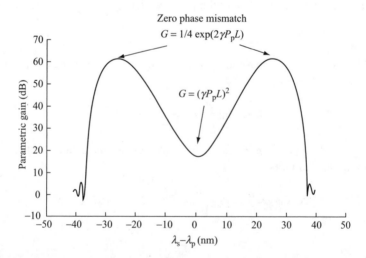

Figure 6.9 Calculated amplifier gain versus the wavelength difference between pump and signal for the same fiber of Fig. 6.8, considering a pump power of 1.4 W. (After Ref. [20]; © 2002 IEEE.)

Considering the approximation given by Eq. (6.22) for the linear phase mismatch, the FOPA bandwidth $\Delta\Omega$ can be obtained from the maximum effective phase mismatch

$$\kappa_m = \beta_2(\Omega_s + \Delta\Omega)^2 + 2\gamma P_p \qquad (6.53)$$

with $\Delta\Omega \ll \Omega_s$ and Ω_s given by Eq. (6.26). This maximum value of the effective phase mismatch occurs when the parametric gain in Eq. (6.46) vanishes, which gives $\kappa_m = 2\gamma P_p$. In these circumstances, the FOPA bandwidth is approximately given by

$$\Delta\Omega \approx \sqrt{\frac{\gamma P_p}{2|\beta_2|}} \qquad (6.54)$$

Equation (6.54) shows that the FOPA bandwidth can be increased by increasing the nonlinear parameter γ and reducing $|\beta_2|$. This is the reason why modern FOPAs generally use HNLFs and the pump wavelengths are chosen near the ZDW of the fiber. Moreover, Eq. (6.48) shows that, for a fixed value of the FOPA gain, the value of $\gamma P_p L$ must be kept constant. In such a case, the amplifier bandwidth will increase with decreasing L. Using an HNLF provides the possibility of simultaneously decreasing L and increasing the nonlinear parameter γ. For example, a value $\gamma P_p L = 10$ can be achieved with a pump power of 1 W and an HNLF with $L = 1$ km and $\gamma = 10\,\mathrm{W^{-1}\,km^{-1}}$. The bandwidth will be increased 5 times using, for instance, a fiber length of 40 m, a pump power of 5 W, and a nonlinear parameter $\gamma = 50\,\mathrm{W^{-1}\,km^{-1}}$. Such a high value of γ can be achieved using some types of HNLFs, such as tapered or photonic fibers, as will be seen in Chapter 13.

As the FOPA bandwidth is proportional to the square root of the pump power, it could be, in principle, arbitrarily increased if enough optical power is available. However, such high values of the pump power are limited in practice by SBS, which shows a reduced threshold in the case of HNLFs due to the small effective core area of this type of fibers.

The limitations imposed by the SBS process can be partially circumvented if the pump power is distributed between two pumps, instead of being concentrated in one pump. Moreover, a two-pump FOPA offers additional degrees of design freedom, which makes it fundamentally different from the conventional one-pump device [22–27]. Figure 6.10 illustrates the operating principle of a two-pump FOPA, with one pump in the anomalous regime and the other pump in the normal dispersion regime. A flat, broadband parametric gain can be generated by controlling three distinct parametric processes. First, the degenerate exchange $2\omega_{1,2} \to \omega_{1-,2-} + \omega_{1+,2+}$, which is recognized as modulation instability. Second, phase conjugation [25], which allows for symmetric, nondegenerate FWM: $\omega_1 + \omega_2 \to \omega_{1-} + \omega_{2+}$ and $\omega_1 + \omega_2 \to \omega_{1+} + \omega_{2-}$. Finally, Bragg scattering [25,28] enables the inherently stable exchanges $\omega_1 + \omega_{2+} \to \omega_{1+} + \omega_2$ and $\omega_{1-} + \omega_2 \to \omega_1 + \omega_{2-}$. When the signal frequency is close to the pump frequency, modulation instability is the dominant parametric process, whereas phase conjugation is responsible for photon creation in spectral regions farther from pump frequencies [25].

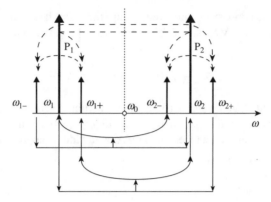

Figure 6.10 Two-pump parametric processes. $P_{1,2}$ are pumps and $1\pm, 2\pm$ are parametric sidebands. Modulation instability is indicated by thin dashed lines, phase conjugation by thick dashed lines, and Bragg scattering by thin solid lines. (After Ref. [26]; © 2003 Elsevier.)

Properly selecting the pump wavelengths can provide a flat, broadband parametric gain. Figure 6.11a illustrates the effect of pump spectral tuning, for the case of a two-pump FOPA constituted by a 1 km long HNLF with a ZDW at 1578 nm [26]. When the pump wavelengths are at 1552 and 1599 nm, the parametric gain between the inner bands is negligible. However, tuning the pumps to 1561 and 1593 nm provides a considerable gain increase across the entire spectral region. After choosing the proper pump wavelengths, the pump powers can be adjusted to achieve an optimum gain profile, as shown in Fig. 6.11b.

The polarization sensitivity is an important issue for any fiber communication device. This is indeed the main problem of a one-pump FOPA, which presents a high degree of polarization sensitivity. In the case of a two-pump FOPA, the gain is maximized when the pump and signal are copolarized along the entire interaction length. Any deviation from such a copolarized state determines a significant reduction of the parametric gain, which can become negligible for a signal polarization orthogonal to two copolarized pumps [29]. A simple and elegant solution to this problem can be achieved using orthogonally multiplexed pumps [30–32]. This configuration results in the near polarization invariance of the parametric process. However, such improvement is achieved at expense of a significant reduction of the parametric gain, compared to the copolarized two-pump scheme [24].

6.6 PARAMETRIC OSCILLATORS

A fiber-optic parametric oscillator (FOPO) can be realized by placing a fiber inside an optical cavity and pumping it with a single pump beam. The presence of the cavity allows for the coherent buildup of the amplified spontaneous emission (ASE), from which the signal and idler waves are created at frequencies determined by the phase matching condition. Both the signal and idler waves are amplified through the FWM process and are subsequently emitted by the FOPO.

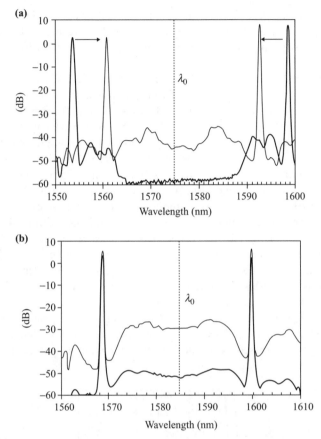

Figure 6.11 (a) Effect of pump spectral tuning in a two-pump FOPA. (b) Effect of pump power variation in a spectrally tuned two-pump FOPA. (After Ref. [26]; © 2003 Elsevier.)

The first FOPO was reported in 1987 [33] and used a cavity of a few meters in length pumped by Q-switched pulses from a 1.06 μm Nd:YAG laser. Synchronous pumping was realized adjusting the cavity length and the laser emitted 1.15 μm pulses of about 65 ps duration. After that experiment, several other FOPOs have been constructed using standard and highly nonlinear dispersion-shifted optical fibers [34–37]. A FOPO was realized in a 1988 experiment using a 100 m long ring cavity pumped synchronously with a mode-locked color-center laser operating in the 1550 nm wavelength region [34]. FOPOs of this type, which are realized by pumping optical fibers in the anomalous dispersion region, are known as modulation instability lasers. This kind of lasers offer the possibility of converting a continuous-wave (CW) pump into a train of short optical pulses, instead of generating a CW output [38].

Tunability is an important feature of any laser source. A tunable range of 40 nm was achieved in a 1999 experiment with a FOPO employing a nonlinear Sagnac interferometer as a FOPA [35]. The laser was pumped by using 7.7 ps mode-locked pulses from a color-center laser operating at 1539 nm. More recently, a CW FOPO that delivers few mW output power over a tunable range from 1510 to 1600 nm was reported [37].

Microstructered Fibers (MFs), which will be discussed in Chapter 13, allow the construction of very short devices and offer the promise of dramatically extending the wavelengths of operation and efficiency of FOPOs [39–42]. The first MF-based OPO was reported in 2002 [39] and utilized 2.1 m of MF within a Fabry–Perot cavity in which a diffraction grating in Littrow configuration was used as a tunable element. A tuning range of 40 nm and output powers of the order of a few hundred microwatts were achieved. In a 2005 experiment, a pulsed oscillator using a 75 cm long MF in a ring configuration was demonstrated [40]. A saturated output power of about 500 μW, a pulse width of 400 fs, and an impressive tuning range of 200 nm were obtained. Similar results in terms of wavelength tunability were reported more recently [41]. The first OPO capable of producing sub-100 fs pulses was reported very recently [42]. The OPO emitted nearly transform-limited 70 fs pulses at a center wavelength of 880 nm and incorporated a reduced diameter 4.2 cm long MF within a Fabry–Perot cavity.

The principal drawback of virtually all MF-based OPOs already realized is their relatively low output power. It was found experimentally that a good operating point for pulsed FOPOs is achieved for a moderate amount of self-phase modulation of the pump ($\gamma P_p L \sim \pi$) [43]. This means that reducing the fiber length the pump power must be increased. Considering an MF with a nonlinear coefficient

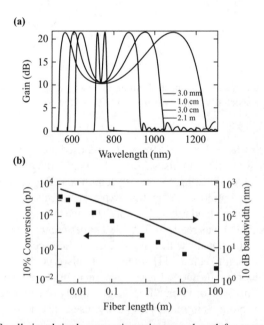

Figure 6.12 (a) Small-signal single-pass gain against wavelength for several fiber lengths. (b) Expected output pulse energy (left axis) and gain bandwidth (right axis) as a function of fiber length. (After Ref. [43]; © 2008 IEEE.)

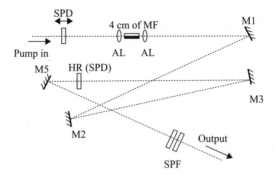

Figure 6.13 Schematic of a typical MF-based FOPO. (After Ref. [43]; © 2008 IEEE.)

$\gamma = 75\,\mathrm{W^{-1}\,km^{-1}}$ and a pump power of 2.5 kW typically delivered by a commercial Ti:sapphire oscillator, the optimal fiber length is found to be about 2 cm.

Figure 6.12a shows the small-signal single-pass gain obtained from Eq. (6.48) as a function of the wavelength for different fiber lengths under the condition $\gamma P_p L = \pi$ [43]. We can observe that reducing the fiber length provides a dramatic increase in the gain bandwidth. This result is also illustrated in the right axis of Fig. 6.12b, which also displays the expected output energy (left axis). For a fiber length of 2 cm, one can expect on the order of 10–20% of the pump energy to be converted to either signal or idler output energy.

Figure 6.13 shows a typical fiber-based FOPO that utilizes a short MF within a Fabry–Perot cavity [43]. Pump coupling into the cavity is accomplished by using a short-pass dielectric mirror (SPD). Aspheric lenses (AL) focus the beam into the MF, and the cavity is folded using broadband metallic mirrors M1–M3. Long wavelengths are reflected by both SPDs and thus oscillate within the cavity, while the idler field at short wavelengths is coupled through the high reflector (HR). The pump is removed before detection using short-pass filters (SPFs).

6.7 NONLINEAR PHASE CONJUGATION WITH FWM

As seen in Section 6.5, the FWM process can be used to amplify a weak signal and to generate simultaneously a new wave at the idler frequency. The idler wave has characteristics similar to the signal wave, except that its phase is conjugated. An important application of the phase-conjugated wave generated by FWM is for distortion compensation in optical transmission systems using a technique called *mid-span spectral inversion*. The use of such a technique for dispersion compensation was suggested in 1979 [44]. However, the first experiments demonstrating its validity occurred only in 1993. In one such experiment, a 6 Gb/s signal centered at 1546 nm was transmitted over 152 km of standard fiber and its phase conjugation was achieved by using FWM in a 23 km long fiber pumped at 1549 nm [45].

It is easy to show that, under certain conditions, nonlinear phase conjugation (NPC) can compensate both the GVD and SPM. Taking the complex conjugate of Eq. (4.49) gives

$$\frac{\partial U^*}{\partial z} - i\frac{1}{2}\beta_2 \frac{\partial^2 U^*}{\partial \tau^2} + i\gamma|U|^2 U^* = 0 \qquad (6.55)$$

A comparison of Eqs. (4.49) and (6.55) shows that the phase-conjugated field U^* propagates with the sign reversed for the GVD and the nonlinear parameters. This observation suggests that if the optical field is phase conjugated in the middle of the fiber link, the dispersion and the SPM-induced chirp acquired along the first part of the link will be exactly compensated in the second part. Moreover, it can be shown that, besides compensating GVD and SPM simultaneously, NPC can also compensate for the stimulated Raman scattering (SRS) and all even-order dispersion terms [46]. The compensation of all effects (odd- and even-order dispersion, SPM, SRS, and self-steepening) can be achieved by combining optical phase conjugation and a suitably chosen dispersion map [47,48].

In practice, fiber losses make difficult the exact compensation of SPM-induced distortions. In fact, due to the power dependence of the phase shift induced by SPM, much larger phase shifts are induced in the first half of the link than in the second half, and NPC cannot compensate completely for the nonlinear effects. For such compensation to be possible, an exact converse amplification distribution is required for the second half of the transmission link.

Using a single pump to realize the mid-span FWM-based phase conjugation, the signal angular frequency changes from ω_s to $\omega_c = 2\omega_p - \omega_s$, where ω_p is the pump angular frequency. As a consequence, the GVD parameter β_2 becomes different in the second half of the link, which means that the ideal location of the phase compensator to achieve an exact compensation will not be the midpoint of the fiber link. Such an ideal location, L_{pc}, can be determined by using the condition

$$\beta_2(\omega_s)L_{pc} = \beta_2(\omega_c)(L - L_{pc}) \qquad (6.56)$$

where L is the total link length.

Another aspect that must be considered while implementing the mid-span NPC technique is the polarization dependence of the FWM process. In practical applications, the output phase-conjugated signal must be insensitive to the polarization of the input signal. As discussed in Section 6.5, polarization-independent FWM can be achieved when two orthogonally polarized pump waves are used [49]. As an additional advantage, if the pump wavelengths are located symmetrically in relation to the zero dispersion wavelength, λ_{ZD}, of the fiber, and the signal wavelength coincides with λ_{ZD}, the phase-conjugated wave is generated at the same frequency of the signal [50].

In general, using NPC for distortion compensation in long-haul transmission systems requires the periodic use of optical amplifiers and phase conjugators. Fiber parametric amplifiers appear as an ideal solution in these circumstances, since they provide simultaneously the generation and the amplification of the phase-conjugated signal.

6.8 SQUEEZING AND PHOTON PAIR SOURCES

Coherent light contains equal minimum noise fluctuations in cophasal and quadrature components. Such noise fluctuations cannot be suppressed altogether, because that would violate the uncertainty principle. However, we can rearrange them between the quadratures. Light that has less noise in one selected quadrature than that dictated by the quantum noise level is called squeezed light.

Light squeezing can be generated taking advantage of correlations between photons introduced by some nonlinear processes, such as the FWM process. In the quantum mechanical description, FWM occurs when two pump photons, with frequencies ω_1 and ω_2, are annihilated to create two new photons: the signal photon, with frequency ω_3, and the idler photon, with frequency ω_4. These new photons are created simultaneously to satisfy the rule of conservation of energy. Because the two modes, signal and idler, initially have no photons in them, the newly created twin photons can be perfectly correlated. This correlation can be exploited for light squeezing [51]. In practice, four-wave mixing increases or decreases the number of specific signal–idler photon pairs, depending on their relative phases. The noise reduction below the quantum noise level can be verified using a phase-sensitive detection scheme in which the phase of the local oscillator is conveniently adjusted. Squeezed light might play a role in enhancing the performance of future optical communication systems limited by the quantum properties of light.

The results of some experiments using squeezed light can be greatly affected by spontaneous and stimulated Brillouin scattering, which is driven by the thermal excitations of the material. Inside the fiber, this drives acoustic waves that have a complex spectrum of resonances. A consequence of these effects is the generation of a broadband spectrum of phase noise, called guided acoustic wave Brillouin scattering (GAWBS) [52]. This adds to the noise in the squeezing quadrature and can mask all squeezing. Several techniques have been proposed to control the GAWBS noise [51]. A simple method consists in cooling the fiber, namely, by immersing the fiber in a liquid helium bath [53]. Another method to suppress the GAWBS noise consists in modulating the pump beam. This approach was used in a 1986 experiment, in which a 12.5% reduction in the quantum noise level was observed [54]. Figure 6.14 shows the noise spectrum obtained in such an experiment when the local oscillator phase was set to record the minimum noise. Squeezing is observed in the spectral regions around 45 and 55 MHz.

The generation of correlated photon pairs has gained much attention in recent years, due to their potential applications in quantum communications, quantum cryptography, and quantum computing. The spontaneous four-wave mixing (SFWM) process occurring in optical fibers provides a simple way of generating such photon pairs from quantum noise. As mentioned above, the signal and idler photons are created simultaneously to satisfy the rule of conservation of energy and they are

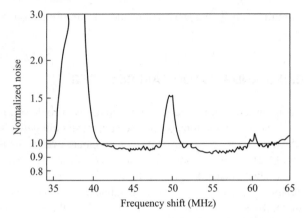

Figure 6.14 Noise spectrum showing light squeezing in optical fibers. The quantum noise level is indicated by the horizontal line. (After Ref. [54]; © 1986 APS.)

perfectly correlated. For cryptography, the signal photon can be sent out as the carrier, while the idler is detected as a trigger to indicate that a signal photon has been generated and sent out. Spontaneous Raman scattering, which typically occurs together with SFWM, is the main source of error that can destroy that perfect correlation.

The generation of correlated two-photon states via SFWM in a single-mode fiber was first proposed in 1999 [55,56]. In a 2001 experiment, correlated photon pairs were generated using a dispersion-shifted fiber and taking advantage of the anomalous dispersion spectral region, where SFWM occurs with a small detuning from the pump frequency and the Raman gain is low [57]. However, wavelength filtering needed to reject the pump light becomes a difficult task using this scheme. Another approach consists in using a frequency detuning greater than that corresponding to the Raman gain peak, such that the Raman gain is significantly reduced and the SFWM bandwidth is broad.

The gain of phase-matched SFWM at low power is proportional to $(\gamma P_p L)^2$. Increasing the SFWM gain by increasing the product $P_p L$ adds unwanted multiphoton emission through stimulated FWM and Raman scattering process. This can be avoided using highly nonlinear fibers, such as MFs, which provide a large value of the nonlinear parameter γ [58]. Due to this higher nonlinearity, lower pump powers are needed to achieve the same two-photon signal, which also reduces the Raman background. MFs also offer the possibility of adjusting the dispersive characteristics, in order to enhance the phase matching conditions [58,59]. By flattening the GVD profile and making β_2 small, higher order GVD terms become important and gain bands can appear in the normal dispersion regime, such that $\beta_2 > 0$ and $\beta_4 < 0$ [60]. Considering Eqs. (6.27) and (6.46), the sideband gain can be written in the form

$$g = \left[-\left(\frac{\beta_2 \Omega^2}{2} + \frac{\beta_4 \Omega^4}{24} \right) \left(\frac{\beta_2 \Omega^2}{2} + \frac{\beta_4 \Omega^4}{24} + 2\gamma P_0 \right) \right]^{1/2} \qquad (6.57)$$

This expression shows that an effective gain is obtained when $\beta_2\Omega^2 + \beta_4\Omega^4/12 < 0$ and $|\beta_2\Omega^2 + \beta_4\Omega^4/12| < 4\gamma P_0$. The sidebands can be widely spaced from the pump frequency, their position and width being controllable by engineering the even-order higher order dispersion terms. In particular, it is possible to arrange that these sidebands appear well beyond the Raman gain peak, thus reducing the Raman noise. In a 2005 experiment, an MF with ZDW at 715 nm was pumped by a Ti: sapphire laser at 708 nm (normal dispersion) [61]. In these circumstances, phase matching was satisfied by signal and idler waves at 587 and 897 nm, and 10 million photon pairs per second were generated and sent via SMF to Si avalanche detectors, producing $\sim 3.2 \times 10^5$ coincidences per second for a pump power of 0.5 mW.

PROBLEMS

6.1 Derive the coupled equations for the pump, signal, and idler waves similar to Eqs. (6.8)–(6.11) for the case of a single pump.

6.2 Explain how FWM affects the performance of a WDM system. Discuss the merits of using unequal channel spacing and/or dispersion management to suppress the impact of FWM.

6.3 Consider a pump and a signal wave with a frequency spacing Δv between them. Show that the linear phase mismatch in such case is given by $\Delta k = \beta_2 (2\pi\Delta v)^2$, where β_2 is the dispersion at the pump frequency.

6.4 Consider a pump beam at a wavelength of 1550 nm carrying a power of 10 W in a single-mode fiber that has a nonlinear parameter $\gamma = 4\,\mathrm{W}^{-1}\,\mathrm{km}^{-1}$ and a GVD $\beta_2 = -20\,\mathrm{ps}^2/\mathrm{km}$. Find the signal wavelength for which a perfect phase matching occurs.

6.5 Two signals in the vicinity of 1.55 μm and with a frequency difference of 100 GHz are launched into a 50 km fiber with $\alpha = 0.2\,\mathrm{dB/km}$, $\beta_2 = -1\,\mathrm{ps}^2/\mathrm{km}$, and negligible dispersion slope. Estimate the value of the FWM efficiency in such case.

6.6 Verify that Eqs. (6.38) and (6.39) are indeed solutions of Eqs. (6.36) and (6.37) under the condition $V_4^*(0) = 0$. Also confirm the result given by Eq. (6.40) for the parametric gain.

REFERENCES

1. N. Shibata, R. P. Braun, and R. G. Waarts, *IEEE J. Quantum Electron.* **23**, 1205 (1987).
2. K. Inoue, *J. Lightwave Technol.* **10**, 1553 (1992).
3. F. Forghieri, R. W. Tkach, and A. R. Chraplyvy, *J. Lightwave Technol.* **13**, 889 (1995).
4. H. Taga, *J. Lightwave Technol.* **14**, 1287 (1996).

5. H. Suzuki, S. Ohteru, and N. Takachio, *IEEE Photon. Technol. Lett.* **11**, 1677 (1999).

6. S. Beti, M. Giaconi, and M. Nardini, *IEEE Photon. Technol. Lett.* **15**, 1079 (2003).

7. R. H. Stolen, J. E. Bjorkhholm, and A. Ashkin, *Appl. Phys. Lett.* **24**, 308 (1974).

8. F. Forghieri, R. W. Tkach, and A. R. Chraplyvy, Fiber nonlinearities and their impact on transmission systems, in I. P. Kaminow and T. L. Koch (Eds.), *Optical Fiber Telecommunications*, Vol. IIIA, Academic Press, San Diego, CA, 1997.

9. K. Stenersen and R. K. Jain, *Opt. Commun.* **51**, 121 (1984).

10. W. J. Wadsworth, N. Joly, J. C. Knight, T. A. Birks, F. Biancalana, and P. St. J. Russell, *Opt. Express* **12**, 299 (2004).

11. G. K. Wong, A. Y. Chen, S. G. Murdoch, R. Leonhardt, J. D. Harvey, N. Y. Joly, J. C. Knight, W. J. Wadsworth, and P. St. J. Russel, *J. Opt. Soc. Am. B* **22**, 2505 (2005).

12. F. Forghieri, R. W. Tkach, and A. R. Chraplyvy, *IEEE Photon. Technol. Lett.* **6**, 754 (1994).

13. R. W. Tkach, A. R. Chraplyvy, F. Forghieri, A. H. Gnauck, and R. M. Derosier, *J. Lightwave Technol.* **5**, 841 (1995).

14. D. F. Grosz, C. Mazzali, S. Celaschi, A. Paradisi, and H. L. Fragnito, *IEEE Photon. Technol. Lett.* **11**, 379 (1999).

15. G. P. Agrawal, *Fiber-Optic Communication Systems*, 3rd ed., Wiley, New York, 2002.

16. M. E. Marhic, N. Kagi, T.-K. Chiang, and L. G. Kazovsky, *Opt. Lett.* **21**, 573 (1996).

17. M. Karlsson, *J. Opt. Soc. Am. B* **15**, 2269 (1998).

18. J. Hansryd and P. A. Andrekson, *IEEE Photon. Technol. Lett.* **13**, 194 (2001).

19. M. Westlund, J. Hansryd, P. A. Andrekson, and S. N. Knudsen, *Electron. Lett.* **38**, 85 (2002).

20. J. Hansryd, P. A. Andrekson, M. Westlund, J. Li, and P. O. Hedekvist, *IEEE J. Sel. Top. Quantum Electron.* **8**, 506 (2002).

21. G. Kalogerakis, M. E. Marhic, K. K. Wong, and L. G. Kazovsky, *J. Lightwave Technol.* **23**, 2945 (2005).

22. M. E. Marhic, Y. Park, F. S. Yang, and L. G. Kazovsky, *Opt. Lett.* **21**, 1354 (1996).

23. K. K. Y. Wong, M. E. Marhic, K. Uesaka, and L. G. Kazovsky, *IEEE Photon. Technol. Lett.* **14**, 911 (2002).

24. C. J. McKinstrie and S. Radic, *Opt. Lett.* **27**, 1138 (2002).

25. C. J. McKinstrie, S. Radic, and A. R. Chraplyvy, *IEEE J. Sel. Top. Quantum Electron.* **8**, 538 (2002).

26. S. Radic and C. J. McKinstrie, *Opt. Fiber Technol.* **9**, 7 (2003).

27. S. Radic, C. J. McKinstrie, A. R. Chraplyvy, G. Raybon, J. C. Centanni, C. G. Jorgensen, K. Brar, and C. Headley, *IEEE Photon. Technol. Lett.* **14**, 1406 (2002).

28. M. Yu, C. J. McKinstrie, and G. P. Agrawal, *Phys. Rev.* **48**, 2178 (1993).

29. R. H. Stolen and J. E. Bjorkholm, *IEEE J. Quantum Electron.* **18**, 1062 (1982).

30. K. Inoue, *J. Lightwave Technol.* **12**, 1916 (1994).

31. R. M. Jopson and R. E. Tench, *Electron. Lett.* **29**, 2216 (1993).

32. O. Leclerc, B. Lavigne, E. Balmefrezol, et al., *J. Lightwave Technol.* **21**, 2779 (2003).

33. W. Margulis and U. Österberg, *Opt. Lett.* **12**, 519 (1987).

34. M. Nakazawa, K. Suzuki, and H. A. Haus, *Phys. Rev. A* **38**, 5193 (1988).

35. D. K. Serkland and P. Kumar, *Opt. Lett.* **24**, 92 (1999).

36. M. E. Marhic, K. K. Y. Wong, L. G. Kazvsky, and T. E. Tsai, *Opt. Lett.* **2**, 1439 (2002).

37. T. Torounnidis and P. Andrekson, *IEEE Photon. Lett.* **19**, 650 (2007).

38. S. Coen, M. Haelterman, P. Emplit, L. Delage, L. M. Simohamed, and F. Reynaud, *J. Opt. Soc. Am. B* **15**, 2283 (1998).

39. J. E. Sharping, M. Fiorentino, P. Kumar, and R. S. Windeler, *Opt. Lett.* **27**, 1675 (2002).

40. Y. Deng, Q. Lin, F. Lu, G. Agrawal, and W. Knox, *Opt. Lett.* **30**, 1234 (2005).

41. J. E. Sharping, M. A. Foster, A. L. Gaeta, J. Lasri, O. Lyngnes, and K. Vogel, *Opt. Express* **15**, 1474 (2007).

42. J. E. Sharping, J. R. Sanborn, M. A. Foster, D. Broaddus, and A. L. Gaeta, *Opt. Express* **16**, 18050 (2008).

43. J. E. Sharping, *J. Lightwave Technol.* **26**, 2184 (2008).

44. A. Yariv, D. Fekete, and D. M. Pepper, *Opt. Lett.* **4**, 52 (1979).

45. S. Watanabe, N. Saito, and T. Chikama, *IEEE Photon. Technol. Lett.* **5**, 92 (1993).

46. S. Watanabe and M. Shirasaki, *J. Lightwave Technol.* **14**, 243 (1996).

47. J. Pina, B. Abueva, and G. Goedde, *Opt. Commun.* **176**, 397 (2000).

48. M. Tsang and D. Psaltis, *Opt. Lett.* **28**, 1558 (2003).

49. T. Schneider and D. Schilder, *J. Opt. Commun.* **19**, 115 (1998).

50. R. M. Jopson and R. E. Tench, *Electron. Lett.* **29**, 2216 (1993).

51. A. Sizmann and G. Leuchs, in E. Wolf (Ed.), *Progress in Optics*, Vol. **39**, Elsevier, New York, 1999, Chapter 5.

52. M. D. Levenson, R. M. Shelby, A. Aspect, M. Reid, and D. F. Walls, *Phys. Rev. A* **32**, 1550 (1985).

53. G. J. Milburn, M. D. Levenson, R. M. Shelby, S. H. Perlmutter, R. G. DeVoe, and D. F. Walls, *J. Opt. Soc. Am. B* **4**, 1476 (1987).

54. R. M. Shelby, M. D. Levenson, S. H. Perlmutter, R. G. DeVoe, and D. F. Walls, *Phys. Rev. Lett.* **57**, 691 (1986).

55. L. J. Wang, C. K. Hong, and S. R. Friberg, NEC Research Institute Technical Report, TR99-106, 1999.

56. L. J. Wang, C. K. Hong, and S. R. Friberg, *J. Opt. B* **3**, 346 (2001).

57. J. E. Sharping, et al., *Opt. Lett.* **26**, 367 (2001).

58. P. S. J. Russel, *J. Lightwave Technol.* **24**, 4729 (2006).

59. K. Saito and M. Koshiba, *J. Lightwave Technol.* **23**, 3580 (2005).

60. W. H. Reeves, D. V. Skryabin, F. Biancalana, J. C. Knight, P. St. J. Russell, F. G. Omenetto, A. Efimov, and A. J. Taylor, *Nature* **424**, 511 (2003).

61. J. Fulconis, O. Alibart, W. J. Wadsworth, P. St. J. Russell, and J. G. Rarity, *Opt. Express* **13**, 7572 (2005).

INTRACHANNEL NONLINEAR EFFECTS

Dispersion-managed transmission links constitute a key technology to realize optical communication systems operating at bit rates of 40 Gb/s or more. These systems employ relatively short optical pulses, which are rapidly dispersed in the fiber, spreading over a very large number of bits. This spreading reduces the peak power and lowers the impact of self-phase modulation (SPM) [1, 2] and interchannel crosstalk in wavelength division multiplexed (WDM) systems [3–5].

There are two main options concerning the design of a dispersion-managed transmission system. In one scheme, dispersion accumulates along most of the link length and is compensated using dispersion compensation fibers (DCFs) only at the transmitter (precompensation) and receiver (postcompensation) ends. In the other option, dispersion is compensated periodically along the link, either completely or partially. In this case, the pulse broadens and recompresses periodically, according to the dispersion characteristics of the fiber segments.

Dispersion management is useful for both pseudolinear systems and soliton transmission systems, which will be discussed in Chapter 8. In the case of dispersion-managed soliton systems, the fiber nonlinearity plays an essential role in preserving the pulse shape, whereas in pseudolinear systems the nonlinearity can be considered as a small perturbation.

Pseudolinear dispersion-managed transmission systems operate in the regime in which the nonlinear length far exceeds the dispersion length in all fiber sections. As the optical pulses spread over a very large number of bits as they propagate along the link, the effect of fiber nonlinearity will be averaged out [6]. This fact, together with the high local dispersion, determines an effective suppression of the *interchannel* nonlinear effects, such as four-wave mixing (FWM) and cross-phase modulation (XPM). However, the same reasons determine an enhanced interaction among the pulses of the same channel, leading to new *intrachannel* nonlinear effects that can significantly impact the system performance [7–11].

Besides the SPM effect, which can be suppressed by an appropriate design of the dispersion map, there are two new intrachannel nonlinear effects that are of concern. One is the intrachannel cross-phase modulation (IXPM), which leads to a timing jitter of the pulses. The other is the intrachannel four-wave mixing (IFWM), which is responsible for two effects that degrade the system performance: the amplitude jitter and the generation of ghost pulses. Both IXPM and IFWM can occur only when the pulses overlap in time, at least partly, during their propagation in the dispersion-managed system. Such overlapping is especially significant in systems with strong dispersion management.

7.1 MATHEMATICAL DESCRIPTION

The propagation of an optical bit stream along a dispersion-managed transmission line is governed by the nonlinear Schrödinger equation (NLSE), which can be written in the form given by Eq. (4.49):

$$i\frac{\partial U}{\partial z} - \frac{1}{2}\beta_2\frac{\partial^2 U}{\partial\tau^2} + \gamma|U|^2 U = 0 \tag{7.1}$$

Equation (7.1) can be solved numerically considering a given bit stream with N bits at the input, such that

$$U(0,\tau) = \sum_{n=1}^{N} U_n(0,\tau-\tau_n) \tag{7.2}$$

where $\tau_n = nt_B$ and t_B is the duration of the bit slot. Substituting this sum in Eq. (7.1), we obtain

$$\sum_{n=1}^{N}\left(i\frac{\partial U_n}{\partial z} - \frac{1}{2}\beta_2\frac{\partial^2 U_n}{\partial\tau^2}\right) = -\gamma\sum_{j,k,l=1}^{N} U_j U_k^* U_l \tag{7.3}$$

The various terms on the right-hand side of Eq. (7.3) correspond to different nonlinear effects as follows: when $j = k = l$ we have SPM, when $j = k \neq l$ or $j \neq k = l$ it is IXPM, and when $j \neq k \neq l$ or $j = l \neq k$ it is IFWM. The location of the nonlinear interaction is given approximately by

$$\tau_{j,k,l} = \tau_j + \tau_l - \tau_k \tag{7.4}$$

This relation is similar to the phase matching condition used to find the frequency of a wave generated by FWM.

Assuming a sequence of only three pulses, such that $U = U_1 + U_2 + U_3$, Eq. (7.3) reduces to the following coupled NLSEs [6]:

$$i\frac{\partial U_1}{\partial z} - \frac{1}{2}\beta_2\frac{\partial^2 U_1}{\partial \tau^2} = -\gamma\left[(|U_1|^2 + 2|U_2|^2 + 2|U_3|^2)U_1 + U_2^2 U_3^*\right] \quad (7.5)$$

$$i\frac{\partial U_2}{\partial z} - \frac{1}{2}\beta_2\frac{\partial^2 U_2}{\partial \tau^2} = -\gamma\left[(|U_2|^2 + 2|U_1|^2 + 2|U_3|^2)U_2 + 2U_1 U_2^* U_3\right] \quad (7.6)$$

$$i\frac{\partial U_3}{\partial z} - \frac{1}{2}\beta_2\frac{\partial^2 U_3}{\partial \tau^2} = -\gamma\left[(|U_3|^2 + 2|U_1|^2 + 2|U_2|^2)U_3 + U_2^2 U_1^*\right] \quad (7.7)$$

The first term on the right-hand side of Eqs. (7.5)–(7.7) describes the SPM, whereas the next two terms represent the modulation of the phase of a given pulse by the neighboring pulses, that is, the IXPM effect. The last term describes the IFWM, a phenomenon that causes the transfer of energy among the interacting pulses and even the creation of new pulses in the time domain, called *ghost* pulses [8].

Assuming a system with strong dispersion management, a perturbative approach can be used to describe the impact of the relatively weak intrachannel nonlinearities. The main idea is to assume that the solution of the NLSE (7.1) can be written in the form

$$U(z,\tau) = \sum_{j=1}^{N} U_j(z,\tau-\tau_j) + \sum_{j,k,l=1}^{N} \Delta U_{jkl}(z,\tau) \equiv U_{\text{lin}}(z,\tau) + U_{\text{p}}(z,\tau) \quad (7.8)$$

where

$$U_{\text{lin}}(z,\tau) = \sum_{j=1}^{N} U_j(z,\tau-\tau_j) \quad (7.9)$$

is the solution of the NLSE for the linear case ($\gamma = 0$) and

$$U_{\text{p}}(z,\tau) = \sum_{j,k,l=1}^{N} \Delta U_{jkl}(z,\tau) \quad (7.10)$$

is the nonlinear perturbation of the linear solution due to the relatively weak nonlinearity. This nonlinear perturbation satisfies the equation

$$i\frac{\partial U_{\text{p}}}{\partial z} - \frac{1}{2}\beta_2\frac{\partial^2 U_{\text{p}}}{\partial \tau^2} = -\gamma|U_{\text{lin}} + U_{\text{p}}|^2(U_{\text{lin}} + U_{\text{p}}) \approx -\gamma|U_{\text{lin}}|^2 U_{\text{lin}} \quad (7.11)$$

where the condition $U_{\text{p}} \ll U_{\text{lin}}$ was assumed on the right-hand side.

Let us consider the simplest case of two bits located at $t_1 = t_B$ and $t_2 = 2t_B$, such that

$$U_{\text{lin}} = \hat{U}_1 e^{i\omega(t-t_B)} + \hat{U}_2 e^{i\omega(t-2t_B)} \tag{7.12}$$

Introducing (7.12) into (7.11), the right-hand side can be written in the form

$$-\gamma|U_{\text{lin}}|^2 U_{\text{lin}} = -\gamma[((|\hat{U}_1|^2 + 2|\hat{U}_2|^2)\hat{U}_1 e^{i\omega(t-t_B)} + (|\hat{U}_2|^2 + 2|\hat{U}_1|^2)\hat{U}_2 e^{i\omega(t-2t_B)}$$
$$+ \hat{U}_1^2 \hat{U}_2^* e^{i\omega t} + \hat{U}_2^2 \hat{U}_1^* e^{i\omega(t-3t_B)}] \tag{7.13}$$

The terms in the first bracket on the right-hand side describe the influence of the nonlinearity on the pulse located at $t_1 = t_B$. The first term in this bracket describes the SPM, whereas the second term describes IXPM. The terms in the second bracket describe the same features for the pulse located at $t_2 = 2t_B$. Finally, the last two terms on the right-hand side of Eq. (7.13) correspond to perturbations created by IFWM at locations 0 and $3t_B$, which do not coincide with the original positions of the input pulses. If the two new locations are occupied by "1" bits, the perturbations beat with them and produce amplitude jitter. However, if they correspond to "0" bits, a ghost pulse is created in such positions.

7.2   INTRACHANNEL XPM

In order to describe the main characteristics of the IXPM, we can consider two isolated bits by setting $U_3 = 0$ in Eqs. (7.5)–(7.7). The optical fields of the two bits satisfy the equations

$$i\frac{\partial U_1}{\partial z} - \frac{1}{2}\beta_2 \frac{\partial^2 U_1}{\partial \tau^2} = -\gamma\left[(|U_1|^2 + 2|U_2|^2)U_1\right] \tag{7.14}$$

$$i\frac{\partial U_2}{\partial z} - \frac{1}{2}\beta_2 \frac{\partial^2 U_2}{\partial \tau^2} = -\gamma\left[(|U_2|^2 + 2|U_1|^2)U_2\right] \tag{7.15}$$

Clearly, the last term in the above equations corresponds to intrachannel XPM. The nonlinear interaction between the pulses leads to an alteration of the phase of both. The nonlinear phase shift resulting from the IXPM after the propagation over a distance z is given by

$$\phi_n = 2\gamma|U_{3-n}(z,\tau)|^2 z \tag{7.16}$$

where $n = 1$ or 2. Since this phase shift varies across the pulse, it is translated into a frequency chirp

$$\Delta\omega_n = -\frac{\partial\phi_n}{\partial z} = -2\gamma\frac{\partial}{\partial z}|U_{3-n}(z,z)|^2 z \tag{7.17}$$

According to (7.16), the frequency shift that one pulse experiences depends on the slope of the power of the other pulse. On the other hand, a mutual interaction due to IXPM can take place only if at least parts of the pulses overlap temporally during their propagation. Such overlapping can occur as a result of the fiber dispersion. This is the case of dispersion-managed transmission systems, in which the pulses are periodically broadened and compressed during propagation.

If the dispersion is sufficiently low, there is an overlapping of the tailing edge of the first pulse with the leading edge of the second (Fig. 7.1a). In this case, the pulses overlap in the regions where the slope of the pulse power is strong, resulting in a large frequency shift. Such an XPM-induced frequency shift determines a shift in the pulse position through changes in the group velocity of the pulse. The time shift varies from bit to bit, since it depends on the pattern surrounding each pulse. As a consequence, IXPM introduces timing jitter that can significantly degrade the system performance.

In the case of high dispersion, the pulses experience a strong broadening and their power slope is low (Fig. 7.1b). As a consequence, the IXPM-induced chirp is weak. In this regime, the IXPM can be effectively suppressed by optimum dispersion management or by suitably chirping input pulses before launching them into the fiber link [12].

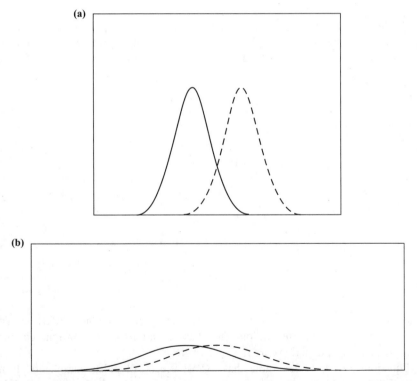

Figure 7.1 Schematic representation of a pair of overlapping pulses in the case of low (a) and high (b) dispersion.

The magnitude of the IXPM-induced chirp can be reduced by controlling the relation between the pulse width, t_0, and the pulse separation, t_B, which is equal to the inverse of the bit rate, R. If $t_B \gg t_0$, the tails of the pulses do not overlap significantly and the IXPM is low. If $t_B \approx t_0$, the IXPM has its maximum and the frequency shift is largest. Finally, if $t_B \ll t_0$, the effectiveness of IXPM will decrease again [8]. This fact seems surprising, since in this case the pulses almost completely overlap. The reason of this behavior is related to the dependence of the IXPM-induced frequency chirp on the slope of the pulse power. In fact, a large pulse width corresponds to the figure for a high dispersion, shown in Fig. 7.1b. The power slope is smaller for wider pulses and also changes sign, resulting in an averaging effect. As a result, the timing jitter caused by IXPM can be effectively reduced by stretching optical pulses over multiple bit slots.

The effects of IXPM can be described analytically by using different methods [13–20]. One of these methods is provided by the variational approach presented in Section 5.1.2, extending its formulation to the case of two interacting pulses. Assuming such an approach, let us consider the following ansatz:

$$U_n(z,\tau) = u_n \exp\left[-\frac{1}{2}(1+iC_n)\left(\frac{\tau-\tau_n}{t_n}\right)^2 - i\Omega_n(\tau-\tau_n) + i\psi_n \right], \quad n = 1,2 \quad (7.18)$$

where the parameters u_n, C_n, t_n, τ_n, Ω_n, and ψ_n represent the nth pulse amplitude, chirp, width, position, frequency, and phase, respectively. Applying the variational method to Eqs. (7.14) and (7.15) with ansatz (7.18) gives the following set of equations [13, 15]:

$$\frac{du_n}{dz} = -\frac{u_n \beta_2(z)C_n}{2 \; t_n^2} \tag{7.19}$$

$$\frac{dt_n}{dz} = \frac{\beta_2(z)C_n}{t_n} \tag{7.20}$$

$$\frac{dC_n}{dz} = \frac{\beta_2(z)(1+C_n^2)}{t_n^2} + \frac{1}{\sqrt{2\pi}}\frac{E_n\gamma}{t_n} - \sqrt{\frac{2}{\pi}}E_{3-n}\gamma t_n^2 P^3 (1-\mu^2)\exp(-\mu^2/2) \tag{7.21}$$

$$\frac{d\Delta\Omega}{dz} = \sqrt{\frac{2}{\pi}}\gamma(E_1+E_2)P^2\mu\exp(-\mu^2/2) \tag{7.22}$$

$$\frac{d\Delta\tau}{dz} = -\beta_2(z)\Delta\Omega \tag{7.23}$$

where $P = \sqrt{2}/\sqrt{t_1^2+t_2^2}$, $\mu = P\Delta\tau$, $\Delta\Omega = \Omega_1-\Omega_2$ is the frequency difference, and $\Delta\tau = \tau_1-\tau_2$ is the pulse separation. The phase equations have been ignored since phase-dependent interactions are neglected here.

Equations (7.19)–(7.23) show that IXPM affects all pulse parameters. From Eqs. (7.19) and (7.20) one can see that the pulse energy $E_n = \sqrt{\pi}u_n^2 t_n$ remains constant. The chirp is affected by both SPM and IXPM, as given by the second and third terms, respectively, on the right-hand side of Eq. (7.21). However, the effects of SPM are generally reduced in the pseudolinear regime. Moreover, since those effects are identical for all pulses, their impact on the system performance can be neglected. Concerning the IXPM-induced pulse chirping, it depends on the presence or absence of neighboring pulses. Equation (7.20) shows that any change in the chirp affects the pulse width and, consequently, its amplitude, since the pulse energy is conserved. As the bit pattern varies randomly in an optical bit stream, such effect results in IXPM-induced amplitude jitter.

A more significant effect of IXPM is the generation of timing jitter through the generation of frequency shifts that are converted into time shifts by dispersion, as described by Eqs. (7.21) and (7.22). However, the suppression of the IXPM-induced timing jitter is possible using symmetric links [19]. This possibility is illustrated in Fig. 7.2, which shows the frequency shift for the case of a lossless 100 km link formed by two 50 km fiber sections with $D = \pm 10$ ps/(km nm) [20]. The two pulses have 5 ps width and are separated by $t_B = 25$ ps, corresponding to the separation between two pulses in a 40 Gb/s pulse train. The almost coinciding solid and dashed curves in

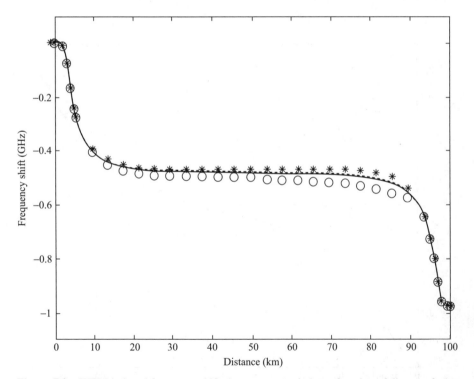

Figure 7.2 IXPM-induced frequency shift along a map period as a function of distance for a lossless system for two launch positions: at the input of the normal fiber (solid curve and asterisks) and at input of the anomalous fiber (dashed curve and open circles). Curves were obtained using the variational approach to the case of two interacting pulses, whereas symbols correspond to numerical results obtained from the NLSE. (After Ref. [20]; © 2001 OSA.)

Fig. 7.2, obtained using the variational approach, correspond to the cases in which pulses are launched at the input of each fiber type. The symbols were obtained by numerically solving the NLSE. The good agreement between the curves and symbols demonstrates the accuracy of the variational formulation.

Figure 7.3 illustrates the time shift in pulse spacing as a function of launch position in the anomalous dispersion fiber segment ($D = 10$ ps/(km nm)) of the same link considered in Fig. 7.2. This figure shows that a launch point in the middle of this fiber leads to zero net time shift. The same result is obtained if the launch point is in the middle of the normal dispersion fiber segment ($D = -10$ ps/(km nm)). Both cases correspond to symmetric dispersion maps. In fact, even if the frequency shift is approximately the same for the symmetric and nonsymmetric maps, the timing shift is significantly different in both cases, as seen from Fig. 7.3. The timing jitter is completely canceled for a symmetric dispersion map only when losses are compensated through distributed amplification. In such a case, timing shifts produced in both sections completely cancel each other. However, even using lumped amplification, a considerable improvement can be achieved with the symmetric dispersion map [19].

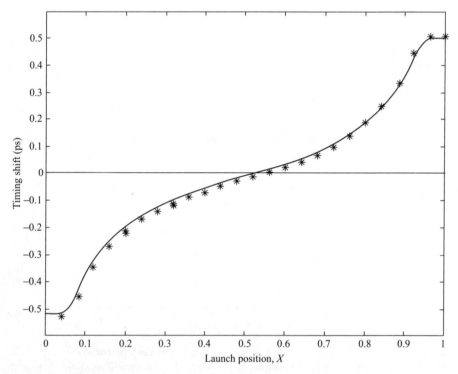

Figure 7.3 Net time shift after one dispersion map as a function of launch position in the anomalous dispersion fiber. The solid curve was obtained using the variational approximation, whereas the asterisks correspond to numerical simulations. (After Ref. [20]; © 2001 OSA.)

7.3 INTRACHANNEL FWM

In contrast to IXPM, IFWM causes the transference of energy among the interacting pulses. In particular, this phenomenon can create new pulses in the time domain, which are usually referred to as *ghost* pulses [8]. These pulses were observed for the first time in 1992 [21] in the context of an experiment involving a pair of ultrashort pulses. Signal perturbations by IFWM were first reported in 1998 in the form of amplitude fluctuations in the "1" bit slots and ghost pulse generation in the "0" bit slots [22, 23]. IFWM-induced phase noise can also degrade the performance of fiber transmission systems using coherent detection of phase-modulated signals [24–26].

Regarding Eq. (7.11), the last two terms correspond to ghost pulses, which are symmetrically arranged relatively to the two original pulses, at locations 0 and $3t_B$. The formation of new pulses at both sides of the original pump pulses is similar to the generation of new mixing frequencies via FWM, as observed in Chapter 6. However, in contrast to FWM, the frequency axis is now replaced by a time axis, and the phase matching for the frequency mixing is now taken over by a time condition. In fact, the pulses can interact with each other only if they overlap temporally, at least partially. This happens particularly for high bit rate systems in strong dispersion maps, in which interchannel FWM is effectively suppressed. Conversely, a good phase matching and consequently a strong interaction between different wavelength channels via FWM occurs in systems with low local dispersion, as discussed in Chapter 6. In such a case, IFWM becomes weak, since the pulses in one channel will not overlap during their propagation.

Various analytical methods have been used to describe IFWM [9, 16–20]. Among them, the so-called small field perturbation method [9, 16, 19] offers the advantage of encompassing all types of nonlinear interactions between pulses. Using this method and considering the case of Gaussian input pulses of width t_0, the following result can be obtained for the perturbation term ΔU_{jkl} in Eq. (7.10) [9, 19]:

$$\Delta U_{jkl}(L, \tau_j + \tau_l - \tau_k) = \gamma U_{in}^3 \exp\left(-\frac{\tau^2}{6t_0^2}\right) e^{i\Delta\phi} \int_0^L \frac{id(z)}{\sqrt{1 + 2id + 3d^2}}$$

$$\times \exp\left(-\frac{3[2\tau/3 + (\tau_j - \tau_k)][2\tau/3 + (\tau_l - \tau_k)]}{t_0^2(1 + 3id)} - \frac{(\tau_j - \tau_l)^2}{t_0^2(1 + 2id + 3d^2)}\right) \tag{7.24}$$

where $\Delta\phi = \phi_j + \phi_l - \phi_k$ is related to the phases of the pulses, $U_{in} = \sqrt{P_0}$ is the pulse peak amplitude at the input of the transmission fiber, and $d(z) = t_0^{-2} \int_0^z \beta_2(z')dz'$ is the chirp due to dispersion. Equation (7.24) is rather general and includes SPM, IXPM, and IFWM effects through different sets of indices [jkl]. Considering a pair of pulses, the perturbations representing SPM correspond to the combinations [111] and [222], whereas the combinations for IXPM are [221], [122], [112], and [211]. Finally, the effect of IFWM is given by the combinations [121] and [212].

The integral in Eq. (7.24) can be calculated analytically in the case of a constant-dispersion fiber, considering a fiber length L much longer than the dispersion distance $L_D = t_0^2/|\beta_2|$. Considering the simplest case of two bits located at $\tau_1 = t_B$ and $\tau_2 = 2t_B$, the peak power of the ghost pulse generated through IFWM at $\tau = 0$ can be obtained from Eq. (7.24) considering $j = l = 1$ and $k = 2$ [9]:

$$P_g(L) = |\Delta Q_{121}(L,0)|^2 = \frac{1}{3}(\gamma P_0 L_D)^2 \exp(-\tau^2/3t_0^2)|E_1(x)|^2 \qquad (7.25)$$

where

$$x = \frac{2it_B^2}{|\beta_2|L} \qquad (7.26)$$

and $E_1(x)$ is the exponential integral function. Note that the temporal shape of the perturbation is Gaussian. To obtain the dependence on fiber length, we can use the approximation $|E_1(x)| \sim \ln|1/x|$. Figure 7.4 shows the variation of the peak power of the ghost pulse with the link length L, considering a 40 Gb/s signal. The solid line represents the analytic result given by Eq. (7.25), whereas the dotted line is the asymptotic logarithmic approximation for E_1 and the squares are results of numerical simulation.

Another important property of the nonlinear interaction is its dependence on the temporal separation between consecutive pulses, which should reflect the dependence on the bit rate in a fiber communication system. Calculations based on Eq. (7.24) show, as expected, that the peak power of the ghost pulse is reduced by increasing the pulse separation t_B [9]. For large values of x, we can consider the approximation $E_1(x)| \approx 1/x$, which predicts that P_g decreases with t_B as t_B^{-4}.

Some simulations have shown that the position of the ghost pulse is not stationary, but it zigzags in time during its propagation along a dispersion-managed link [27].

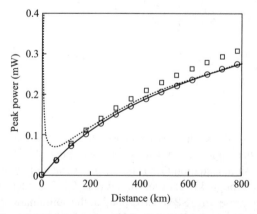

Figure 7.4   Variation of the peak power with the link length L, considering a 40 Gb/s signal. Symbols were obtained by numerical simulations. (After Ref. [9]; © 2000 IEEE.)

However, both the shape and the position are recovered after every dispersion map period. Moreover, the peak power of the ghost pulse shows a resonant growth as a result of periodic forcing and temporal phase matching with the signal pulses [17]. As a result, the total peak power is found to experience a quadratic increase with the number of dispersion-managed cells, such that

$$P_{g,tot}(L) = P_g(L_{map})(L/L_{map})^2 \qquad (7.27)$$

where L_{map} is the map period. This quadratic growth of the ghost pulse peak power can seriously impact the performance of long-haul dispersion-managed transmission systems.

Considering the more general case of interaction among N pulses, the number of possible mixing products is given by Eq. (6.2), as happens for interchannel FWM. Besides the generation of ghost pulses in time slots originally carrying a "zero", some of the mixing products fall in the same time slots of the original pulses. In such cases, the power is transferred between the pulses. The power of each original pulse can be increased or decreased, depending on the relative phase between this pulse and the newly generated one. The power variation in the "ones" is maximum (minimum) when they are in phase (in quadrature) with the generated pulses. In high-dispersion fibers, where many pulses interact with each other, this effect leads to stochastic amplitude fluctuations of the "one" bits. In the case of a periodic dispersion-managed system, energy fluctuations grow linearly with the length of the fiber link [9].

When the "ones" from a pulse train have identical phases, the difference of phase between the perturbation ΔU_{jkl} and each pulse at the end of the transmission line is given by the phase of the integral on the right-hand side of Eq. (7.24). It can be verified that the real part of ΔU_{jkl} (which determines the in-phase interference) vanishes for a symmetric dispersion map. Using such a symmetric map will then be very effective to reduce the amplitude jitter created by the in-phase interference, which is the dominant source of distortion due to IFWM. This possibility was confirmed experimentally [19].

Concerning the imaginary part of ΔU_{jkl}, it is not possible to vanish it with any specific dispersion map. Such imaginary part is responsible for the IFWM-induced phase shifts of the original light pulse [24], which are pattern dependent and cause the degradation of phase-modulated optical communication systems [25, 26].

7.4 CONTROL OF INTRACHANNEL NONLINEAR EFFECTS

As observed in previous sections, both IXPM and IFWM can have a significant impact on the performance of quasilinear transmission systems. As a consequence, it is important to find possible methods for the suppression of these effects.

Concerning the IXPM, it can be suppressed using a symmetric dispersion map, as discussed in Section 7.2. In particular, the XPM-induced timing jitter would vanish under these conditions if a distributed amplification scheme is used. Using a symmetric dispersion map together with a lumped amplification scheme causes some residual timing jitter due to variations in the average power of the transmitted signal.

In the case of a periodic dispersion map employing two fiber sections of equal length but opposite dispersions, both the timing and the amplitude jitters can be reduced simply by reversing the two fibers in alternate map periods. This is illustrated in Fig. 7.5 at three power levels: 3, 6, and 9 dBm [28]. Both the timing and amplitude jitters increase linearly with the link length in the case of the asymmetric map, whereas for the symmetric one they increase much more slowly. The insertion of an optical phase conjugator in the middle of the transmission link was also found to be effective in reducing the impact of intrachannel nonlinear effects for any dispersion map [29, 30].

The advantage of using a symmetric dispersion map is verified not only in the case of return-to-zero on-off keying (RZ-OOK), but also in the case of return-to-zero differential-phase-shift-keyed (RZ-DPSK) system. This fact is illustrated in Fig. 7.6, which shows the calculated bit error rate (BER) results for an RZ-DPSK system considering (a) a nonsymmetric and (b) a symmetric dispersion map [31]. The BER for the symmetric map was obtained by removing the amplitude fluctuations from the corresponding results for the nonsymmetric map. These results clearly show that

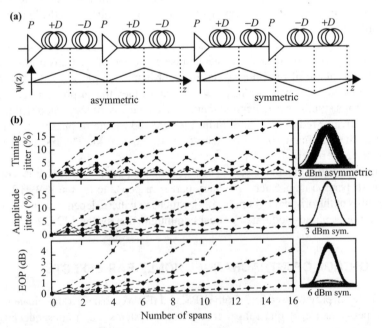

Figure 7.5 (a) Symmetric and asymmetric fiber links. (b) Timing jitter, amplitude jitter, and eye-opening penalty over 16 periods for symmetric (solid curves) and asymmetric (dashed curves) links. Launched power is 3 (diamonds), 6 (circles), and 9 dBm (squares). (After Ref. 28; © 2004 IEEE.)

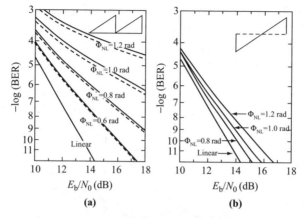

Figure 7.6 Calculated BER against the SNR for an RZ-DPSK system with (a) a nonsymmetric and (b) a symmetric map. The different values of Φ_{NL} indicate different power levels used for the calculation. The dashed curves in (a) were obtained neglecting the IFWM-induced phase fluctuations. (After Ref. [31]; © 2006 OSA.)

the nonlinear penalties with the nonsymmetric map are truly dominated by the nonlinear amplitude fluctuations, which overtake the effect of the nonlinear phase fluctuations.

The ghost pulse generation due to IFWM is similar to the frequency generation in the case of FWM. Using an unequal time spacing between the bits of a bit stream, it is possible to effectively suppress the ghost pulses [32]. In practice, this can be done by conveniently adjusting the time delay between the pulses. Another method to reduce the power of the ghosts consists in splitting the pulses into two subchannels spaced by a frequency shift $\Delta\Omega$ [33]. A hybrid optical time and WDM scheme is used to realize the subchannel multiplexing of the signals. In practice, for a channel operating at a given bit rate, two subchannels operating at half that bit rate are created. At the receiver, the two subchannels are treated as a single channel with the original bit rate. The efficiency of the IFWM generation depends on this frequency spacing and it can be particularly reduced if some resonances are avoided.

Equation (7.24) shows that the perturbation resulting from the nonlinear intrachannel effects depends on the phase of pulses generating it. This fact suggests that those nonlinear effects could be controlled introducing a relative phase shift between any two consecutive pulses forming the bit stream. Such kind of modulation is known as the alternate-phase return-to-zero (AP-RZ) format. The carrier-suppressed return-to-zero (CSRZ) format is an example of the AP-RZ format for which the relative phase difference between two adjacent bits is $\delta\phi = \pi$. However, other values can be chosen for $\delta\phi$. Some numerical and experimental results have shown that the growth of ghost pulses generated through IFWM is suppressed by a large amount for $\delta\phi = \pi/2$ [34].

Data-dependent phase encoding schemes such as alternate-mark-inversion RZ (AMI-RZ) have been proposed to suppress the IFWM-induced ghost pulses in

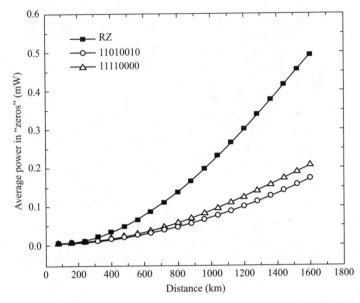

Figure 7.7 Calculated growth of ghost pulses as the average power in "zero" bit slots against transmission distance. (After Ref. [38]; © 2005 OSA.)

symmetric patterns where the one "0" bit is surrounded by several "1" bits, such as 1110111 [35–37]. However, such schemes need a data-dependent electrical encoder, which becomes complex and expensive when the bit rate is up to 40 Gb/s. Alternatively, it was shown that phase modulation with some particular fixed 8-bit patterns such as $\pi\pi0\pi00\pi0$ or $\pi\pi\pi\pi0000$ is effective in suppressing the ghost pulses. This kind of fixed pattern can be constructed by using the exhaustive search method. In such a case, because the pattern of the phase modulation is independent of the input data stream, the phase encoder is not necessary. The efficiency of this method is illustrated in Fig. 7.7, where the average energy in "zeros" is shown against the transmission distance for a 40 Gb/s system [38]. The transmission link is composed of twenty 80 km spans of single-mode fiber, each of which is followed by a two-stage optical amplifier containing a DCF module, used to compensate the loss and the dispersion. With the fixed-pattern phase modulation scheme adopted, the average energy in "zeros" is greatly decreased to one-third of that in normal RZ systems.

An alternative technique to control the intrachannel nonlinear effects consists in the alternation of the polarization of consecutive bits in an RZ signal. This technique is suggested by the fact that both IXPM and IFWM processes depend on the state of polarization of the interacting pulses. In practice, neighboring bits of a single channel are orthogonally polarized through time-domain interleaving [39, 40]. Some experimental results confirm that the use of this technique helps considerably to reduce the intrachannel nonlinear effects in high-speed transmission systems operating at bit rates of 40 Gb/s and above [40].

PROBLEMS

7.1 Obtain an explicit expression for the right-hand side of Eq. (7.11) when Eq. (7.12) is generalized for the case of three pulses located at τ_j, τ_l, and τ_k.

7.2 Explain how IXPM among pulses in a pseudolinear system produces timing jitter. Why is this effect enhanced when the pulse duration is similar to the bit slot? Describe one technique that can be used in practice to reduce the impact of IXPM.

7.3 Explain how IFWM affects the performance of a pseudolinear system. Point out the differences and the similarities between intrachannel and interchannel FWM.

7.4 Verify that the real part of ΔU_{jkl} given by Eq. (7.24) vanishes for a symmetric dispersion map. Discuss the importance of such a fact concerning the reduction of the amplitude jitter.

REFERENCES

1. M. J. Ablowitz, T. Hirroka, and G. Blodini, *Opt. Lett.* **26**, 459 (2001).
2. M. J. Ablowitz and T. Hirroka, *J. Opt. Soc. Am. B* **11**, 425 (2002).
3. C. Kurtzke, *IEEE Photon. Technol. Lett.* **5**, 1250 (1993).
4. R. W. Tkach, A. R. Chraplyvy, F. Forghieri, A. H. Gnauck, and R. M. Derosier, *J. Lightwave Technol.* **13**, 841 (1995).
5. G. P. Agrawal, *Fiber-Optic Communication Systems*, 3rd ed., Wiley, New York, 2002.
6. R.-J. Essiambre, G. Raybon, and B. Mikkelsen, in I. P. Kaminow and T. Li (Eds.), *Optical Fiber Telecommunications*, Vol. 4B, Academic Press, San Diego, CA, 2002, Chapter 6.
7. R. J. Essiambre, B. Mikkelsen, and G. Raybon, *Electron. Lett.* **35**, 1576 (1999).
8. P. V. Mamyshev and N. A. Mamysheva, *Opt. Lett.* **24**, 1454 (1999).
9. A. Mecozzi, C. B. Clausen, and A. H. Gnauck, *IEEE Photon. Technol. Lett.* **12**, 392 (2000).
10. J. Martensson, A. Berntson, M. Westlund, A. Danielsson, P. Johannisson, D. Anderson, and M. Lisak, *Opt. Lett.* **26**, 55 (2001).
11. M. J. Ablowitz and T. Hirooka, *Opt. Lett.* **27**, 203 (2002).
12. R. I. Killey, H. J. Thiele, V. Mikhailov, and P. Bayvel, *IEEE Photon. Technol. Lett.* **12**, 1624 (2000).
13. M. Matsumoto, *IEEE Photon. Technol. Lett.* **10**, 373 (1998).
14. S. Kumar, M. Wald, F. Lederer, and A. Hasegawa, *Opt. Lett.* **23**, 1019 (1998).
15. T. Inoue, H. Sugahara, A. Maruta, and U. Kodama, *IEEE Photon. Technol. Lett.* **12**, 299 (2000).
16. A. Mecozzi, C. B. Clausen, and M. Shtaif, *IEEE Photon. Technol. Lett.* **12**, 1633 (2000).
17. M. J. Ablowitz and T. Horooka, *Opt. Lett.* **25**, 1750 (2000).
18. F. Merlaud and S. K. Turitsyn, *ECOC'00*, Munich, Germany, 2000, Paper 7.2.4, p. 39.

19. A. Mecozzi, C. B. Clausen, M. Shtaif, S. G. Park, and A. H. Gnauck, *IEEE Photon. Technol. Lett.* **13**, 445 (2001).

20. J. Märtensson, A. Berntson, M. Westlund, A. Danielsson, P. Johannisson, D. Anderson, and M. Lisak, *Opt. Lett.* **26**, 55 (2001).

21. M. K. Jackson, G. R. Boyer, J. Paye, M. A. Franco, and A. Mysyrowicz, *Opt. Lett.* **17**, 1770 (1992).

22. I. Shake, H. Takara, K. Mori, S. Kawanishi, and Y. Yamabayashi, *Electron. Lett.* **34**, 1600 (1998).

23. R.-J. Essiambre, B. Mikkelsen, and G. Raybon, *Electron. Lett.* **18**, 1576 (1999).

24. X. Wei and X. Liu, *Opt. Lett.* **28**, 2300 (2003).

25. K. P. Ho, *IEEE Photon. Technol. Lett.* **17**, 789 (2005).

26. A. P. Lau, S. Rabani, and J. M. Kahan, *J. Lightwave Technol.* **26**, 2128 (2008).

27. P. Johannisson, D. Anderson, A. Berntson, and J. Märtensson, *Opt. Lett.* **26**, 1227 (2001).

28. A. G. Striegler and B. Schmauss, *J. Lightwave Technol.* **22**, 1877 (2004).

29. A. Chowdhury and R. J. Essiambre, *Opt. Lett.* **29**, 1105 (2004).

30. A. Chowdhury, G. Raybon, R. J. Essiambre, J. H. Sinsky, A. Adamiecki, J. Leuthold, C. R. Roerr, and S. Chandrasekhar, *J. Lightwave Technol.* **23**, 172 (2005).

31. X. Wei, X. Liu, S. H. Simon, and C. J. McKinstrie, *Opt. Lett.* **31**, 29 (2006).

32. S. Kumar, *IEEE Photon. Technol. Lett.* **13**, 800 (2001).

33. J. Zweck and C. R. Menyuk, *Opt. Lett.* **27**, 1235 (2002).

34. S. Appathurai, V. Mikhailov, R. I. Killey, and P. Bayvel, *J. Lightwave Technol.* **22**, 239 (2004).

35. N. Alic and Y. Fainman, *IEEE Photon. Technol. Lett.* **16**, 1212 (2004).

36. X. Liu, X. Wei, A. H. Gnauck, C. Xu, and L. K. Wickham, *Opt. Lett.* **27**, 1177 (2002).

37. P. J. Winzer, A. H. Gnaick, G. Raybon, S. Chandrasekhar, Y. Su, and J. Leuthold, *IEEE J. Sel. Top. Quantum Electron.* **15**, 766 (2003).

38. M. Zou, M. Chen, and S. Xie, *Opt. Express* **13**, 2251 (2005).

39. X. Liu, C. Xu, and X. Wei, *IEEE Photon. Technol. Lett.* **16**, 302 (2004).

40. C. Xie, I. Kang, A. H. Gnauck, L. Möller, L. F. Mollenauer, and A. R. Grant, *J. Lightwave Technol.* **22**, 806 (2004).

8

SOLITON LIGHTWAVE SYSTEMS

As described in Chapter 4, the formation of solitons in optical fibers is the result of a balance between the negative (anomalous) group velocity dispersion (GVD) of the glass fiber, which occurs for wavelengths longer than 1.3 μm in a standard fiber, and the Kerr nonlinearity. The existence of fiber solitons was suggested for the first time by Hasegawa and Tappert [1] and it soon appeared as an ideal solution to the problem of pulse spreading caused by fiber dispersion.

After the first experimental observation by Mollenauer et al. [2], Hasegawa [3] made the imaginative proposal that solitons could be used in all-optical trans-mission systems based on optical amplifiers instead of regenerative repeaters, which were considered standard until 1990. In particular, he suggested the use of the Raman effect of transmission fiber itself for optical amplification. The distrib-uted Raman amplification was replaced by erbium-doped fiber amplifiers (EDFAs) during the 1990s.

Meanwhile, Gordon and Haus [4] anticipated that the transmission of a signal made of optical solitons could not be extended to an unlimited distance when optical amplification is used. In fact, the amplifiers needed to compensate for the fiber loss generate amplified spontaneous emission (ASE), and this noise is, in part, incorpo-rated by the soliton, whose mean frequency is then shifted. Due to GVD, the arrival time of the soliton then becomes a random variable, whose variance is proportional to the cube of the propagation distance. This is the so-called *Gordon–Haus effect*.

At the end of 1991, some research groups suggested the use of frequency filters to extend the limit set by the Gordon–Haus effect [5–7]. However, using this technique, the transmission distance was still limited by the growth of narrowband noise at the center frequency of the filters. This problem was beautifully solved by Mollenauer and coworkers, who developed the sliding-guiding filter concept [8–10]. Other proposals to achieve a stable soliton propagation in fiber systems made use of amplitude modulators [11] or nonlinear optical amplifiers [12–14].

The nonlinear pulses propagating in the presence of some of the control techniques referred above are the result of the double balance between nonlinearity and dispersion and between gain and loss. Such nonlinear pulses are known as *dissipative solitons* and their properties are completely determined by the external parameters of the optical system [15–17].

Attempts to solve much of the intrinsic problems of optical soliton transmission by proper control of the fiber group dispersion—a technique usually referred to as *dispersion management*—have emerged in the latter half of the 1990s. In particular, Smith et al. [18] showed in 1996 that a nonlinear soliton-like pulse can exist in a fiber having a periodic variation of the dispersion, even if the dispersion is almost zero. The nonlinear pulse that can propagate in such a system is usually called the *dispersion-managed (DM) soliton* and presents several remarkable characteristics, namely, an enhanced pulse energy, reduced Gordon–Haus timing jitter, longer collision lengths, and greater robustness to PMD [18–22]. Due to their characteristics, dispersion-managed solitons are the preferred option for use in new ultrahigh-speed multiplexed systems.

8.1 SOLITON PROPERTIES

As shown in Chapter 4, the nonlinear Schrödinger equation (NLSE) has an exact and stable solution, called the fundamental soliton solution, given by

$$q(T, Z) = \eta \, \text{sech}[\eta(T - T_0)] \exp\left(i \frac{\eta^2}{2} Z + i\sigma \right) \qquad (8.1)$$

If the central frequency deviates from ω_0 by κ/t_0, the Galilean transformation of Eq. (8.1) gives

$$q(T, Z) = \eta \, \text{sech}[\eta(T + \kappa Z - T_0)] \exp\left(-i\kappa T + \frac{i}{2}(\eta^2 - \kappa^2)Z + i\sigma \right) \qquad (8.2)$$

where κ is the normalized frequency shift. Thus, the fundamental soliton solution of the NLSE is characterized by four parameters: the amplitude η (also the pulse width), the frequency κ (also the pulse speed), the time position T_0, and the phase σ. The soliton amplitude and width are coupled in Eq. (8.2), such that there is an inverse relationship between these two quantities. This is a very important property of fundamental solitons in fibers. The canonical form of the fundamental soliton can be obtained from Eq. (8.2) considering $\eta = 1$, $T_0 = 0$, $\kappa = 0$, and $\sigma = 0$. With these values, Eq. (8.2) reduces to

$$q(T, Z) = \text{sech}(T) \exp(iZ/2) \qquad (8.3)$$

An interesting feature of the fundamental soliton solution, which demonstrates its notable robustness, is that it can be obtained starting from an arbitrary input pulse, since it has an area that lies in the range between $\pi/2$ and $3\pi/2$ [23]. Figure 4.5b shows, for example, the formation of a soliton solution obtained numerically from the NLSE using an input rectangular pulse with an area $A = 3.05$.

According to Eq. (4.66), the power of a fundamental soliton ($N = 1$) in an optical fiber is given by

$$P_0 = \frac{|\beta_2|}{\gamma t_0^2} \approx \frac{3.11|\beta_2|}{\gamma t_{\text{FWHM}}^2} \tag{8.4}$$

where $t_{\text{FWHM}} = 1.763 t_0$ is the full width at half maximum (FWHM) of the soliton. Using typical values $\beta_2 = -20\,\text{ps}^2/\text{km}$ and $\gamma = 1.3\,\text{W}^{-1}\,\text{km}^{-1}$ for standard fibers near the $1.55\,\mu\text{m}$ wavelength and considering pulses with a temporal width $t_0 = 25\,\text{ps}$, we obtain a soliton power of 25 mW. This value is significantly reduced using dispersion-shifted fibers in the same wavelength region.

The required power for the generation of higher order solitons is N^2 times that for the fundamental soliton, according to Eq. (4.66). Moreover, higher order solitons compress periodically (see Fig. 4.6), resulting in soliton chirping and spectral broadening. In contrast, fundamental solitons preserve their shape during propagation. This fact, together with the lower power required for their generation, makes fundamental solitons the preferred option in soliton communication systems.

The soliton solution (8.2) exists in the range $-\infty < T < \infty$ and remains valid for a train of solitons only if individual solitons are well separated. This means that the soliton width must be a small fraction of the bit slot. The bit rate R is related to the soliton width t_0 as

$$R = \frac{1}{t_B} = \frac{1}{s_0 t_0} \tag{8.5}$$

where t_B is the duration of the bit slot and $s_0 = t_B/t_0$ is the separation between neighboring solitons in normalized units.

8.1.1 Soliton Interaction

Since a soliton solution ideally allows only one soliton in $-\infty < T < \infty$, the existence of an adjacent soliton must modify the ideal soliton solution. This modification appears as an interaction force between two neighboring solitons. One can numerically study the interaction between two solitons by solving the NLSE with an input field of the form

$$q(0, T) = \text{sech}(T - s_0/2) + r\,\text{sech}(T + s_0/2)\exp(i\Delta\phi) \tag{8.6}$$

where s_0 is the normalized separation between the two solitons, r is the ratio of their amplitudes, and $\Delta\phi$ is their relative phase difference. Numerical simulations show that the soliton separation evolves from its initial value in a way that depends on r and $\Delta\phi$. Figure 8.1 shows the contour plots corresponding to a pair of solitons with equal

(a)

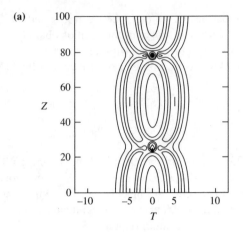

(b)

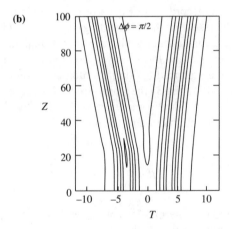

Figure 8.1 Interaction between two solitons with equal amplitudes $(r = 1)$, an initial separation $S_0 = 7$, and a phase difference $\Delta\phi = 0$ (a) or $\Delta\phi = \pi/2$ (b).

amplitudes $(r = 1)$, an initial separation $s_0 = 7$, and a phase difference $\Delta\phi = 0$ (a) or $\Delta\phi = \pi/2$ (b). Figure 8.1a shows that two in-phase solitons with equal amplitudes collide periodically. The separation between the solitons varies with the propagation distance as [24]

$$s = \ln\left[\frac{1}{2}(1 + \cos(4Ze^{-s_0/2}))e^{s_0}\right] \tag{8.7}$$

Equation (8.7) shows that the separation s evolves periodically along the fiber, with a period

$$Z_p = \frac{\pi}{2}e^{s_0/2} \tag{8.8}$$

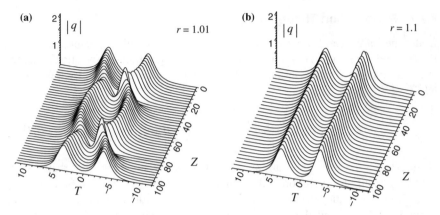

Figure 8.2   Interaction between two solitons when $S_0 = 7$, $\Delta\phi = 0$, $r = 1.01$ (a), and $r = 1.1$ (b).

The periodic evolution observed in Fig. 8.1a does not occur for other values of the phase difference between the two interacting solitons. For example, Fig. 8.1b shows that introducing a phase difference $\Delta\phi = \pi/2$, the two solitons will separate monotonously during the propagation.

In order to avoid soliton interaction, two adjacent solitons should be separated enough (typically more than six times the pulse width in time). This requirement is a severe drawback for soliton-based communications because it requires a bandwidth at least three times wider than linear systems. However, the soliton interaction can be controlled and the tolerable spacing can be reduced by using adjacent solitons with slightly different amplitudes, as illustrated in Fig. 8.2. Since small changes in the peak power are not detrimental for the soliton nature of pulse propagation, this scheme is feasible in practice and can be useful for increasing the system capacity.

8.2 PERTURBATION OF SOLITONS

High-speed fiber-optic communication systems are generally limited by both the fiber nonlinearity and the group dispersion that causes the pulse broadening. However, since fundamental solitons are obtained by the balance between the group dispersion and Kerr nonlinearity, their width can be maintained over long distances. In other words, fundamental solitons are free from either the dispersive distortion or self-phase modulation. Thus, it is quite natural to use solitons as information carrier in fibers since all other format will face distortion from either dispersion or nonlinearity.

In a realistic communication system, the soliton propagation is affected by many perturbations related to the input pulse shape, chirp and power, amplifier noise, optical filters, modulators, and so on. However, because of its particle-like nature, it turns out that the soliton remains stable under most of these perturbations.

8.2.1 Perturbation Theory

Under the influence of arbitrary perturbations, the four soliton parameters used in Eq. (8.2) do not remain constant but evolve during the soliton propagation along the fiber. Soliton propagation in these circumstances can be described by a perturbed NLSE of the form

$$i\frac{\partial q}{\partial Z} + \frac{1}{2}\frac{\partial^2 q}{\partial T^2} + |q|^2 q = iP(q) \tag{8.9}$$

where $P(q)$ represents the various perturbations. When these perturbations are relatively small, the evolution of the four soliton parameters can be described using the soliton perturbation theory [25]. Considering the variational approach discussed in Section 5.1.2, we find that Eq. (8.9) can be obtained from the Euler–Lagrange equation using the Lagrangian density [26]

$$L_d = \frac{i}{2}\left(q\frac{\partial q^*}{\partial Z} - q^*\frac{\partial q}{\partial Z}\right) + \frac{1}{2}\left|\frac{\partial q}{\partial T}\right|^2 - \frac{1}{2}|q|^4 + i(Pq^* - P^*q) \tag{8.10}$$

Integrating the Lagrangian density over T and using the reduced Euler–Lagrange equation, we obtain the following evolution equations for the soliton parameters:

$$\frac{d\eta}{dZ} = \mathrm{Re}\int_{-\infty}^{\infty} P(q)q^*\,dT \tag{8.11}$$

$$\frac{d\kappa}{dZ} = -\mathrm{Im}\int_{-\infty}^{\infty} P(q)\tanh(\eta(T-T_0))q^*\,dT \tag{8.12}$$

$$\frac{dT_0}{dZ} = -\kappa + \frac{1}{\eta^2}\mathrm{Re}\int_{-\infty}^{\infty} P(q)(T-T_0)q^*\,dT \tag{8.13}$$

$$\frac{d\sigma}{dZ} = \frac{1}{2}(\eta^2-\kappa^2) + T_0\frac{d\kappa}{dZ} + \frac{1}{\eta}\mathrm{Im}\int_{-\infty}^{\infty} P(q)(1-\eta(T-T_0))\tanh[\eta(T-T_0)]q^*\,dT \tag{8.14}$$

where Re and Im stand for the real and imaginary parts, respectively. Equations (8.11)–(8.14) are used extensively in the theory of soliton communication systems [27–29].

8.2.2 Fiber Losses

Fiber loss is the first perturbation affecting the soliton propagation and one of the main causes of signal degradation in long-distance fiber-optic communication systems.

This limitation can be minimized by operating near $\lambda = 1.55\,\mu$m. However, even with fiber losses as low as $0.2\,$dB/km, the signal power is reduced by $20\,$dB after transmission over $100\,$km of fiber.

Fiber loss can be taken into account theoretically by adding a loss term to the NLSE, which becomes

$$i\frac{\partial q}{\partial Z} + \frac{1}{2}\frac{\partial^2 q}{\partial T^2} + |q|^2 q = -\frac{i}{2}\Gamma q \qquad (8.15)$$

where $\Gamma = \alpha_0 L_D$ is the loss rate per dispersion distance and α_0 is the fiber loss coefficient. Assuming that $\Gamma \ll 1$ and using $P(q) = -\Gamma q/2$ in Eqs. (8.11)–(8.14), we find the following results:

$$\eta(Z) = \eta(0)\exp(-\Gamma Z) \equiv \eta_0 \exp(-\Gamma Z) \qquad (8.16a)$$

$$\kappa(Z) = \kappa(0) \equiv \kappa_0 \qquad (8.16b)$$

$$T_0(Z) = -\kappa_0 Z + T_0(0) \qquad (8.16c)$$

$$\sigma(Z) = -\frac{1}{2}\kappa_0^2 Z + \frac{\eta_0^2}{4\Gamma}(1-\exp(-2\Gamma Z)) + \sigma(0) \qquad (8.16d)$$

Equation (8.16) indicates that the soliton amplitude decreases along the fiber length at the same rate as the power amplitude. The decrease in the amplitude at a rate twice as fast as a linear pulse is a consequence of the nonlinear property of a soliton. Since the amplitude and width are inversely related, the soliton width will increase exponentially according to

$$t_p(Z) = t_0 \exp(\Gamma Z) \qquad (8.17)$$

In fact, the exponential increase of the soliton width predicted by the perturbation theory occurs only for relatively short propagation distances. Figure 8.3 shows the variation of the broadening factor $t_p(Z)/t_0$ with distance for the fundamental soliton when $\Gamma = 0.07$ [30]. Clearly, the result predicted by the perturbation theory is valid only for distances up to $Z \approx \Gamma^{-1}$. The numerical result displayed in Fig. 8.3 shows that for large distances the soliton width increases linearly with distance, but at a rate slower than the linear pulse.

Loss Compensation   In the context of communication systems, the fiber loss problem was solved prior to 1991 using repeaters periodically installed in the transmission line. A repeater consists of a light detector and light pulse generators. In fact, it was the most expensive unit in the transmission system and its use also represented the bottleneck to increase the transmission speed.

A repeaterless soliton transmission system using distributed Raman gain provided by the fiber itself to compensate for the fiber loss was suggested by Hasegawa [3] in 1983. Its use requires periodic injection of the pump power into the transmission fiber.

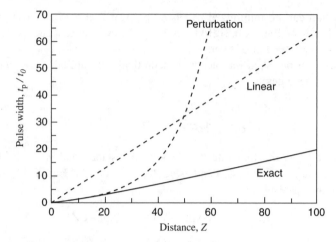

Figure 8.3 Broadening factor $t_p(Z)/t_0$ against propagation distance obtained numerically and predicted by the perturbation theory (dotted curve). The dashed curve corresponds to the case of a linear pulse. (After Ref. [30]; © 1985 Elsvier.)

The most significant feature of the Raman gain in silica fibers is that it extends over a large frequency range (up to 40 THz, with a typical FWHM of 6 THz), presenting a broad dominant peak near 13 THz below the pump frequency [31].

The concept of all-optical transmission using the distributed Raman amplification was replaced by lumped amplification during the 1990s, through the use of EDFAs. The EDFA can be pumped by laser diodes and presents some remarkable properties [32]. Using lumped EDFAs, the amplification occurs over a very short distance (~10 m), which compensates for the loss occurring over 40–50 km. However, by lightly doping the transmission fiber, a distributed amplification configuration can also be achieved.

8.3   PATH-AVERAGED SOLITONS

The behavior of a soliton in the presence of fiber losses and optical amplification depends strongly on the relative magnitudes of dispersion length, L_D, and amplifier spacing, L_A. If the conditions $L_D \ll L_A$ and $\Gamma = \alpha_0 L_D \ll 1$ are satisfied, we have the so-called *quasi-adiabatic regime*. In such a regime, the soliton can adapt adiabatically to losses by increasing the width and decreasing its peak power. Conversely, if the amplifier spacing is much smaller than the dispersion distance ($L_A \ll L_D$), the soliton shape is not distorted significantly by fiber loss between successive amplifications. In this case, solitons can be amplified hundreds of times while preserving their shape and width. Such solitons are referred to as *path-averaged solitons*, since their properties are given by the soliton energy averaged over one amplifier spacing.

8.3.1 Lumped Amplification

Taking into account the fiber loss, periodic lumped amplification, and possible variations in fiber dispersion, Eq. (8.15) is generalized and becomes

$$i\frac{\partial q}{\partial Z} + \frac{d(Z)}{2}\frac{\partial^2 q}{\partial T^2} + |q|^2 q = -i\frac{\Gamma}{2}q + iG(Z)q \tag{8.18}$$

where

$$G(Z) = g\sum_{n=1}^{M}\delta(Z-nZ_A) \tag{8.19}$$

and $d(Z)$ is the dispersion normalized so that its average becomes unity. In Eq. (8.19), it is assumed that the fiber link presents M amplifiers spaced apart by $Z_A = L_A/L_D$ and $g = \exp(\Gamma Z_A/2)-1$ gives the proper compensation of the fiber loss.

Equation (8.18) may be transformed to a Hamiltonian form by introducing a new amplitude u through

$$u(Z,T) = \frac{q(Z,T)}{a(Z)} \tag{8.20}$$

where $a(Z)$ contains rapid amplitude variations and $u(Z,T)$ is a slowly varying function of Z. Substituting (8.20) into (8.18), one obtains that $a(Z)$ satisfies

$$\frac{da}{dZ} = \left[-\frac{\Gamma}{2} + g\sum_{n=1}^{M}\delta(Z-nZ_A)\right]a \tag{8.21}$$

while the new amplitude u satisfies

$$i\frac{\partial u}{\partial Z} + \frac{d(Z)}{2}\frac{\partial^2 u}{\partial T^2} + a^2(Z)|u|^2 u = 0 \tag{8.22}$$

In this section we will consider the case with no variation in dispersion, $d(Z) = 1$. The concept of the average soliton makes use of the fact that $a^2(Z)$ in Eq. (8.22) varies rapidly with a period $Z_A \ll 1$. Since solitons evolve little over a short distance Z_A, one can replace $a^2(Z)$ by its average value $\langle a^2\rangle$. The resulting equation is then satisfied by the average soliton $\bar{u}$ such that $u = \bar{u} + \delta u$. The perturbation δu is relatively small, since the leading-order correction varies as Z_A^2 rather than Z_A [33]. The average soliton description proves to be accurate even for $Z_A = 0.2$.

The normalization $\langle a^2\rangle = 1$ is achieved by choosing the integration constant a_0 in (8.21), leading to

$$a_0 = \left(\frac{\Gamma Z_A}{1-\exp(-\Gamma Z_A)}\right)^{1/2} = \left(\frac{G\ln G}{G-1}\right)^{1/2} \tag{8.23}$$

where $G = \exp(\Gamma Z_A)$ is the amplifier gain. Thus, soliton evolution in lossy fibers with periodic lumped amplification is identical to that in lossless fibers provided that the amplifier spacing L_A is much less than the dispersion distance L_D and the initial amplitude is enhanced by a factor a_0 given in Eq. (8.23).

Using $L_D = t_0^2/|\beta_2|$, the condition $L_A \ll L_D$ can be expressed in terms of the soliton width t_0 as

$$t_0 \gg \sqrt{|\beta_2|L_A} \qquad (8.24)$$

Considering typical values, $\beta_2 = -0.5 \text{ ps}^2/\text{km}$, $L_A = 50 \text{ km}$, and $s = 10$, we obtain from Eqs. (8.24) and (8.5) that $t_0 \gg 5$ ps and $R \ll 20$ Gbit/s. This limitation on the bit rate for soliton communication systems can be alleviated considerably using distributed amplification instead of lumped amplification.

8.3.2 Distributed Amplification

In comparison to the lumped amplification scheme, the distributed amplification configuration appears as a better approach since it can provide a nearly lossless fiber by compensating losses locally at every point along the fiber link. In the presence of a distributed gain with $G \ll 1$, such as that provided by the Raman amplification or a distributed EDFA, the soliton amplitude given by Eq. (8.16) is simply modified to

$$\eta(Z) = \eta_0 \exp\left(\int_0^Z [2G(Z) - \Gamma] dZ\right) \qquad (8.25)$$

By designing the gain so that the exponent of Eq. (8.25) vanishes—namely, making $G(Z)$ constant and equal to $\Gamma/2$ for all Z—one can achieve a system in which a soliton propagates without any distortion.

In practice, the effective gain cannot be made constant along the fiber, since the pump power also suffers from fiber loss. The evolution of the soliton energy is given in physical units as [34]

$$\frac{dE_s}{dz} = [g(z) - \alpha_0]E_s \qquad (8.26)$$

Assuming a bidirectional pumping scheme and neglecting the gain saturation, the gain coefficient $g(z)$ can be approximated as

$$g(z) = g_0\{\exp(-\alpha_p z) + \exp[-\alpha_p(L_A - z)]\} \qquad (8.27)$$

where α_p is the fiber loss at the pump wavelength and L_A now stands for the distributed amplifier spacing or the spacing between the pump stations. The gain constant g_0 is related to the pump power injected at both ends. Using Eq. (8.27), the integration of Eq. (8.26) gives the following result for the soliton energy:

$$\ln\left(\frac{E_s(z)}{E_{in}}\right) = \alpha_0 L_A \left\{ \frac{\sinh[\alpha_p(z - L_A/2)] + \sinh(\alpha_p L_A/2)}{2\sinh(\alpha F L_A/2)} - \frac{z}{L_A} \right\} \qquad (8.28)$$

where E_{in} is the soliton energy at the fiber input. Equation (8.28) shows that the range of energy variations increases with L_A. However, it remains much smaller than that occurring in the lumped amplification scheme.

Like in the case of lumped amplification, the effects of energy variations on solitons depend on the relative magnitudes of dispersion length, L_D, and amplifier spacing, L_A. When $L_D \ll L_A$, we have the so-called *quasi-adiabatic regime*, in which solitons evolve adiabatically with some emission of dispersive waves. On the other hand, for $L_A \ll L_D$ little soliton reshaping is observed. A more complicated behavior occurs when the values of L_A and L_D become of the same order. In particular, for $L_A \approx 4\pi L_D$ dispersive waves and solitons are resonantly amplified, leading to unstable and chaotic evolution [34].

The propagation of short solitons ($t_0 < 5\,\text{ps}$) in distributed amplifiers can be described by a generalized NLSE that, besides the addition of a gain term, must also include the effects of finite gain bandwidth, third-order dispersion (TOD), and intrapulse Raman scattering (IRS). The inclusion of these effects provides a generalization of Eq. (8.18) in the form [35]

$$i\frac{\partial q}{\partial Z} + \frac{1}{2}\frac{\partial^2 q}{\partial T^2} + |q|^2 q = -\frac{i}{2}\Gamma q + \frac{i}{2}g(Z)L_D\left(q + \tau_2^2\frac{\partial^2 q}{\partial T^2}\right) + i\delta_3\frac{\partial^3 q}{\partial T^3} + \tau_R q\frac{\partial |q|^2}{T}$$

(8.29)

where τ_2 is inversely related to the gain bandwidth. The TOD and Raman effects are represented, respectively, by the parameters δ_3 and τ_R, given by

$$\delta_3 = \frac{\beta_3}{6|\beta_2|t_0}, \qquad \tau_R = \frac{t_R}{t_0}$$

(8.30)

where t_R is a constant that depends on the Raman gain slope. The TOD term becomes important only when $|\beta_2|$ is sufficiently small. On the other hand, since $t_R \approx 5\,\text{fs}$, the Raman term in Eq. (8.29) can generally be treated as a small perturbation. Considering only this term and applying the perturbation theory discussed in Section 8.2.1, it is found that the soliton amplitude remains constant, whereas its frequency varies with distance as

$$\kappa(Z) = -\frac{8}{15}\tau_R\eta^4 Z$$

(8.31)

The IRS effect leads to a continuous downshift of the carrier frequency, an effect known as the *soliton self-frequency shift* (SSFS) [36]. This effect was observed for the first time by Mitschke and Mollenauer in 1986 using 0.5 ps pulses obtained from a passively mode-locked color-center laser [37]. The origin of SSFS can be understood by noting that for ultrashort solitons the pulse spectrum becomes so broad that the high-frequency components of the pulse can transfer energy through Raman amplification to the low-frequency components of the same pulse. Such an energy transfer

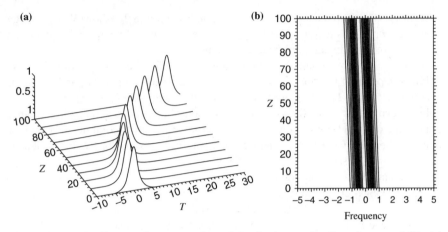

Figure 8.4 IRS effect on the propagation of the fundamental soliton for $\tau_R = 0.01$: (a) evolution of the amplitude and (b) contour plot of the soliton spectrum.

appears as a red shift of the soliton spectrum, with shift increasing with distance. Since the amplitude and width are inversely related, the frequency shift scales as t_0^4, indicating that it can become quite large for short pulses.

The IRS effect on the propagation of the fundamental soliton is illustrated in Fig. 8.4 for the case $\delta_3 = \tau_2 = 0, g(Z)L_D = \Gamma$, and $\tau_R = 0.01$. The linear variation of the frequency shift with distance is clearly illustrated by the contour plot of the soliton spectrum in Fig. 8.4b. As a consequence, there is a continuous variation of the soliton velocity and position, as observed in Fig. 8.4a. In practice, the limited bandwidth of the gain spectrum ($\tau_2 \neq 0$) can provide in some circumstances the SSFS compensation and adiabatic soliton trapping [35].

8.3.3 Timing Jitter

When one soliton is used to represent one digit in a communication system, the major cause of information loss is the timing jitter, which corresponds to the random variation of the soliton position T_0. This situation is quite different from a nonsoliton pulse, where the major cause of information loss is deformation of the waveform itself. The soliton timing jitter can be induced by several physical mechanisms. Among them, the ASE noise added by the optical amplifiers is often dominant in practice. The amplifier noise normally deteriorates the signal-to-noise ratio (SNR) for a linear signal due to the superposition of noise with the signal. For soliton transmission, this problem is generally not serious because the signal amplitude can be made large with proper choice of the fiber dispersion. The variation of the soliton position T_0 originates from the frequency shift κ due to the amplifier noise and the finite fiber dispersion. When the soliton frequency shifts, it modifies the group velocity through the group dispersion, resulting in jitter in arrival times of the soliton pulses. This is known as the *Gordon–Haus timing jitter* [38–40].

Neglecting other perturbations, the soliton position variation at the end of one amplifier spacing due to the frequency shift κ is given from Eq. (8.13) as $T_0 = -\kappa Z_A$. Considering a link with M amplifiers, the total timing jitter is given by

$$T_0 = -Z_A \sum_{j=1}^{M} \sum_{i=1}^{j} \kappa_i \tag{8.32}$$

where κ_i is the frequency shift induced by the ith amplifier. When N is sufficiently large, the summations in Eq. (8.32) can be replaced by an integral and the variance of the timing jitter becomes [38]

$$\sigma_{GH}^2 = \frac{\sigma_\kappa^2}{3} \frac{Z^3}{Z_A} \tag{8.33}$$

In Eq. (8.33), $Z = MZ_A$ is the total transmission distance and σ_κ^2 is the normalized variance of frequency fluctuations, given by [40]

$$\sigma_\kappa^2 = 2\eta n_{sp} F(G)/(3N_s) \tag{8.34}$$

where n_{sp} is the spontaneous emission factor, N_s is the number of photons in the soliton, G is the amplifier gain, and $F(G) = (G-1)^2/(G \ln G)$.

As observed from Eq. (8.33), the Gordon–Haus timing jitter increases with the cube of distance and, in practice, it sets a limit on the bit rate–distance product of a communication system. Substituting Eq. (8.5), $z = MZ_A L_D$, $N_s = 2P_0 t_0/(h\nu_0)$, and $P_0 = 1/(\gamma L_D)$ in Eq. (8.33), we find that the total bit rate–distance product Rz is limited by

$$(Rz)^3 < \frac{18\pi f_b^2 L_A}{n_{sp} F(G) S_0 \lambda h \gamma D} \tag{8.35}$$

where D is the dispersion parameter and f_b is the tolerable fraction of the bit slot corresponding to the arrival time jitter. Considering a 10 Gb/s soliton communication system operating at 1.55 μm, with typical parameter values $L_A = 50$ km, $D = 0.5$ ps/(km nm), $\alpha_0 = 0.2$ dB/km, $\gamma = 3$ W^{-1} km^{-1}, $n_{sp} = 2$, $S_0 = 12$, and $f_b = 0.1$, the transmission distance is limited to $z = 6000$ km.

Besides the frequency, the ASE noise added by the amplifiers also affects the other three soliton parameters (amplitude, position, and phase), as well as its polarization. In particular, ASE noise-induced amplitude fluctuations give rise to a timing jitter that, in the case of ultrashort solitons, becomes more important than that due to the Gordon–Haus effect. Such amplitude fluctuations are converted into frequency fluctuations by the IRS effect (see Eq. (8.31)) and give rise to a timing jitter that increases with the fifth power of distance [41].

Another contribution to the timing jitter, which was observed in the earliest long-distance soliton transmission experiments [42], arises from an acoustic interaction among the solitons. As a soliton propagates down the fiber, an acoustic wave is generated through electrostriction. This acoustic wave induces a variation in the refractive index, which affects the speed of other pulses following in the wake of the soliton. In practice, since a bit stream consists of a random string of 1 and 0 bits, changes in the speed of a given soliton depend on the presence or absence of solitons in the preceding bit slots. The timing jitter arises because, in these circumstances, different solitons experience different changes of speed.

Considering a fiber with $A_{eff} = 50\,\mu m^2$, the acoustic jitter can be approximated by [43]

$$\sigma_{acou} \approx 4.3 \frac{D^2}{t_{FWHM}} z^2 \sqrt{R-0.99} \tag{8.36}$$

where σ_{acou} is in ps, D is in ps/(km nm), t_{FWHM} is in ps, R is in Gbit/s, and z is in thousands of km. For example, if $D = 0.5$ ps/(km nm), $t_{FWHM} = 10$ ps, $z = 9000$ km (trans-Pacific distance), and $R = 10$ Gbit/s, we have $\sigma_{acou} = 26$ ps. Unlike the Gordon–Haus jitter, the acoustic jitter increases with bit rate and also increases as a higher power of the distance. As a result, the acoustic jitter becomes particularly important in high bit rate long-distance communication systems.

Polarization mode dispersion (PMD) is an additional cause of timing jitter in soliton transmission systems [44]. As the solitons are periodically amplified, their state of polarization is scattered by the ASE noise added by optical amplifiers. The polarization scattering is transformed in a random transit time, since the two orthogonally polarized components propagate with slightly different speeds in birefringent fibers. The variance of the timing jitter introduced by the combination of ASE and PMD is given by [44]

$$\sigma_P^2 = \frac{\pi}{16} \frac{F(G)n_{sp}}{N_s} \frac{D_p^2 z^2}{L_A} \tag{8.37}$$

where D_p is the so-called PMD parameter. It must be noted that σ_P increases linearly with both the transmission distance z and the PMD parameter D_p. Generally, the timing jitter due to PMD is much less than the Gordon–Haus jitter. However, the PMD-induced timing jitter becomes significant for fibers having large values of the PMD parameter and for soliton communication systems with bit rates of 40 Gbit/s or above.

8.4   SOLITON TRANSMISSION CONTROL

Although solitons are clearly a better choice as the information carrier in optical fibers because of their robust nature, their interaction and the timing jitter can lead to the loss of information in transmission systems. These problems can be suppressed using some techniques to control the soliton parameters, namely, its amplitude (or width), time position, velocity (or frequency), and phase. In fact, while the original wave equation has infinite-dimensional parameters, control of finite-dimensional parameters is sufficient to control the soliton transmission system. This is an important characteristic of a soliton system.

8.4.1 Fixed-Frequency Filters

The use of narrowband filters was early suggested to control the Gordon–Haus jitter and other noise effects [5–7]. The basic idea is that any soliton whose central frequency has strayed from the filter peak will be returned to the peak by virtue of the differential loss the filters induce across its spectrum. Suppression of the timing jitter is a consequence of the frequency jitter damping. In practice, etalon guiding filters are used. The multiple peaks present in the response of these filters provide the possibility of their use in wavelength division multiplexed (WDM) soliton transmission systems.

The control of timing jitter and/or soliton interaction by means of fixed-frequency filters can be demonstrated using the adiabatic soliton perturbation theory presented in Section 8.2.1. If we designate the filter strength by β ($\beta > 0$) and the excess gain to compensate for the filter-induced loss by δ, the perturbation term $P(q)$ in Eqs. (8.11)–(8.14) reads

$$P(q) = \delta q + \beta \frac{\partial^2 q}{\partial T^2} \qquad (8.38)$$

In particular, we obtain the following evolution equations for the soliton amplitude η and the frequency κ:

$$\frac{\partial \eta}{\partial Z} = 2\delta\eta - 2\beta\eta \left(\frac{1}{3}\eta^2 + \kappa^2 \right) \qquad (8.39)$$

$$\frac{\partial \kappa}{\partial Z} = -\frac{4}{3}\beta\eta^2\kappa \qquad (8.40)$$

From Eq. (8.40) it can be seen that the soliton frequency approaches asymptotically to $\kappa = 0$ (stable fixed point) if $\eta \neq 0$. In such a limit, Eq. (39) provides a stationary amplitude $\eta = 1$ if the condition $\beta = 3\delta$ is satisfied.

Figure 8.5 shows the phase portrait of Eqs. (8.39) and (8.40) for the case $\beta = 3\delta$ and $\beta = 0.15$. We observe that solitons having an initial range of amplitudes and frequencies emerge as solitons with an identical amplitude and frequency imposed by the attractor at $\eta = 1$ and $\kappa = 0$ after repeated amplifications. This process may be interpreted as soliton cooling and can be used to overcome the Gordon–Haus effect. In fact, the use of fixed-frequency guiding filters provides an effective suppression of the soliton timing jitter, whose variance is reduced relatively to the uncontrolled case, given by Eq. (8.33), by the factor

$$F_r(x) = \frac{3}{2}\frac{1}{x^3}\left[2x - 3 + 4\exp(-x) - \exp(-2x)\right] \qquad (8.41)$$

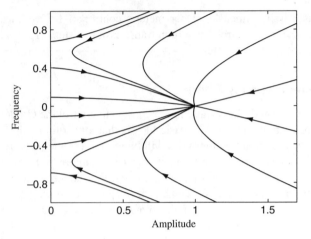

Figure 8.5   Phase portrait of Eqs. (8.39) and (8.40) for the case $\beta = 3\delta$ and $\beta = 0.15$.

where $x = 4\delta Z$. If $x \gg 1, f(x) \sim 3x^{-2}$ and the variance of the timing jitter increases linearly with distance, instead of the cubic dependence shown in the uncontrolled case [6, 7, 28].

As mentioned above, to compensate for the filter-induced loss some excess gain must be provided to the soliton. However, this excess gain also amplifies the small-amplitude waves coexistent with soliton. Such amplification results in a background instability that can significantly affect and even destroy the soliton itself. This process is illustrated in Fig. 8.6 for $\beta = 3\delta$ and $\delta = 0.05$. In this case, we observe that the soliton is severely disturbed for transmission distances $Z > 80$.

8.4.2   Sliding-Frequency Filters

An approach to avoid the background instability was suggested by Mollenauer et al. [8] and consists in using filters whose central frequency is gradually shifted along the transmission line. In this scheme, the linear waves that initially grow near $\kappa = 0$ eventually fall into negative gain region of the filter and are dissipated while the soliton central frequency is shifted, following the central frequency of the filter. This example represents a remarkable property of solitons, since the signal carried by them can be effectively separated from noise that has the same frequency components as the signal.

The evolution equations for the soliton amplitude η and the frequency κ in the presence of sliding-frequency filters can be obtained from Eqs. (8.11) and (8.12) with

$$P(q) = -i\alpha_s Tq + \delta q + \beta \frac{\partial^2 q}{\partial T^2} \tag{8.42}$$

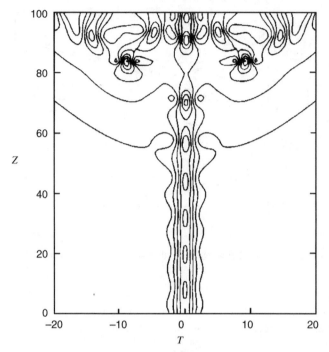

Figure 8.6 Soliton propagation under the effect of background instability for the case $\beta = 3\delta$ and $\beta = 0.15$.

where α_s is a real constant representing the sliding rate. It comes out that the evolution equation for the soliton amplitude is again given by Eq. (8.39), whereas the evolution equation for the frequency becomes

$$\frac{\partial \kappa}{\partial Z} = \alpha_s - \frac{4}{3} \beta \eta^2 \kappa \qquad (8.43)$$

The equilibrium points of Eqs. (8.39) and (8.43) are given by the conditions

$$\frac{\eta^2}{3} + \kappa^2 = \frac{\delta}{\beta} \quad \text{and} \quad \alpha_s = \frac{4}{3} \beta \eta^2 \kappa \qquad (8.44)$$

There exists a critical value for the sliding rate, α_c, such that the number of stationary points is two for $|\alpha_s| < \alpha_c$, one for $|\alpha_s| = \alpha_c$, and none for $|\alpha_s| > \alpha_c$. This means that for strong sliding the soliton is pushed outside the filter and is destroyed. In the case $|\alpha_s| < \alpha_c$, we have one stable point, (κ_s, η_s), and one unstable point, (κ_u, η_u), such that $|\kappa_s| \leq |\kappa_u|$ and $|\eta_s| \geq |\eta_u|$. This case is illustrated in Fig. 8.7, which shows the phase portrait of Eqs. (8.39) and (8.43) considering the values $\alpha_s = 0.075$, $\beta = 0.3$, and $\delta = 0.11$. These values provide a stable point ($\kappa_s = 0.1875, \eta_s = 1$).

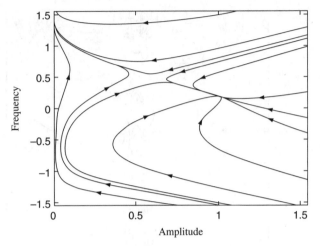

Figure 8.7 Phase portrait of Eqs. (8.39) and (8.43) for the case $\alpha_s = 0.075$, $\beta = 0.3$, and $\delta = 0.111$.

In the presence of sliding-frequency filters, the variance of the Gordon–Haus timing jitter is reduced relatively to the uncontrolled case by a factor given by [45]

$$F_r(x_1, x_2) = \frac{3}{4} [g(x_1, x_1) + g(x_2, x_2)] - \frac{1}{2} g(x_1, x_2) \qquad (8.45a)$$

where

$$g(x_1, x_2) = \frac{3}{x_1 x_2} [1 - h(x_1) - h(x_2) + h(x_1 + x_2)] \qquad (8.45b)$$

$$h(x) = \frac{1 - \exp(-x)}{x} \qquad (8.45c)$$

$$x_{1,2} = \frac{4}{3} \beta Z \left(1 \pm \frac{3\alpha_s}{4\beta} \right) \qquad (8.45d)$$

Figure 8.8 shows the reduction factor F_r of the variance of timing jitter, against the propagation distance for $\beta = 0.1$ and several values of the sliding rate α_s, namely, $\alpha_s = 0$ (curve a), $\alpha_s = 0.03$ (curve b), $\alpha_s = 0.04$ (curve c), and $\alpha_s = 0.05$ (curve d). Besides the positive effect in controlling the background instability, Fig. 8.8 shows that the use of sliding-frequency filters determines an enhancement of the timing jitter compared to the use of fixed-frequency filters with the same strength. This enhancement was verified experimentally by Mollenauer et al. [43] and it is due to the coupling between amplitude and frequency fluctuations.

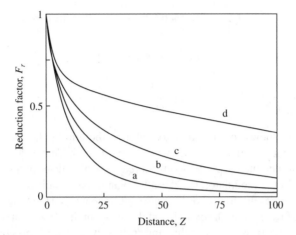

Figure 8.8 Reduction factor of the Gordon–Haus timing jitter, given by Eq. (8.45), for several values of the sliding rate.

8.4.3 Synchronous Modulators

An alternative way to control the timing jitter due to the Gordon–Haus effect was suggested by Nakazawa et al. [11] and consisted of the use of a modulator that is timed to pass solitons at the peak of its transmission. Synchronous modulators work by forcing the soliton to move toward their transmission peak, where loss is minimum, and such forcing reduces timing jitter considerably.

The evolution equations for the soliton amplitude η, frequency κ, and soliton position T_0 in the presence of amplitude modulators and fixed-frequency filters can be obtained from Eqs. (8.11)–(8.13) with

$$P(q) = \delta q + \beta \frac{\partial^2 q}{\partial T^2} + \mu_R \left\{ \cos\left(\frac{2\pi}{T_M} T\right) - 1 \right\} q \qquad (8.46)$$

where μ_R is the average extinction ratio and T_M is the modulator period. The resulting evolution equations are [28]

$$\frac{\partial \eta}{\partial Z} = 2(\delta - \mu_R)\eta - 2\beta\eta\left(\frac{1}{3}\eta^2 + \kappa^2\right) + \frac{2\pi^2 \mu_R}{T_M} \cos\left(2\pi \frac{T_0}{T_M}\right) \operatorname{cosech}\left(\frac{\pi^2}{\eta T_M}\right) \qquad (8.47)$$

$$\frac{\partial \kappa}{\partial Z} = -\frac{4}{3}\beta\eta^2\kappa \qquad (8.48)$$

$$\frac{\partial T_0}{\partial Z} = -\kappa + \frac{\pi \mu_R}{\eta} \sin\left(2\pi \frac{T_0}{T_M}\right) \operatorname{cosech}\left(\frac{\pi^2}{\eta T_M}\right) \left\{ 1 - \frac{\pi^2}{\eta T_M} \coth\left(\frac{\pi^2}{\eta T_M}\right) \right\} \qquad (8.49)$$

The suppression of the Gordon–Haus timing jitter results due to the fact that both κ and T_0 tend exponentially to zero. Considering the case $T_M \gg \eta^{-1} = 1$, a stable fixed point ($\eta = 1, \kappa = 0, T_0 = 0$) is achieved if the following conditions are satisfied:

$$\delta - \frac{1}{3}\beta - \frac{\pi^4}{6T_M^2}\mu_R = 0 \tag{8.50}$$

$$0 < \frac{\pi^4}{2T_M^2}\mu_R < \beta \tag{8.51}$$

Equation (8.51) shows that in order to stabilize the amplitude fluctuations of the soliton, a filter must be used together with the modulator. This is due to the fact that when the amplitude increases, its width decreases, which determines a reduced loss of the soliton in the modulator. On the other hand, the combined use of filters and modulators provides the control of solitons simultaneously in the time and frequency domains. This fact explains the experimental results obtained by Nakazawa et al. [46], which indicated the possibility of achieving arbitrarily large transmission distances by this combination. However, this scheme of soliton control requires active devices, having drawbacks of complexity, reduced reliability, and high cost. On the other hand, it is not immediately compatible with WDM, because the pulses in different channels drift apart and must be retimed separately.

8.4.4 Amplifier with Nonlinear Gain

An alternative approach to avoid the background instability consists in the use of an amplifier having a nonlinear property of gain, or gain and saturable absorption in combination, such as the nonlinear loop mirror or nonlinear polarization rotation in combination with a polarization-dependent loss element [13]. The key property of the nonlinearity in gain is to give an effective gain to the soliton and a suppression (or very small gain) to the noise. This method of nonlinear gain may be particularly useful for transmission of solitons with subpicosecond or femtosecond durations, where the gain bandwidth of amplifiers will not be wide enough for the sliding of the filter frequency to be allowed [47].

The pulse propagation in optical fibers where linear and nonlinear amplifiers and narrowband filters are periodically inserted may be described by the following modified NLSE [48–51]:

$$i\frac{\partial q}{\partial Z} + \frac{1}{2}\frac{\partial^2 q}{\partial T^2} + |q|^2 q = i\delta q + i\beta\frac{\partial^2 q}{\partial T^2} + i\varepsilon|q|^2 q + i\mu|q|^4 q + v|q|^4 q \tag{8.52}$$

where β stands for spectral filtering, δ is the linear gain or loss coefficient, ε accounts for nonlinear gain–absorption processes, μ represents a higher order correction to the nonlinear gain–absorption, and v is a higher order correction term to the nonlinear refractive index.

Equation (8.52) is known as the complex Ginzburg–Landau equation (CGLE), so-called cubic for $\mu = v = 0$ and quintic for $\mu, v \neq 0$. It describes to a good approximation the soliton behavior in both optical transmission lines and mode-locked fiber lasers [52–56].

When all the coefficients on the right-hand side of Eq. (8.52) are small and $v = 0$, the dynamical evolution of the soliton parameters can be obtained using Eqs. (8.11)–(8.14) with

$$P(q) = \delta q + \beta \frac{\partial^2 q}{\partial T^2} + \varepsilon |q|^2 q + \mu |q|^4 q \qquad (8.53)$$

In particular, Eq. (8.40) is again derived for the soliton frequency κ, whereas the soliton amplitude η satisfies

$$\frac{\partial \eta}{\partial Z} = 2\delta\eta - 2\beta\eta \left(\frac{1}{3}\eta^2 + \kappa^2 \right) + \frac{4}{3}\varepsilon\eta^3 + \frac{16}{15}\mu\eta^5 \qquad (8.54)$$

The stable fixed points for the soliton amplitude are given by minimums of the potential function ϕ defined by

$$\phi(\eta) = -\delta\eta^2 + \frac{1}{6}(\beta - 2\varepsilon)\eta^4 - \frac{8}{45}\mu\eta^6 \qquad (8.55)$$

For the zero-amplitude state to be stable, the potential function must have a minimum at $\eta = 0$, in addition to a minimum at $\eta = \eta_s \neq 0$. These objectives can be achieved if the following conditions are verified [14]:

$$\delta < 0, \ \mu < 0, \ \varepsilon > \beta/2, \ 15\delta > 8\mu\eta_s^4 \qquad (8.56)$$

where η_s is the stationary value for the soliton amplitude. The above conditions show that the inclusion of the quintic term in Eq. (8.52) is necessary to guarantee the stability of the whole solution: pulse and background.

Figure 8.9 illustrates the stability characteristics of a system with spectral filtering and linear and nonlinear gains, considering the following parameter values: $\beta = 0.15$,

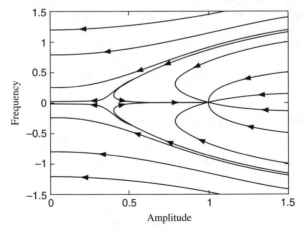

Figure 8.9 Phase portrait of Eqs. (8.40) and (8.54) for the case $\beta = 0.15$, $\delta = -0.011$, $\varepsilon = 0.2$, and $\mu = -0.137$.

$\delta = -0.01$, $\varepsilon = 0.2$, and $\mu = -0.1375$. These values satisfy the stability conditions given by Eq. (8.56). We confirm from Fig. 8.9 that the stationary point ($\kappa = 0, \eta = 1$) is indeed a stable point. Moreover, the background instability is avoided in this case since the small-amplitude waves are attenuated, irrespective of their frequency κ.

The use of spectral filtering and linear and nonlinear gains is also effective in controlling the propagation of ultrashort solitons. This technique allows to control both the self-frequency shift and the background instability, as well as to suppress the soliton phase jitter [57–59].

8.5 DISSIPATIVE SOLITONS

Besides the propagation of nonlinear pulses in fiber systems with gain and loss, the CGLE given by Eq. (8.52) has been used to describe many nonequilibrium phenomena, such as convection instabilities, binary fluid convection, and phase transitions [60–62]. In particular, the formation of solitons in systems far from equilibrium has emerged in the past few years as an active field of research [15]. These solitons are termed *dissipative solitons* and they emerge as a result of a double balance between nonlinearity and dispersion and between gain and loss [63]. The properties of dissipative solitons are completely determined by the external parameters of the system and they can exist indefinitely in time, as long as these parameters stay constant. However, they cease to exist when the source of energy is switched off, or if the parameters of the system move outside the range that provides their existence.

Even if it is a stationary object, a dissipative soliton shows nontrivial energy flows with the environment and between different parts of the pulse. Hence, this kind of soliton is an object that is far from equilibrium and that presents some characteristics similar to a living thing. Conversely, we can consider animal species in nature as elaborate forms of dissipative solitons. An animal is a localized and persistent "structure" that has material and energy inputs and outputs and complicated internal dynamics. Moreover, it exists only for a certain range of parameters (pressure, temperature, humidity, etc.) and dies if the supply of energy is switched off. These ideas can be applied to various fiber-optic devices such as passively mode-locked fiber lasers [64, 65] and high-density optical transmission lines [66].

8.5.1 Analytical Results of the CGLE

Several types of exact analytical solutions of the CGLE can be obtained considering a particular ansatz [67–69]. However, due to restrictions imposed by the ansatz, these solutions do not cover the whole range of parameters. In the following, we will look for stationary solutions of Eq. (8.52) in the form

$$q(T,Z) = a(T)\exp\{ib\ln[a(T)]-i\omega Z\} \qquad (8.57)$$

where $a(T)$ is a real function and b and ω are real constants.

Solutions of the Cubic CGLE We will first consider the cubic CGLE, which is given by Eq. (8.52) with $\mu = \nu = 0$. Inserting Eq. (8.57) in this equation, the following solution for $a(T)$ is obtained:

$$a(T) = A \, \text{sech}(BT) \tag{8.58}$$

where

$$A = \sqrt{\frac{B^2(2-b^2)}{2} + 3b\beta B^2}, \; B = \sqrt{\frac{\delta}{\beta b^2 + b - \beta}} \tag{8.59}$$

and b is given by

$$b = \frac{3(1+2\varepsilon\beta) \pm \sqrt{9(1+2\varepsilon\beta)^2 + 8(\varepsilon-2\beta)^2}}{2(\varepsilon-2\beta)} \tag{8.60}$$

On the other hand, we have

$$\omega = -\frac{\delta(1-b^2+4\beta b)}{2(b-\beta+\beta b^2)} \tag{8.61}$$

The solution (8.57)–(8.61) is known as the solution of Pereira and Stenflo [70]. Although the amplitude profile of the solution (8.57)–(8.61) is a hyperbolic secant as in the case of the NLSE solitons, two important differences exist between the CGLE and the NLSE solitons. First, for CGLE pulses the amplitude and width are independently fixed by the parameters of Eq. (8.52), whereas for NLSE solitons we have $A = B$. Second, the CGLE solitons are chirped.

The solution given by Eqs. (8.58)–(8.60) has a singularity at $b-\beta+\beta b^2 = 0$, which takes place on the line $\varepsilon_s(\beta)$ in the plane (β, ε) given by

$$\varepsilon_s = \frac{\beta}{2} \frac{3\sqrt{1+4\beta^2}-1}{2+9\beta^2} \tag{8.62}$$

For a given value of β, the denominator in the expression for B in Eq. (8.59) is positive for $\varepsilon < \varepsilon_s$ and negative for $\varepsilon > \varepsilon_s$. Hence, for solution (8.58)–(8.60) to exist, the excess linear gain δ must be positive for $\varepsilon < \varepsilon_s$ and negative for $\varepsilon > \varepsilon_s$. In the latter case, both numerical simulations and the soliton perturbation theory show that the soliton is unstable relatively to any small amplitude fluctuation [67–69]. On the other hand, for $\delta > 0$ and $\varepsilon < \varepsilon_s$ the solution (8.58)–(8.60) is stable, since after any small perturbation it approaches the stationary state. However, the background state is unstable in this case, since the positive excess gain also amplifies the linear waves coexistent with the soliton trains. The general conclusion is that either the soliton itself or the background state is unstable at any point in the plane (β, ε), which means that the total solution is always unstable.

If β and ε satisfy Eq. (8.62) and $\delta = 0$, a solution of the cubic CGLE with arbitrary amplitude exists, given by [68]

$$a(T) = C \operatorname{sech}(DT) \tag{8.63}$$

where C is an arbitrary positive parameter and C/D is given by

$$\frac{C}{D} = \sqrt{\frac{(2+9\beta^2)\sqrt{1+4\beta^2}(\sqrt{1+4\beta^2}-1)}{2\beta^2(3\sqrt{1+4\beta^2}-1)}} \tag{8.64}$$

We also have

$$b = \frac{\sqrt{1+4\beta^2}-1}{2\beta} \tag{8.65}$$

$$\omega = -b\frac{1+4\beta^2}{2\beta}D^2 \tag{8.66}$$

Arbitrary-amplitude solitons are stable pulses, which propagate in a stable background because $\delta = 0$. This feature is illustrated in Fig. 8.10, which shows the simultaneous propagation of four stable solitons with amplitudes 2, 1.5, 1, and 0.5, for $\delta = 0$, $\beta = 0.2$, and $\varepsilon = \varepsilon_s$.

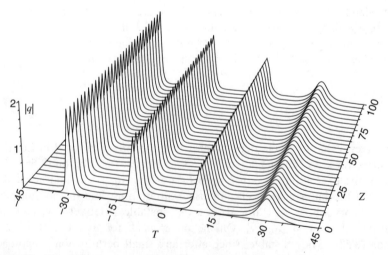

Figure 8.10   Simultaneous propagation of four solitons with amplitudes 2, 1.5, 1, and 0.5, for $\delta = \mu = 0$, $\beta = 0.2$, and $\varepsilon = \varepsilon_s$ (given by Eq. (8.62)).

Solutions of the Quintic CGLE   Considering the quintic CGLE and inserting Eq. (8.57) in Eq. (8.52), the following general solution can be obtained for $f = a^2$ [68]:

$$f(T) = \frac{2f_1 f_2}{(f_1 + f_2) - (f_1 - f_2)\cosh(2\alpha \sqrt{f_1 |f_2|} T)} \tag{8.67}$$

where

$$\alpha = \sqrt{\left| \frac{\mu}{3\beta - 2d - \beta b^2} \right|} \tag{8.68}$$

and b is given by Eq. (8.60). The parameters f_1 and f_1 are the roots of the equation

$$\frac{2\nu}{8\beta b - b^2 + 3} f^2 + \frac{2(2\beta - \varepsilon)}{3b(1 + 4\beta^2)} f - \frac{\delta}{b - \beta + \beta b^2} = 0 \tag{8.69}$$

and the coefficients are connected by the relation

$$\nu \left[\frac{12\varepsilon\beta^2 + 4\varepsilon - 2\beta}{\varepsilon - 2\beta} b - 2\beta \right] + \mu \left[\frac{2\varepsilon\beta - 16\beta^2 - 3}{\varepsilon - 2\beta} b + 1 \right] = 0 \tag{8.70}$$

One of the roots of Eq. (8.69) must be positive for the solution (8.67) to exist, while the other can have either sign.

When the two roots are both positive, the general solution given by Eq. (8.67) becomes wider and flatter as they approach each other, as can be seen in Fig. 8.11.

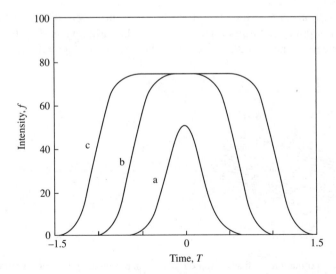

Figure 8.11   The pulse shapes given by Eq. (8.67) when the two roots f_1 and f_2 are close to each other.

These flat-top solitons correspond to stable pulses, whereas the solution (8.67) is generally unstable for arbitrary choice of parameters. If $f_1 = f_2$, the width of the flat-top soliton tends to infinity and the soliton splits into two fronts. The formation and stable propagation of a flat-top soliton will be demonstrated numerically in Section 8.5.2.

If β and ε satisfy Eq. (8.62) and $\delta = 0$, a solution of the quintic CGLE with arbitrary amplitude exists, given by [68]

$$f(T) = [a(T)]^2 = \frac{3b(1+4\beta^2)P}{(2\beta-\varepsilon) + S\cosh(2\sqrt{P}T)} \qquad (8.71)$$

where P is an arbitrary positive parameter and

$$S = \sqrt{(2\beta-\varepsilon)^2 + \frac{9b^2\mu(1+4\beta^2)^2}{3\beta-2b-\beta b^2}P} \qquad (8.72)$$

The parameters b and ω are given by Eqs. (8.65) and (8.66). When $\mu \to 0$, Eq. (8.76) transforms to the arbitrary-amplitude solution of the cubic CGLE, given by Eqs. (8.63) and (8.64).

8.5.2   Numerical Solutions of the CGLE

Due to restrictions imposed by the ansatz, the analytic solutions of the quintic CGLE presented above do not cover the whole range of parameters and almost all of them are unstable. To find stable solutions in other regions of the parameters, different approximate methods [71, 72], a variational approach [73–77], or numerical techniques must be used.

Stable solitons can be found numerically from the propagation equation (8.52) taking as the initial condition a pulse of somewhat arbitrary profile. In fact, such profile appears to be of little importance. For example, Fig. 8.12 illustrates the

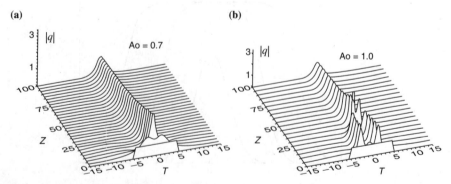

Figure 8.12   Formation of a fixed-amplitude soliton of the cubic CGLE starting from an initial pulse with a rectangular profile of amplitude $A_0 = 0.7$ (a) and $A_0 = 1.0$ (b), $\delta = -0.003$, $\beta = 0.2$, and $\varepsilon = 0.09$.

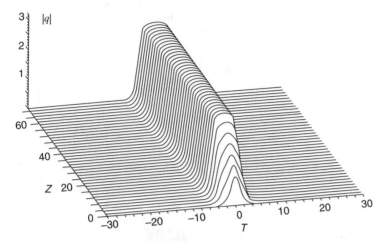

Figure 8.13 Formation and stable propagation of a flat-top soliton, starting from an initial pulse with a sech profile for $\delta = -0.1$, $\beta = 0.5$, $\varepsilon = 0.66$, and $\mu = v = -0.01$.

formation of a fixed-amplitude soliton of the cubic CGLE starting from an initial pulse with a rectangular profile. In this case, the linear gain is positive but relatively small ($\delta = 0.003$) and the soliton propagation remains stable within the displayed distance.

In general, if the result of the numerical calculation converges to a stationary solution, it can be considered as a stable one, and the chosen set of parameters can be deemed to belong to the class of those that permit the existence of solitons. For small values of the parameters on the right-hand side of Eq. (8.52), the stable soliton solutions of the CGLE have a sech profile, similar to the soliton solutions of the NLSE, and correspond to the so-called plain pulses (PPs). However, rather different pulse profiles can be obtained for nonsmall values of those parameters. For example, Fig. 8.13 illustrates the formation and stable propagation of a flat-top soliton, starting from an initial pulse with a sech profile and assuming the following parameter values: $\delta = -0.1$, $\beta = 0.5$, $\varepsilon = 0.66$, and $\mu = v = -0.01$.

Another example is given in Fig. 8.14, which shows (a) the amplitude profiles and (b) the spectra of a plain pulse, as well as of two composite pulses (CPs) [78]. The following parameter values were considered in this case: $\delta = -0.01$, $\beta = 0.5$, $\mu = -0.03$, $v = 0$, $\varepsilon = 1.5$ (plain pulse), $\varepsilon = 2.0$ (narrow composite pulse), and $\varepsilon = 2.5$ (wide composite pulse). Figure 8.14c illustrates the formation and propagation of the wide composite pulse starting from the plain pulse solution represented in (a) and (b). A composite pulse exhibits a dual-frequency but symmetric spectrum (Fig. 8.14b) and can be considered as a bound state of a plain pulse and two fronts attached to it from both sides [67]. The "hill" between the two fronts should be counted as a source, because it follows from the phase profile that energy flows from the center to the CP wings.

If one of the fronts of a CP is missing, one has a moving soliton (MS) [67]. The MS always moves with a velocity smaller than the velocity of the front for the same set of parameters. In fact, the front tends to move with its own velocity but the soliton

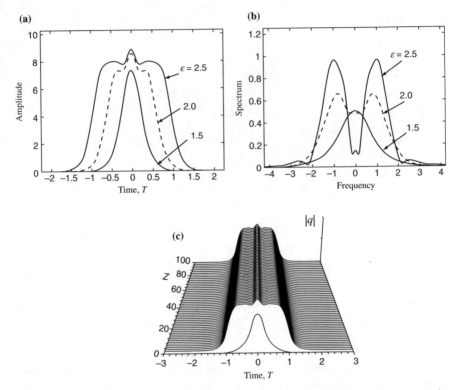

Figure 8.14 Amplitude profiles (a) and spectra of a plain pulse, as well as of two composite pulses (b) for $\delta = -0.01$, $\beta = 0.5$, $\mu = -0.03$, $\nu = 0$, $\varepsilon = 1.5$ (plain pulse), $\varepsilon = 2.0$ (narrow composite pulse), and $\varepsilon = 2.5$ (wide composite pulse). Part (c) illustrates the formation and propagation of the wide composite pulse starting from the plain pulse solution represented in (a) and (b).

tends to be stationary, due to the spectral filtering. The resulting velocity of the MS is determined by competition between these two processes.

Pulsating and exploding soliton solutions of the CGLE were also observed recently [63, 79, 80]. Pulsating solitons correspond to fixed solutions in the same way as the stationary pulses and can be found when the parameters of the CGLE are far enough from the NLSE limit. On the other hand, exploding solitons appear for a wide range of parameters of the CGLE and originate from soliton solutions that remain stationary only for a limited period of time. An example of an exploding soliton is shown in Fig. 8.15. Following the explosion, there is a "cooling" period, after which the solution becomes "stationary" again. This is a periodic phenomenon, like other phenomena occurring in the nature [15].

It was observed that different stable stationary solutions of the quintic CGLE can exist simultaneously for the same set of parameters [15]. This can be understood considering that solitons, fronts, and sources are elementary building units that can be combined to form more complicated structures. In more complex systems, the number

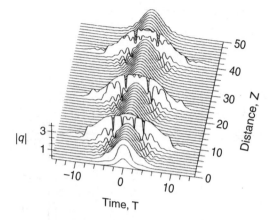

Figure 8.15 Propagation on an exploding soliton for $\delta = -0.1$, $\beta = 0.125$, $\varepsilon = 1.0$, $\mu = -0.1$, and $\nu = -0.6$.

of solutions may be very high. This reality again resembles the world of biology, where the number of species is truly impressive.

8.6 DISPERSION-MANAGED SOLITONS

Various dispersion profiles (or dispersion maps) have been tried in linear transmission systems to minimize the nonlinear effects. Basically, the idea is to introduce locally relatively large dispersion to make nonlinear effects relatively less important and to compensate for the accumulated dispersion at the end of the line so that the integrated dispersion becomes zero. A method of programming dispersion is called *dispersion management*. Dispersion management was also found to be effective in soliton transmission systems since it can significantly reduce the timing jitter, which is the major cause of bit error in such systems.

Smith et al. [18] proposed the use of a periodic map using both anomalous dispersion and normal dispersion fibers alternately. The nonlinear stationary pulse that propagates in such a fiber has a peak power larger than the soliton that propagates in a fiber with a constant dispersion equal to the average dispersion of the period map. Such a nonlinear stationary pulse is commonly called a *dispersion-managed soliton* and it has been introduced simultaneous by many authors [22].

8.6.1 The True DM Soliton

To illustrate the effects of programmed dispersion, let us consider the propagation of an optical pulse in a transmission line in which the group velocity dispersion is alternately positive and negative, as shown in Fig. 8.16. Soliton evolution in such a fiber is governed by Eq. (8.22), which can be solved numerically using the split-step Fourier method. In early numerical simulations, the exact shape of the periodic pulse

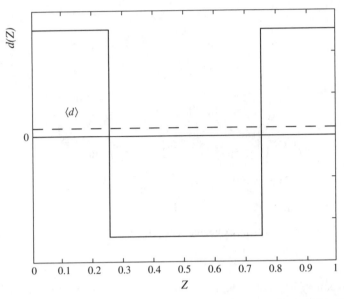

Figure 8.16   Dispersion map with anomalous and normal dispersion fibers of equal length and anomalous average dispersion.

solutions was unknown and an iterative method has been used to obtain a soliton solution. Given a Gaussian initial guess, of approximately the correct power, it is found that it converges upon successive iterations to a DM soliton solution. These simulations showed the existence of almost periodic solutions, but with some remaining oscillations. Subsequently, using a numerical averaging algorithm, it was shown that, indeed, there exist exactly periodic DM solitons [81–83]. The used algorithm requires memorizing the pulses on one oscillation period (typically a few dispersion maps) and detects the extrema in pulse width. Both pulse widths, maximum and minimum, are used to define the pulses u_{max} and u_{min}, respectively, then being stored. The difference of phase between u_{max} and u_{min} is adjusted, and then follows an energy modification so that the energy is conserved. The iterations can be stopped when the variation of the extremum pulse width is smaller than the precision wanted.

After the averaging method is applied, a long-term stable pulse can be obtained. Figure 8.17 shows an example of a long-term stable stationary solution, represented in a logarithmic scale. This result was obtained considering the lossless case ($a^2(Z) = 1$), which occurs in practice when distributed amplification is used so that fiber losses are nearly compensated by the local gain all along the fiber. Figure 8.17 shows that the true DM soliton develops oscillatory tails in the wings.

DM solitons are entities whose shape varies substantially over one period of the dispersion map but return to the same shape at the end of each period. Figure 8.18a and b shows the evolution of the DM soliton during one period of the dispersion map, in normal and logarithmic scales, respectively. The DM pulse alternately spreads and compresses as the sign of the dispersion is switched. The pulse peak power varies

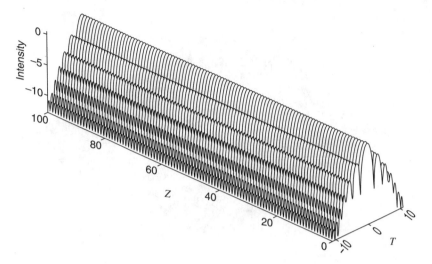

Figure 8.17 Long-term stable stationary solution, represented in a logarithmic scale.

rapidly and the pulse width becomes minimum at the center of each fiber where frequency chirp vanishes.

The DM soliton is quite different from a standard fundamental soliton, since its shape, width, and peak power are not constant. However, since these parameters repeat from period to period at any location within the map, DM solitons can be used in optical communications. Moreover, such solitons present some remarkable characteristics that make them the preferred option for use in high-capacity transmission systems.

8.6.2 The Variational Approach to DM Solitons

The variational approach is one of the most powerful techniques capable of providing insight into the main characteristics of the DM soliton [84–90]. As discussed in Section 5.1.2, this approximation involves choosing an initial ansatz for the shape of solution sought, but leaves in the ansatz a set of free parameters, which may evolve with Z.

The Lagrangian density corresponding to Eq. (8.22) is

$$L_d = \frac{i}{2}\left(u\frac{\partial u^*}{\partial Z} - u^*\frac{\partial u}{\partial Z}\right) + \frac{d(Z)}{2}\left|\frac{\partial u}{\partial T}\right|^2 - \frac{a^2(Z)}{2}|u|^4 \qquad (8.73)$$

Unlike a standard soliton in fibers with constant dispersion, the DM soliton inevitably has a frequency chirp $C(Z)$. Hence, we introduce the ansatz $u(T, Z)$:

$$u(T, Z) = u(Z)f[T/T_p(Z)]\exp[i\psi(Z) + iC(Z)T^2] \qquad (8.74)$$

where $u(Z)$ is the amplitude, $T_p(Z) = t_p(Z)/t_0$ is the normalized width, $\psi(Z)$ is the phase, and $C(Z)$ is the chirp parameter. The form f of the pulse is still arbitrary. In fact,

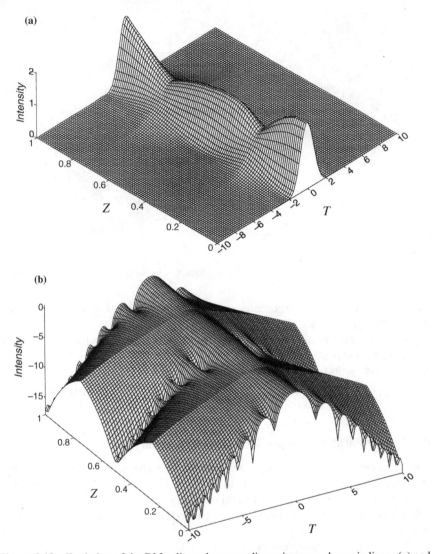

Figure 8.18   Evolution of the DM soliton along one dispersion map shown in linear (a) and logarithmic (b) scales.

an important feature of DM solitons is that their profile in general is not given by a sech, as it happens for the NLSE fundamental soliton, but varies from sech to a Gaussian function, and even more flat distributions.

Inserting the trial function (8.74) into Eq. (8.73) and integrating the result over T, we obtain the Lagrangian

$$L(Z) = u^2 T_p F_1 \psi_Z + u^2 T_p^3 F_2 C_Z + \frac{d(Z)}{2} \left[u^2 T_p^{-1} F_3 + 4C^2 u^2 T_p^3 F_2 \right] - \frac{a^2(Z)}{2} u^4 T_p F_4$$

$$(8.75)$$

where

$$F_1 = \int_{-\infty}^{\infty} |f(x)|^2 \, dx, \quad F_2 = \int_{-\infty}^{\infty} x^2 |f(x)|^2 \, dx$$

$$F_3 = \int_{-\infty}^{\infty} |f_x(x)|^2 \, dx, \quad F_4 = \int_{-\infty}^{\infty} |f(x)|^4 \, dx \qquad (8.76)$$

Now, variation of the Lagrangian gives the Lagrangian equation of motion

$$\frac{\partial L}{\partial v} - \frac{d}{dz} \frac{\partial L}{\partial v_Z} = 0 \qquad (8.77)$$

for $v(Z) = \psi(Z)$, $C(Z)$, $u(Z)$, $T_p(Z)$. The results are

$$u^2 T_p = 1 \qquad (8.78)$$

$$\frac{dT_p}{dZ} = 2d(Z)T_p C \qquad (8.79)$$

$$\frac{dC}{dZ} = \frac{d(Z)K_1}{2T_p^4} - \frac{a^2(Z)K_2}{T_p^3} - 2d(Z)C^2 \qquad (8.80)$$

where $K_1 = F_1/F_2$ and $K_2 = F_4/F_2$. The constants K_1 and K_2 are related to the structural function $f(x)$. In particular, K_2 is related to the pulse energy. Introducing $I = CT_p$, we obtain from Eqs. (8.79) and (8.80)

$$\frac{dT_p}{dZ} = 2d(Z)I(Z) \qquad (8.81)$$

$$\frac{dI}{dZ} = \frac{d(Z)K_1}{2T_p^3} - \frac{a^2(Z)K_2}{T_p^2} \qquad (8.82)$$

To describe periodic breathing pulses, solutions of Eqs. (8.81) and (8.82) must satisfy two conditions of periodicity: $T_p(1) = T_p(0)$ and $I(1) = I(0)$. From Eqs. (8.81) and (8.82), we obtain the following condition for the existence of the periodic solution:

$$\left\langle \frac{d(Z)}{T_p^2} \right\rangle > 0 \qquad (8.83)$$

This condition can be satisfied even if the average dispersion is normal [86–88, 91]. For example, Fig. 8.19 shows the evolution of the pulse width (a) and chirp (b) in the case of normal average dispersion ($\langle d(Z) \rangle = -0.02$), for a pulse energy $K_2 = 1.485$. The pulse width achieves a minimum value in the middle of each fiber segment,

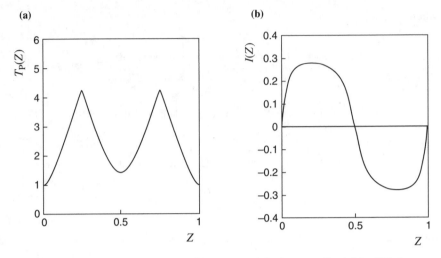

(a)

(b)

Figure 8.19 (a) Variation of the pulse width $T_p(Z)$ and (b) the normalized chirp $I(Z)$ along one dispersion map for an average dispersion $\langle d(Z) \rangle = -0.02$ and $K_2 = 1.485$.

where the chirp becomes zero. However, the shortest pulse occurs in the middle of the anomalous dispersion fiber.

Figure 8.20 shows the phase space diagram for the variables T_p and I, corresponding to the same situation of Fig. 8.19. The gap between the two C-shaped curves is due to the frequency chirp produced by the self-induced phase shift. In fact, the pulse width in the middle of the normal dispersion fiber is higher than that in the middle of the anomalous dispersion fiber (see Fig. 8.19a) because of this phase shift. This makes $\langle d(Z)/T_p^2 \rangle > 0$ even if the average dispersion is zero or normal. In contrast, the linear stationary pulse is possible only if the average dispersion $\langle d(Z) \rangle$ is zero. As a consequence, DM solitons have a much larger tolerance relative to the fiber dispersion in comparison to the linear pulses. Stable transmission of DM solitons in a system with average normal dispersion was demonstrated by Jacob et al. [92].

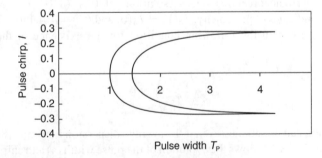

Figure 8.20 Trajectory of parameters T_p and I in the T_p–I plane for the case represented in Fig. 8.19.

Since a DM soliton can exist even when the average dispersion is zero, the timing jitter can be significantly reduced [93–95]. Furthermore, the amplitude of the DM soliton can be made sufficiently large, by the so-called soliton enhancement factor [18], with a proper choice of the magnitude of the difference Δd between the dispersions of the two fibers to maintain sufficiently large SNR. As a consequence, a DM soliton can exhibit very low amplitude and timing jitter. The performance of DM soliton transmission can still be improved by using additional control devices, such as filters and modulators [96–101].

8.7 WDM SOLITON SYSTEMS

The capacity of a lightwave transmission system can be increased considerably by using the WDM technique. A WDM soliton system transmits over the same fiber several soliton bit streams, distinguishable through their different carrier frequencies. The new feature that becomes important for WDM soliton systems is the possibility of collisions among solitons belonging to different channels because of their different group velocities. In the collisions, the solitons experience small temporal and phase shifts, but otherwise they recover fully. A collision between two solitons is illustrated in Fig. 8.21. When the two pulse envelops overlap, beating between the two carrier frequencies is clearly discernible.

An important parameter in this context is the collision length, z_l, which is the distance the solitons must travel down the fiber together before the fast-moving

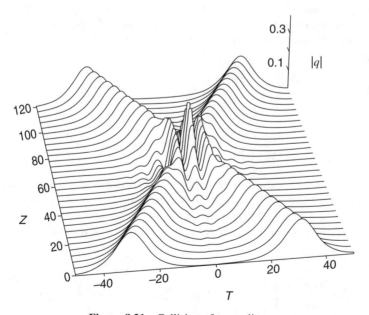

Figure 8.21 Collision of two solitons.

soliton overtakes the slower one. Assuming that a collision begins and ends with overlap at the half-power points, z_I is given by

$$z_I = \frac{2t_{\text{FWHM}}}{D\Delta\lambda} \tag{8.84}$$

where $\Delta\lambda$ is the wavelength spacing between the two pulses. For example, if $t_{\text{FWHM}} = 5\,\text{ps}$, $D = 0.5\,\text{ps/(km\,nm)}$, and $\Delta\lambda = 0.2\,\text{nm}$, $z_I = 100\,\text{km}$.

Considering the importance of dispersion management in the design of WDM systems, we use Eq. (8.22) to study the effects of collisions between solitons. Introducing a new propagation variable Z', related to Z in the form $dZ' = a^2(Z)dZ$, such equation becomes

$$i\frac{\partial u}{\partial Z'} + \frac{1}{2}\frac{\partial^2 u}{\partial T^2} + b|u|^2 u = 0 \tag{8.85}$$

where $b(Z) = a^2(z)/d(Z)$. To simplify the notation, the prime in Z' will be omitted in the following.

Consider the collision of two solitons of equal amplitude in different channels separated in frequency by $\omega_{\text{ch}} = 2\pi f_{\text{ch}} t_0$. Using Eq. (8.2) with $\eta = 1$ and $T_0 = 0$, we have

$$u_m(T, Z) = \text{sech}[T + \kappa_m Z]\exp\left(-i\kappa_m T + \frac{i}{2}(1 - \kappa_m^2)Z + i\sigma_m\right) \tag{8.86}$$

where $\kappa_m = \pm\frac{1}{2}\omega_{\text{ch}}$ for $m = 1, 2$. Now insert $u = u_1 + u_2$ into Eq. (8.85), expand, and group terms according to their frequency dependences. Neglecting the four-wave mixing (FWM) terms, the following two coupled equations for u_1 and u_2 are obtained:

$$i\frac{\partial u_1}{\partial Z} + \frac{1}{2}\frac{\partial^2 u_1}{\partial T^2} + b(|u_1|^2 + 2|u_2|^2)u_1 = 0 \tag{8.87}$$

$$i\frac{\partial u_2}{\partial Z} + \frac{1}{2}\frac{\partial^2 u_2}{\partial T^2} + b(|u_2|^2 + 2|u_1|^2)u_2 = 0 \tag{8.88}$$

These equations are identical to the coupled NLSEs that describe the interaction of two pulses through cross-phase modulation. The second contribution in the nonlinear terms of Eqs. (8.87) and (8.88) corresponds to the cross-phase modulation and is zero except when the pulses overlap. It produces a frequency shift in the faster moving soliton given by [102]

$$\frac{d\kappa_1}{dZ} = \frac{b(Z)}{\omega_{\text{ch}}}\frac{d}{dZ}\left[\int_{-\infty}^{\infty} \text{sech}^2(T - \omega_{\text{ch}}Z/2)\text{sech}^2(T + \omega_{\text{ch}}Z/2)dT\right] \tag{8.89}$$

The change in κ_2 is similar to the change in κ_1. For a constant-dispersion and lossless fiber ($b = 1$), the integration of Eq. (8.89) can be performed analytically and the frequency shift is given by

$$\delta\kappa_1(Z) = -\delta\kappa_2(Z) = \frac{4}{\omega_{ch}} \frac{[\omega_{ch}Z\cosh(\omega_{ch}Z)-\sinh(\omega_{ch}Z)]}{[\sinh(\omega_{ch}Z)]^3} \qquad (8.90)$$

The peak frequency shift, at $Z = 0$, is

$$\delta\kappa_{max} = 4/(3\omega_{ch}) \qquad (8.91)$$

Equation (8.90) can be integrated to give the net temporal shift of the colliding soliton:

$$\delta T = 4/\omega_{ch}^2 \qquad (8.92)$$

Figure 8.22a shows changes in the soliton frequency for the slower moving soliton during collision. The maximum frequency shift occurs at the point of maximum overlap and has a value of about 0.6 GHz. The completed collision in a lossless fiber leaves the soliton intact, with the same frequency and amplitude it had before the collision. The only changes concern their timings and phases.

Since these temporal shifts depend on the sequence of 1 and 0 bits, which occur randomly in real bit streams, different solitons of a channel shift by different amounts. As a consequence, soliton collisions cause some timing jitter even in lossless fibers.

The situation is worse in a system with real fiber and lumped amplifiers, in which case collisions become asymmetric and solitons do not always recover their original frequency and velocity after the collision is over. In fact, the residual frequency shift increases rapidly as the collision length, z_I, approaches the amplifier spacing, L_A. However, the effects of gain–loss variations begin to average out and the residual frequency shift decreases when collisions occur over several amplifier spacings.

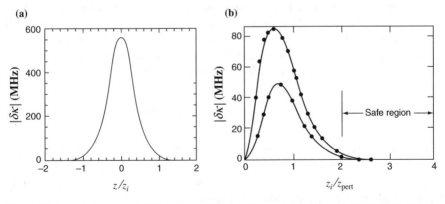

Figure8.22 (a) Frequency shift during collision of two 50 ps solitons with 75 GHz channel spacing in a lossless fiber. (b) Residual frequency shift after collision because of lumped amplification. The amplifier spacing is $L_A = 20$ and 40 km for lower and upper curves, respectively. (After Ref. [102]; © 1991 IEEE.)

Figure 8.22b shows the residual frequency shift after a complete collision of initially well-separated solitons, given as a function of the ratio z_I/z_{pert}, where z_{pert} is equal to the amplifier spacing L_A. We observe that the residual frequency shift increases as z_I approaches L_A and can become ~ 0.1 GHz. However, it becomes essentially zero as long as the condition

$$z_I > 2L_A \tag{8.93}$$

is satisfied.

When combined with Eq. (8.84), condition (8.93) puts an upper bound on the maximum separation between the two outermost channels of a WDM system:

$$\Delta\lambda_{max} = \frac{t_{FWHM}}{DL_A} \tag{8.94}$$

For example, if $t_{FWHM} = 10$ ps, $D = 0.5$ ps/(km nm), and $L_A = 20$ km, Eq. (8.94) gives $\Delta\lambda_{max} = 1$ nm. Assuming a channel wavelength spacing of 0.5 nm, the maximum allowable number of channels is just three.

The limitations referred above can be greatly attenuated through the use of dispersion management and eventually some of the control techniques discussed in Section 8.4. The design optimization of periodic dispersion maps like those considered in Section 8.6 has resulted in WDM soliton systems capable of operating at bit rates close to 1 Tb/s [103–106]. In such systems, colliding solitons move in a zigzag fashion and pass through each other many times before they separate. Since the effective collision length is much larger than the map period, the condition $z_I > 2L_A$ is satisfied even when soliton wavelengths differ by 20 nm or more, which allows a large number of channels.

Several experiments of applying DM solitons for WDM transmissions have been realized during the recent years. In one experiment, Fukuchi et al. [107] have achieved error-free transmission of 1.1 Tbit/s (55×20 Gbit/s) over 3000 km by using both L and C bands of the EDFA. Other experiments confirm that DM solitons are, in fact, suitable for high bit rate and long-haul WDM transmission [108–110]. In particular, it is expected that DM soliton-based WDM systems will be superior to the "quasilinear" WDM systems for bit rates per channel above 40 Gb/s because solitons are more robust against the polarization mode dispersion [111].

PROBLEMS

8.1   Determine the required input power for (i) a fundamental fiber soliton and (ii) a second-order soliton in a dispersion-shifted fiber near the 1.55 μm wavelength using typical values $\beta_2 = -1$ ps^2/km and $\gamma = 3$ W^{-1} km^{-1} and considering pulses with a temporal width $t_0 = 10$ ps.

8.2   Determine the fiber length corresponding to the collision period in a 10 Gbit/s soliton system. The temporal width of the solitons is $t_0 = 20$ ps and the fiber GVD is $\beta_2 = -20$ ps^2/km.

8.3 Consider a soliton communication system with a fiber loss of 0.2 dB/km and an amplifier spacing of 40 km. What should the input power be in order to have the propagation of a fundamental soliton in such a system? What should the amplifier gain be?

8.4 Considering a trans-Pacific soliton transmission system ($z = 9000$ km) operating at 1.55 µm, with typical values $\alpha_0 = 0.2$ dB/km, $D = 0.5$ ps/(km nm), $\gamma = 3\,\text{W}^{-1}\,\text{km}^{-1}$, $n_{sp} = 2$, $s = 10$, $f_b = 0.1$, and $L_A = 50$ km, find the maximum transmission rate imposed by the Gordon–Haus effect.

8.5 Determine the PMD-induced timing jitter in a trans-Pacific soliton transmission system ($z = 9000$ km) operating at 1.55 µm, considering the typical values $\alpha_0 = 0.2$ dB/km, $n_{sp} = 2$, $N_s = 500,000$, $D_p = 0.1$ ps km$^{1/2}$, and $L_A = 50$ km. Discuss the impact of this type of timing jitter considering different values of the PMD parameter and of the system transmission rate.

8.6 Explain the intrapulse Raman scattering effect and the reason of the carrier frequency shift of solitons. Derive an expression for the stationary frequency shift of the fundamental soliton in the presence of distributed spectral filtering with strength β.

8.7 Derive Eq. (8.28) by integrating Eq. (8.26) in the case of bidirectional pumping. Plot $E_s(z)/E_{in}$ for $L_A = 40$ and 80 km using $\alpha_0 = 0.2$ dB/km and $\alpha_p = 0.3$ dB/km.

8.8 Starting from the potential function given by Eq. (8.55), derive conditions (8.56) to guarantee the stability of the pulse and background in a system with spectral filtering and linear and nonlinear gains.

8.9 Starting from the NLSE (8.22), derive the variational equations (8.79) and (8.80) for the pulse width T_p and chirp C using the general ansatz given in Eq. (8.74).

8.10 Find the peak value of the collision-induced frequency and temporal shifts by integrating (8.89) with $b = 1$.

REFERENCES

1. A. Hasegawa and F. D. Tappert, *Appl. Phys. Lett.* **23**, 142 (1973).
2. L. F. Mollenauer, R. H. Stolen, and J. P. Gordon, *Phys. Rev. Lett.* **45**, 1095 (1980).
3. A. Hasegawa, *Opt. Lett.* **8**, 650 (1983).
4. J. P. Gordon and H. A. Haus, *Opt. Lett.* **11**, 665 (1986).
5. L. F. Mollenauer, M. J. Neubelt, M. Haner, E. Lichtman, S. G. Evangelides, and B. M. Nyman, *Electron. Lett.* **27**, 2055 (1991).
6. A. Mecozzi, J. D. Moores, H. A. Haus, and Y. Lai, *Opt. Lett.* **16**, 1841 (1991).
7. Y. Kodama and A. Hasegawa, *Opt. Lett.* **17**, 31 (1992).

8. L. F. Mollenauer, J. P. Gordon, and S. G. Evangelides, *Opt. Lett.* **17**, 1575 (1992).

9. L. F. Mollenauer, E. Lichtman, M. J. Neubelt, and G. T. Harvey, *Electron. Lett.* **29**, 910 (1993).

10. Y. Kodama and S. Wabnitz, *Opt. Lett.* **19**, 162 (1994).

11. M. Nakazawa, Y. Kamada, H. Kubota, and E. Suzuki, *Electron. Lett.* **27**, 1270 (1991).

12. Y. Kodama, M. Romagnoli, and S. Wabnitz, *Electron. Lett.* **28**, 1981 (1992).

13. M. Matsumoto, H. Ikeda, T. Uda, and A. Hasegawa, *J. Lightwave Technol.* **13**, 658 (1995).

14. M. F. Ferreira, M. V. Facão, and S. V. Latas, *Fiber Integr. Opt.* **19**, 31 (2000).

15. N. Akhmediev and A. Ankiewicz, *Dissipative Solitons*, Springer, Berlin, 2005.

16. A. A. Ankiewicz, N. N. Akhmediev, and N. Devine, *Opt. Fiber Technol.*, **13**, 91 (2007).

17. M. F. Ferreira and S. V. Latas, in J. C. Schlesinger (Ed.), *Optical Fibers Research Advances*, Nova Science Publishers, 2008, Chapter 10.

18. N. J. Smith, F. M. Knox, N. J. Doran, K. J. Blow, and I. Bennion, *Electron. Lett.* **32**, 54 (1996).

19. W. Forysiak, F. M. Knox, and N. J. Doran, *Opt. Lett.* **19**, 174 (1994).

20. A. Hasegawa, S. Kumar, and Y. Kodama, *Opt. Lett.* **22**, 39 (1996).

21. M. Suzuki, I. Morita, N. Edagawa, S. Yamamoto, H. Toga, and S. Akiba, *Electron. Lett.* **31**, 2027 (1995).

22. A. Hasegawa (Ed.), *New Trends in Optical Soliton Transmission Systems*, Kluwer, Dordrecht, The Netherlands, 1998.

23. J. Satsuma and N. Yajima, *Prog. Theor. Phys. Suppl.* **55**, 284 (1974).

24. J. P. Gordon, *Opt. Lett.* **8**, 596 (1983).

25. V. I. Karpman and E. M. Maslov, *Zh. Eksp. Teor. Fiz.* **73**, 537 (1977) (*Sov. Phys. JETP* **46**, 281 (1977)).

26. D. Anderson and M. Lisak, *Phys. Rev. A* **27**, 1393 (1983).

27. A. Hasegawa, *Pure Appl. Opt.* **4**, 265 (1995).

28. A. Hasegawa and Y. Kodama, *Solitons in Optical Communications*, Oxford University Press, Oxford, UK, 1995.

29. M. F. Ferreira, M. V. Facão, S. V. Latas, and M. H. Sousa, *Fiber Integr. Opt.* **24**, 287 (2005).

30. K. J. Blow and N. J. Doran, *Opt. Commun.* **52**, 367 (1985).

31. M. F. Ferreira, in B.Guenther (Ed.), *Encyclopedia of Modern Optics*, Elsevier, 2004.

32. M. F. Ferreira, in R. Driggers (Ed.), *Encyclopedia of Optical Engineering*, Marcel Dekker, Inc., 2004.

33. A. Hasegawa and Y. Kodama, *Phys. Rev. Lett.* **66**, 161 (1991).

34. L. F. Mollenauer, J. P. Gordon, and M. N. Islam, *IEEE J. Quantum Electron.* **22**, 157 (1986).

35. M. F. Ferreira, *Opt. Commun.* **107**, 365 (1994).

36. J. P. Gordon, *Opt. Lett.* **11**, 662 (1986).

37. F. M. Mitschke and L. F. Mollenauer, *Opt. Lett.* **11**, 659 (1986).

38. J. P. Gordon and H. A. Haus, *Opt. Lett.* **11**, 665 (1986).

39. D. Marcuse, *J. Lightwave Technol.* **10**, 273 (1992).

40. T. Georges and F. Favre, *J. Opt. Soc. Am. B* **10**, 1880 (1993).

41. M. V. Facão and M. F. Ferreira, *J. Nonlinear Math. Phys.* **8**, 112 (2001).

42. K. Smith and L. F. Mollenauer, *Opt. Lett.* **14**, 1284 (1989).

43. L. F. Mollenauer, P. V. Mamyshev, and M. J. Neubelt, *Opt. Lett.* **19**, 704 (1994).

44. L. F. Mollenauer and J. P. Gordon, *Opt. Lett.* **19**, 375 (1994).

45. M. F. Ferreira and S. V. Latas, *J. Lightwave Technol.* **19**, 332 (2001).

46. M. Nakazawa, K. Suzuki, E. Yamada, H. Kubota, Y. Kimura, and M. Takaya, *Electron. Lett.* **29**, 729 (1993).

47. M. F. Ferreira, in G. Lampropoulos and R. Lessard (Eds.), *Applications of Photonic Technology*, Vol. **2**, Plenum Press, New York, 1997, p. 249.

48. M. F. Ferreira and S. V. Latas, *Opt. Eng.* **41**, 1696 (2002).

49. N. N. Akhmediev, A. Ankiewicz, and J. M. Soto-Crespo, *J. Opt. Soc. Am. B* **15**, 515 (1998).

50. N. N. Akhmediev, V. V. Afanasjev, and J. M. Soto-Crespo, *Phys. Rev. E* **53**, 1190 (1996).

51. J. M. Soto-Crespo, N. N. Akhmediev, and V. V. Afanasjev, *J. Opt. Soc. Am. B* **13**, 1439 (1996).

52. M. F. Ferreira, M. V. Facão, and S. V. Latas, *Photon. Optoelectron.* **5**, 147 (1999).

53. N. N. Akhmediev, A. Rodrigues, and G. Townes, *Opt. Commun.* **187**, 419 (2001).

54. V. V. Afanasjev and N. N. Akhmediev, *Phys. Rev. E* **53**, 6471 (1996).

55. M. F. Ferreira and S. V. Latas, in R. Lessard, G. Lampropoulos, and G. Schinn, (Eds.), *Applications of Photonic Technology; SPIE Proc.* **4833**, 845 (2002).

56. N. N. Akhmediev, A. A. Ankiewicz, and J. M. Soto-Crespo, *J. Opt. Soc. Am. B* **15**, 515 (1998).

57. S. V. Latas and M. F. Ferreira, *Opt. Commun.* **251**, 415 (2005).

58. S. V. Latas and M. F. Ferreira, *J. Math. Comput. Simul.* **74**, 379 (2007).

59. Y. J. He and H. Z. Wang, *Opt. Fiber Technol.* **13**, 67 (2007).

60. C. Normand and Y. Pomeau, *Rev. Mod. Phys.* **49**, 581 (1977).

61. P. Kolodner, *Phys. Rev. A* **44**, 6466 (1991).

62. R. Graham, *Fluctuations, Instabilities and Phase Transactions*, Springer, Berlin, 1975.

63. N. Akhmediev and A. Ankiewicz, in N. Akhmediev and A. Ankiewicz (Eds.), *Dissipative Solitons*, Springer, Heidelberg, 2005.

64. N. Akhmediev, J. M. Soto-Crespo, M. Grapinet, and Ph. Grelu, *Opt. Fiber Technol.* **11**, 209 (2005).

65. J. N. Kutz, in N. Akhmediev and A. Ankiewicz (Eds.), *Dissipative Solitons,* Springer, Heidelberg, 2005.

66. U. Peschel, D. Michaelis, Z. Bakonyi, G. Onishchukov, and F. Lederer, in N. Akhmediev and A. Ankiewicz (Eds.), *Dissipative Solitons*, Springer, Heidelberg, 2005.

67. N. N. Akhmediev and A. A. Ankiewicz, *Solitons: Nonlinear Pulses and Beams*, Chapman & Hall, London, 1997.

68. N. N. Akhmediev, V. V. Afanasjev, and J. M. Soto-Crespo, *Phys. Rev. E* **53**, 1190 (1996).

69. J. M. Soto-Crespo, N. N. Akhmediev, and V. V. Afanasjev, *J. Opt. Soc. Am. B* **13**, 1439 (1996).

70. N. R. Pereira and L. Stenflo, *Phys. Fluids* **20**, 1733 (1977).

71. K. Nozaki and N. Bekki, *Phys. Soc. Jpn.* **53**, 1581 (1984).

72. E. Tsoy, A. Ankiewicz, and N. Akhmediev, *Phys. Rev. E* **73**, 036621 (2006).

73. R. Conte and M. Musette, *Physica D* **69**, 1 (1993).

74. O. Thual and S. Fauvre, *J. Phys.* **49**, 1829 (1988).

75. P. Marcq, H. Chaté, and R. Conte, *Physica D* **73**, 305 (1994).

76. A. D. Boardman and L. Velasco, *IEEE J. Sel. Top. Quantum Electron.* **12**, 388 (2006).

77. A. Ankiewicz, N. Akhmediev, and N. Devine, *Opt. Fiber Technol.* **13**, 91 (2007).

78. S. C. Latas, M. F. Ferreira, and A. S. Rodrigues, *Opt. Fiber Technol.* **11**, 292 (2005).

79. N. N. Akhmediev, J. M. Soto-Crespo, and G. Town, *Phys. Rev. E* **63**, 056602 (2001).

80. L. Song, L. Li, Z. Li, and G. Zhou, *Opt. Commun.* **249**, 301 (2005).

81. J. H. Nijhof, W. Forysiak, and N. J. Doran, *IEEE J. Sel. Top. Quantum Electron.* **6**, 330 (2000).

82. V. Cautaerts, A. Maruta, and Y. Kodama, *Chaos* **10**, 515 (2000).

83. M. H. Sousa, M. F. Ferreira, and E. M. Panameño, *SPIE Proc.* **5622** 1002 (2004).

84. I. R. Gabitov and S. K. Turitsyn, *Opt. Lett.* **21**, 327 (1996).

85. A. Berntson, N. J. Doran, W. Forysiak, and J. H. B. Nijhof, *Opt. Lett.* **23**, 900 (1998).

86. S. K. Turitsyn and E. G. Shapiro, *Opt. Fiber Technol.* **4**, 151 (1998).

87. S. K. Turitsyn, I. Gabitov, E. W. Laedke, V. K. Mezentsev, S. L. Musher, E. G. Shapiro, T. Shafer, and K. H. Spatschek, *Opt. Commun.* **151**, 117 (1998).

88. M. H. Sousa, M. F. Ferreira, and E. M. Panameño, *SPIE Proc.* **5622**, 944 (2004).

89. R. Jackson, C. Jones, and V. Zharnitsky, *Physica D* **190**, 63 (2004).

90. S. Konar, M. Mishra, and S. Jana, *Chaos Solitons Fractals* **29**, 823 (2006).

91. J. H. B. Nijhof, N. J. Doran, W. Forysiak, and A. Berntson, *Electron. Lett.* **33**, 1726 (1998).

92. J. M. Jacob, E. A. Golovchenko, A. N. Pilipetskii, G. M. Carter, and C. R. Menyuk, *IEEE Photon. Technol. Lett.* **9**, 130 (1997).

93. N. J. Smith, W. Forysiak, and N. J. Doran, *Electron. Lett.* **32**, 2085 (1996).

94. G. M. Carter, J. M. Jacob, C. R. Menyuk, E. A. Golovchenko, and A. N. Pilipetskii, *Opt. Lett.* **22**, 513 (1997).

95. T. Okamawari, A. Maruta, and Y. Kodama, *Opt. Commun.* **149**, 261 (1998).

96. S. Waiyapot and M. Matsumoto, *IEEE Photon. Technol. Lett.* **11**, 1408 (1999).

97. J. Kumasako, M. Matsumoto, and S. Waiyapot, *J. Lightwave Technol.* **18**, 1064 (2000).

98. M. F. Ferreira and M. H. Sousa, *Electron. Lett.* **37**, 1184 (2001).

99. M. H. Sousa and M. F. Ferreira, *Laser Phys. Lett.* **1**, 491 (2004).

100. M. H. Sousa and M. F. Ferreira, *Laser Phys. Lett.* **1**, 602 (2004).

101. M. F. Ferreira and M. H. Sousa, *Nonlinear Opt. Quantum Opt.*, **33**, 51 (2005).

102. L. F. Mollenauer, S. G. Evangelides, and J. P. Gordon, *J. Lightwave Technol.* **9**, 362 (1991).

103. F. Favre, D. Le Guen, M. L. Moulinard, M. Henry, and T. Georges, *Electron. Lett.* **33**, 2135 (1997).

104. I. Morita, M. Suzuki, N. Edagawa, K. Tanaka, and S. Yamamoto, *J. Lightwave Technol.* **17**, 80 (1999).

105. M. Nakazawa, H. Kubota, K. Suzuki, E. Yamada, and A. Sahara, *IEEE J. Sel. Top. Quantum Electron.* **6**, 363 (2000).

106. L. F. Mollenauer, P. V. Mamyshev, J. Gripp, M. J. Neubelt, N. Mamysheva, L. Grüner-Nielsen, and T. Veng, *Opt. Lett.* **25**, 704 (2000).

107. K. Fukuchi, M. Kakui, A. Sasaki, T. Ito, Y. Inada, T. Tsuzaki, T. Shitomi, K. Fujii, I. S. Shikii, H. Sugahara, and A. Hasegawa, *ECOC'99*, Nice, France, 1999, Postdeadline Paper PD2-10.

108. M. Suzuki and N. Edagawa, *J. Lightwave Technol.* **21**, 916 (2003).

109. L. F. Mollenauer, A. Grant, X. Liu, X. Wei, C. Xie, and I. Kang, *Opt. Lett.* **28**, 2043 (2003).

110. L. F. Mollenauer and J. P. Gordon, *Solitons in Optical Fibers: Fundamentals and Applications*, Elsevier/Academic Press, San Diego, CA, 2006.

111. M. F. Ferreira, *Fiber Integr. Opt.* **27**, 113 (2008).

9

OTHER APPLICATIONS OF OPTICAL SOLITONS

Optical fiber solitons are leading candidates for ultrahigh-speed long-haul lightwave transmission links because of the unique property that they can propagate over a long distance without either attenuation or change of shape due to the balance between the dispersion and nonlinear effects. One of the key elements of such ultrahigh-speed systems is a source of ultrashort pulses. Besides the optical communications area, such ultrashort pulses also have a great potential in several areas of research and application, such as measurements of ultrafast processes, time-resolved spectroscopy, and sampling systems. Ultrashort pulses can be generated directly by optical fiber soliton lasers using different mode-locking techniques [1–8]. An alternative is provided by the optical pulse compression technique [9] based on, for example, the combination of single-mode fiber (SMF) and grating pair [10]. A more attractive method for producing ultrashort pulses is to make use of the nonlinear properties of the optical fiber itself in order to compress pulses generated from semiconductor lasers. Two widely adopted techniques are the soliton-effect [11–13] and adiabatic pulse compression techniques [14–16]. This chapter provides a description of both soliton fiber lasers (Section 9.1) and pulse compression techniques (Section 9.2). Section 9.3 is dedicated to the fiber gratings, which became a standard component of modern lightwave technology. Such fiber gratings can be used as mirrors for providing optical feedback in a fiber laser or in place of the bulk grating pair in a grating-fiber compressor. The main properties of optical solitons in fiber gratings are also discussed in this section.

9.1 SOLITON FIBER LASERS

One way to generate short optical pulses is to use mode-locked lasers [1,2]. Actively mode-locked operation of lasers is achieved by inserting an amplitude or phase

Nonlinear Effects in Optical Fibers. By Mário F. S. Ferreira.
Copyright © 2011 John Wiley & Sons, Inc. Published 2011 by John Wiley & Sons, Inc.

modulator in the cavity, where the modulation frequency is an integral multiple of the inverse round-trip time in the cavity [3,4]. Passive mode locking can be realized by inserting a fast saturable absorber in the laser cavity [5–8]. Such a fast saturable absorber rejects low-intensity light and can be readily implemented in optical fiber using one of the two methods: a nonlinear loop mirror [17,18] or nonlinear polarization rotation and a polarizer [19].

9.1.1 The First Soliton Laser

The first soliton laser used the passive mode-locking technique and was demonstrated by Mollenauer and Stolen [20] in 1984. Figure 9.1 shows the schematic of such a soliton laser. The output signal of the mode-locked laser is injected into the fiber with anomalous dispersion through the beam splitter S. The light that is reflected by the mirror M_3 is fed back through the same fiber into the laser cavity. After propagating back and forth in this feedback loop, the pulse comes out as a short soliton, which corresponds generally to an $N = 2$ soliton rather than a fundamental soliton. A minimum pulse width of 60 fs was obtained with this system. However, the soliton laser action started and stopped sporadically as vibration and thermal drift caused the relative lengths of the two cavities to vary in and out of the proper interference condition. Servo stabilization of the cavity lengths enabled the laser to achieve a very stable mode of operation [21].

A theoretical interpretation of the soliton laser has been given in terms of the soliton theory [22]. Shortening of the laser output pulses using fibers with normal dispersion instead of anomalous dispersion has also been demonstrated [23,24]. In such an operation, generally called additive-pulse mode locking [25], self-phase modulation (SPM) from a purely nonlinear element in the external cavity causes the returned pulses to interfere constructively with the main laser pulses near their peaks, while interfering destructively with them in their wings, thereby producing a

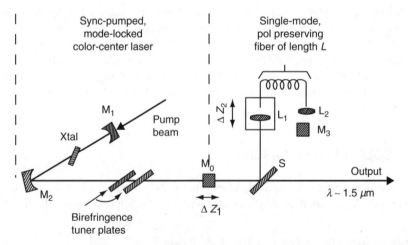

Figure 9.1 Schematic of the first soliton laser. (After Ref. [20]; © 1984 OSA.)

narrower pulse. The pulse shortening process continues until it is balanced by temporal broadening mechanisms, such as bandwidth limiting and group velocity dispersion (GVD).

9.1.2 Figure-Eight Fiber Laser

The advent and rapid progress of the optical fiber amplifiers have enabled one to use them as a gain medium to fabricate fiber lasers. Such lasers are compact, robust, and require minimum bulk optical components. In particular, soliton fiber lasers have been developed with different configurations, such as linear [26], ring [3], and figure-eight [6,27]. Figure 9.2 shows a schematic view of a figure-eight laser [14], which derives its name from the cavity geometry. It can be regarded as a ring laser (to the left of the 50:50 coupler), with a nonlinear amplifying loop mirror (NALM) on the right. The NALM, consisting of a 3 dB fiber coupler, an erbium-doped fiber, and a wavelength division multiplexer for pumping, causes additive-pulse mode locking of the laser.

9.1.3 Nonlinear Loop Mirrors

The NALM is a variant of the nonlinear optical loop mirror (NOLM) [28]; both are antiresonant nonlinear Sagnac interferometers, with intensity-dependent transmission. The key property of devices such as the NOLM and NALM is an asymmetry between the clockwise and anticlockwise waves propagating around the loop. In the NOLM, such asymmetry is created by using a coupler with an uneven coupling ratio $x : (1-x)$ $(x \neq 0.5)$. The incoming signal is split into two counterpropagating parts propagating in clockwise and anticlockwise directions in the loop. In the NALM, the asymmetry is provided by a fiber amplifier that is placed closer to one of the coupler arms than the other. In either case, light traveling clockwise around the loop has a different intensity from light traveling anticlockwise, and hence experiences different

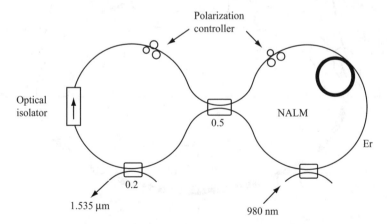

Figure 9.2 Schematic of a figure-eight laser. (After Ref. [6]; © 1991 OSA.)

amounts of self-phase modulation in the loop due to fiber nonlinearity. The interference at the coupler, constructive or destructive, depends on the phase difference between the two waves, which is proportional to the power of the incoming signal. Apart from providing short pulse generation in optical fiber lasers, nonlinear loop mirrors may also be used for ultrafast all-optical switching [28,29] and pulse regeneration in soliton transmission systems [30].

9.1.4 Stretched-Pulse Fiber Lasers

The sideband instability is a problem usually affecting the fiber soliton lasers operating with pulses in the subpicosecond regime [31]. One way to suppress the sideband instability is to use a dispersion-managed (DM) cavity, consisting of fibers with anomalous and normal GVD, such that the total cavity dispersion is close to zero [32]. Figure 9.3 shows an experimental configuration of a fiber laser of this type, also called a stretched-pulse mode-locked laser [33]. All the characteristics of dispersion-managed soliton transmission systems, as discussed in Chapter 8, can also be observed in a stretched-pulse laser. In particular, the enhanced pulse energy and the reduced timing jitter were also reported concerning an actively mode-locked DM fiber ring laser [35].

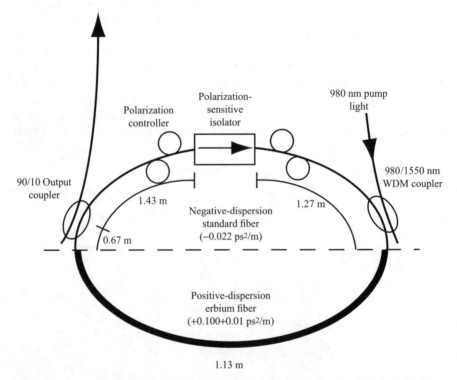

Figure 9.3   Schematic of a stretched-pulse mode-locked laser. (After Ref. [34]; © 1993 OSA.)

A fiber laser with a dispersion close to zero would normally generate very short pulses, but with very low energy in each pulse. In the case of stretched-pulse fiber lasers, however, this is not the case. As seen in Chapter 8, dispersion-managed solitons experience large changes in their temporal width, returning to their minimum temporal width (and maximum peak power) at the midpoint of each fiber section. Since these pulses have a high peak power at only two points in the laser cavity, the average nonlinearity is reduced, and the pulse energy is consequently increased.

9.1.5 Modeling Fiber Soliton Lasers

In analogy with soliton transmission systems, soliton fiber lasers usually require several control mechanisms to limit the effects of soliton interactions [36], soliton self-frequency shift [37], noise and Gordon–Haus jitter [38], and interactions with dispersive waves [39]. For example, the principle of operation of the sliding-frequency soliton laser is very similar to the sliding-filter soliton transmission described in Section 8.4 [40,41]. The main difference between the two cases is that, instead of several filters with different peak wavelengths that are used in the second case, a frequency-shifting device (e.g., an acousto-optic modulator) is generally used in the soliton laser. In such a case, only solitons are able to propagate continuously in the laser cavity, whereas noise is effectively suppressed by the combined action of the frequency shifter and filter [42,43]. By incorporating a broadband Fabry–Perot-like Bragg grating resonator in the cavity, a multiwavelength sliding-frequency soliton laser can be realized [44,45].

Other control techniques are based on the use of spectral filtering and nonlinear gain, as discussed in Section 8.5. Passively mode-locked soliton lasers using these elements can be accurately described by the following quintic complex Ginzburg–Landau equation (CGLE) [46–49], already presented in Chapter 8 in the context of soliton transmission:

$$i\frac{\partial q}{\partial Z} + \frac{D}{2}\frac{\partial^2 q}{\partial T^2} + |q|^2 q = i\delta q + i\beta\frac{\partial^2 q}{\partial T^2} + i\varepsilon|q|^2 q + i\mu|q|^4 q + v|q|^4 q \qquad (9.1)$$

In Eq. (9.1), Z is the cavity round-trip number, T is the retarded time, q is the normalized envelope of the field, D is the GVD coefficient, with $D = +1$ ($D = -1$) for anomalous (normal) GVD, β stands for spectral filtering, δ is the linear gain or loss coefficient, ε accounts for nonlinear gain–absorption processes, μ represents a higher order correction to the nonlinear gain–absorption, and v is a higher order correction term to the nonlinear refractive index. As shown in Chapter 8, quintic terms in the CGLE are necessary to guarantee the stability of the pulse solutions [47,50].

Equation (9.1) models the fiber laser as a distributed system, which is reasonable if the pulse shape changes only slightly during each round-trip. There are many advantages in using such a distributed model, governed by a continuous equation, since it allows, to some extent, an analytic study. However, when the discrete nature of the laser cavity cannot be ignored, the above CGLE can be used such that the parameters D, β, δ, ε, μ, and ν vary periodically with Z, the period corresponding to a cavity round-trip.

As referred in Chapter 8, Eq. (9.1) has a multitude of solutions, which exhibit a variety of shapes and stability properties. Stable pulses can be generated in a very narrow range of the laser parameters, which must be carefully adjusted for such purpose. In general, the pulses may change their shape from one round-trip to another and have complicated dynamics in time [51]. In some circumstances, the pulse shape evolution in time may even become chaotic.

9.2 PULSE COMPRESSION

Pulse compression can be achieved using different techniques. Such techniques utilize dispersion, possibly from gratings or prisms, simultaneously to compensate for chirp and achieve pulse compression. As seen in Section 3.5, when an optical pulse propagates in a linear dispersive medium, it acquires a dispersion-induced chirp. If the pulse has an initial chirp in the opposite direction to that imposed by the dispersive medium, the two tend to cancel each other, resulting in the compression of the optical pulse.

Early work on optical pulse compression did not make use of any nonlinear optical effect. Only during the 1980s, after having understood the evolution of optical pulses in silica fibers, the SPM effect was used to achieve pulse compression. Using such an approach, an optical pulse of 50 fs width from a colliding pulse mode-locked dye laser oscillating at 620 nm was compressed down to 6 fs [10].

9.2.1 Grating-Fiber Compressors

One scheme for pulse compression uses the so-called *grating-fiber compressor* [52,53]. In this scheme, which is generally used at wavelengths $\lambda < 1.3\,\mu m$ [10,54–58], the input pulse is propagated in the normal dispersion regime of the fiber, which imposes a nearly linear, positive chirp on the pulse through a combination of SPM and GVD. The output pulse is then sent through a grating pair, which provides the anomalous (or negative) GVD required to get the pulse compression.

Different spectral components of an optical pulse incident at one grating are diffracted at slightly different angles. As a consequence, they experience different time delays during their passage through the grating pair, the blue-shifted components arriving early than the red-shifted ones. In the case of an optical pulse with a positive chirp, the blue-shifted (red-shifted) components occur near the trailing (leading) edge of the pulse and the passage through the grating pair then provides its compression.

The phase shift acquired by the spectral component of the pulse at the frequency ω passing through the grating pair can be written in the form of a Taylor expansion:

$$\phi_g = \phi_0 + \phi_1(\omega-\omega_0) + \frac{1}{2}\phi_2(\omega-\omega_0)^2 + \frac{1}{6}\phi_3(\omega-\omega_0)^3 + \cdots \qquad (9.2)$$

where ω_0 is the pulse center frequency, ϕ_0 is a constant, ϕ_1 is a constant delay, and ϕ_2 and ϕ_3 are parameters that take into account the GVD effects associated with the grating pair. These parameters depend on the grating period (line spacing), as well as on the orientation and separation of the two gratings. In most cases of practical interest, the spectral width of the pulse satisfies the condition $\Delta\omega \ll \omega_0$ and the cubic and higher order terms in the expansion (9.2) can be neglected.

A limitation of the grating pair is that the spectral components of the pulse are dispersed not only temporally but also spatially. As a consequence, the optical beam becomes deformed, which is undesirable. This problem can be avoided simply using a mirror to reflect the pulse back through the grating pair. Reversing the direction of propagation not only allows the beam to recover its original cross section, but also doubles the amount of GVD, thereby reducing the grating separation by a factor of 2. Such a double-pass configuration was used in a 1984 experiment, in which 33 ps input pulses at 532 nm from a frequency-doubled Nd:YAG laser were propagated through a 105 m long fiber and were compressed to 0.42 ps after passing through the grating pair [52]. A schematic of the optical pulse compressor used in this experiment is shown in Fig. 9.4. In this configuration, the mirror M_3 is used to send the beam back through the grating pair, whereas the mirror M_4 is used to deflect the compressed pulse out of the compressor.

To achieve optimum performance from a grating-fiber compressor, it is necessary to optimize the fiber length, as well as the grating separation. Concerning the first aspect, the effects of both GVD and SPM during the propagation of the pulse inside

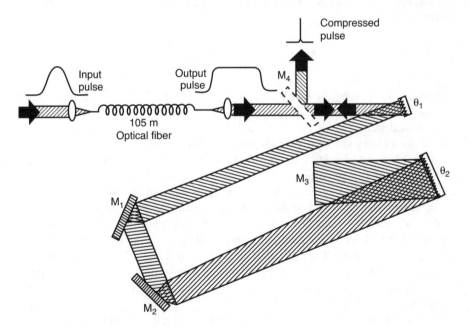

Figure 9.4 Schematic representation of a grating-fiber compressor in the double-pass configuration. (After Ref. [52]; © 1984 AIP.)

the fiber must be considered. SPM alone determines a linear chirp only over the central part of an optical pulse. Since the grating pair compresses only this region, while a significant amount of energy remains in the wings, the compressed pulse is not of high quality in this case. The effect of GVD turns out to be important, since it broadens and reshapes the pulse, which develops a nearly linear chirp across its entire width. In these circumstances, the grating pair can compress most of the pulse energy into a narrow pulse.

The balanced effects of both GVD and SPM explain the existence of an optimum fiber length, z_{opt}, for maximum pulse compression [53]. When the fiber length is less than z_{opt}, the SPM-induced chirp has not yet been linearized. On the other hand, when it is longer than z_{opt}, the SPM effects become negligible due to the GVD-induced pulse broadening.

If the input pulse is unchirped and presents a hyperbolic secant shape, the optimum fiber length for maximum pulse compression is well approximated by [53]

$$z_{opt} \approx (6L_D L_{NL})^{1/2} \tag{9.3}$$

where $L_D = t_0^2/|\beta_2|$ is the dispersion length and $L_{NL} = 1/\gamma P_0$ is the nonlinear length, where γ is the nonlinearity parameter of the fiber and P_0 is the peak power of the input pulse. If the input pulse presents a negative chirp, this must be compensated by the positive chirp provided by the fiber. As a result, the optimum fiber length increases. The opposite occurs in the case of an input pulse with a positive chirp.

The optimum compression factor, F_{opt}, in the case of a grating pair compressor is given by [53]

$$F_{opt} = \frac{t_{FWHM}}{t_{comp}} \approx 0.625 \left(\frac{L_D}{L_{NL}}\right)^{1/2} \tag{9.4}$$

where t_{FWHM} is the full width at half maximum (FWHM) of the input pulse ($t_{FWHM} \approx 1.76 t_0$ for a hyperbolic secant pulse) and t_{comp} is the FWHM of the compressed pulse. Equations (9.3) and (9.4) provide a good estimate even for pulse shapes other than a hyperbolic secant as long as $L_D \gg L_{NL}$.

In a 1987 experiment, a compression factor of 110 was achieved using 60 ps input pulses at 1.06 µm after passing an 880 m long fiber [58]. However, achieving compression factors larger than 100 at 1.06 µm is generally a difficult task. In fact, even though the compression factor increases with the peak power of the input pulse (see Eq. (9.4) and the definition of the nonlinear length L_{NL}), such peak power must be kept below the Raman threshold to avoid the loss of pulse energy through the stimulated Raman scattering process. If this condition is not satisfied, besides the energy loss problem, the generated Raman pulse can interact with the input pulse through cross-phase modulation and deform its frequency chirp characteristics [59,60].

The result given by Eq. (9.4) for the compression factor was obtained neglecting the higher order nonlinear and dispersive effects, which is acceptable for pulse widths $t_0 > 0.1$ ps. However, when shorter pulses are used at the input, their spectral width

$\Delta\omega$ is large enough that the cubic term in the expansion (9.2) must be taken into account. In this case, the compression factor turns out to be smaller than that given by Eq. (9.4). The third-order dispersion (TOD) of the grating pair resulting from cubic term in Eq. (9.2) can be compensated in some cases using a proper combination of gratings and prisms [10]. However, it is difficult to find the right solution for pulse widths in the range of 4–5 fs, as obtained in some experiments [61,62]. In a 1997 experiment, for example, a grating pair and a combination of four prisms were used to compress 13 fs input pulses to 4.9 fs [61].

9.2.2 Soliton-Effect Compressors

In another pulse compression scheme, known as *soliton-effect compressor*, the fiber itself acts as a compressor without the need of an external grating pair [63,64]. The input pulse propagates in the anomalous GVD regime of the fiber and is compressed through an interplay between SPM and GVD. This compression mechanism is related to a fundamental property of the higher order solitons. As seen in Chapter 4, these solitons follow a periodic evolution pattern such that they go through an initial narrowing phase at the beginning of each period. If the fiber length is suitably chosen, the input pulses can be compressed by a factor that depends on the soliton order. Obviously, this compression technique is applicable only in the case of optical pulses with wavelengths exceeding 1.3 μm when propagating in standard SMFs.

The evolution of a soliton of order N inside an optical fiber is governed by the nonlinear Schrödinger equation (NLSE), which can be written in the following form:

$$i\frac{\partial Q}{\partial Z} + \frac{1}{2}\frac{\partial^2 Q}{\partial T^2} + N^2|Q|^2 Q = 0 \tag{9.5}$$

where Q is the normalized amplitude, $Z = z/L_D$, $T = \tau/t_0$, and the parameter N is the soliton order, given by

$$N^2 = \frac{L_D}{L_{NL}} = \frac{\gamma P_0 t_0^2}{|\beta_2|} \tag{9.6}$$

The soliton period, z_0, is given by

$$z_0 = \frac{\pi}{2}L_D = \frac{\pi t_0^2}{2|\beta_2|} \tag{9.7}$$

In writing Eq. (9.5), the fiber losses were neglected, since fiber lengths employed in practice are relatively small. Even though higher order solitons follow an exact periodic evolution only for integer values of N, Eq. (9.4) can be used to describe pulse evolution for arbitrary values of N.

In practice, soliton-effect compression is carried out by initially amplifying optical pulses up to the power level required for the formation of higher order solitons. The peak optical power of the initial pulse required for the formation of the Nth-order soliton can be obtained from Eq. (9.6) and is given by

$$P_0 = 3.11 \frac{|\beta_2|N^2}{\gamma t_{\text{FWHM}}^2} \qquad (9.8)$$

These Nth-order solitons are then passed through the appropriate length of optical fiber to achieve a highly compressed pulse. The optimum fiber length, z_{opt}, and the optimum pulse compression factor, F_{opt}, of a soliton-effect compressor can be estimated from the following empirical relations [64]:

$$z_{\text{opt}} \approx z_0 \left(\frac{0.32}{N} + \frac{1.1}{N^2} \right) \qquad (9.9)$$

$$F_{\text{opt}} \approx 4.1N \qquad (9.10)$$

These relations are accurate to within a few percent for $N > 10$. Figure 9.5 shows the numerically and theoretically obtained compression factor and the fiber length

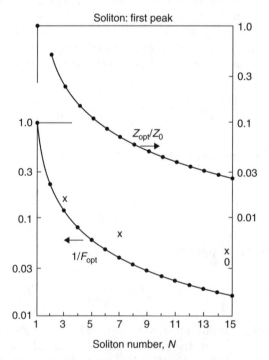

Figure 9.5 Inverse of the compression factor and fiber length for which the compression is maximum in the higher order soliton compression. The fiber length is expressed as a fraction of the soliton period. Data points correspond to experimental results. (After Ref. [63]; © 1983 OSA.)

for which the compression is maximum as a function of the initial pulse amplitude [63].

Compression factors as large as 500 have been attained using soliton-effect compressors [65]. In fact, considering the same input pulse, the soliton-effect compressor provides a pulse compression that is larger and using a shorter fiber, than using a grating-fiber compressor. However, the pulse quality is poorer, since the compressed pulse carries only a fraction of the input energy, while the remaining energy appears in the form of a broad pedestal. From a practical point of view, the pedestal is deleterious since it causes the compressed pulse to be unstable, making it unsuitable for some applications. Despite this, soliton-effect compressed pulses can still be useful because there are some techniques that can eliminate the pedestal. One of them is the intensity discriminator technique [66,67], in which the fiber nonlinear birefringence induced by an intense pulse is used to modify the shape of the same pulse. As a result, the low-intensity tails of a pulse are blocked while the peak is allowed to pass. Another technique for pedestal suppression consists in the use of a NOLM [68,69], described in Section 9.1. Since the transmission of a NOLM is intensity dependent, the loop length can be chosen such that the low-intensity pedestal is reflected while the higher intensity pulse peak is transmitted, resulting in a pedestal-free transmitted pulse.

One difficulty faced when using the soliton-effect compressor is that pulses with high peak power are required for the formation of high-order solitons in conventional fibers. As suggested by Eq. (9.8), the use of dispersion-shifted fibers (DSFs) with small values of β_2 at the operating wavelength can reduce the peak power required for soliton generation by an order of magnitude. However, because the soliton period z_0 is inversely proportional to $|\beta_2|$, Eq. (9.9) indicates that longer lengths of DSFs will be required for solitons to achieve optimum compression. As a result, the total fiber loss experienced by those solitons will be larger and the loss-induced pulse broadening will have a significant impact on the compressor global performance.

Another problem introduced by the use of DSFs in soliton-effect compressors is the third-order dispersion, which generally degrades the quality of the compressed pulse [70]. In the case of soliton-effect compressors using conventional optical fibers, the effects of TOD become significant only when the widths of the compressed pulse become very short ($\ll 1$ ps). In DSFs, however, the relative importance of TOD is increased and those effects become more pronounced because the GVD parameter β_2 is small. In this case, TOD is detrimental in the compression process even for pulse widths of a few picoseconds, resulting in serious degradation of the optimum compression factor F_{opt}.

The result given by Eq. (9.10) for the optimum compression factor holds only for the case of an ideal soliton-effect compressor, when high-order nonlinear and dispersive effects, such as intrapulse Raman scattering (IRS), TOD, and self-steepening, are neglected. However, these effects cannot be neglected for the highly compressed pulses because of their subpicosecond widths and high intensities.

For optical pulses propagating not too close to the zero-dispersion wavelength of the fiber, the dominant higher order effect is IRS, which manifests as a shift of the pulse spectrum toward the red side, the so-called soliton self-frequency shift, which was discussed in Chapter 8. Such effect can be used with advantage to improve the quality of the compressed pulse, by removing the pedestal mentioned above [71]. As a consequence of the change in the group velocity induced by the IRS effect, the sharp narrow spike corresponding to the compressed pulse travels more slowly than the pedestal and separates from it. Moreover, the pedestal can be removed by spectral filtering and a red-shifted, pedestal-free, and highly compressed pulse is then produced. Other studies revealed that the combined effects of negative TOD and IRS, together with the use of dispersion decreasing fibers (DDFs), can further improve the performance of a soliton-effect compressor [72,73].

9.2.3 Compression of Fundamental Solitons

As described in the previous section, the propagation of higher order solitons provides a rapid compression method, but it suffers from the existence of residual pedestals. Some techniques can help to reduce or even eliminate those pedestals, but energy is always wasted in this process. A less rapid technique that provides better pulse quality is the adiabatic compression of fundamental solitons, which is of primary importance in the domain of optical communications. Such adiabatic compression can be achieved using slow amplification [74–76], a proper selection of the initial conditions [77], or by slowly decreasing dispersion [78] along the fiber.

An optical pulse in a fiber having distributed gain obeys the perturbed NLSE

$$i\frac{\partial q}{\partial Z} + \frac{1}{2}\frac{\partial^2 q}{\partial T^2} + |q|^2 q = igq \qquad (9.11)$$

The gain term with a gain coefficient g on the right-hand side leads to the amplification of the pulse energy. Solitons have a fixed area, so the increased energy from amplification is accommodated by an increase in power and a decrease in width. To avoid distortion, the amplification per soliton period cannot be too great. According to the perturbation theory presented in Section 8.2, an approximate solution of Eq. (9.11) is given by

$$q(T,Z) = \eta(Z)\mathrm{sech}[\eta(Z)T]\exp[i\sigma(Z)] \qquad (9.12)$$

where

$$\eta(Z) = \eta_0 \exp(2gZ) \qquad (9.13)$$

and

$$\sigma(Z) = \frac{\eta_0^2}{8g}[\exp(4gZ)-1] \qquad (9.14)$$

From Eqs. (9.12) and (9.13) we verify that the pulse width decreases exponentially as the pulse amplitude increases.

The same effect as adiabatic amplification can be achieved using an optical fiber with dispersion that decreases along the length of the fiber. In fact, for optical solitons, a small variation in the dispersion has a similar perturbative effect as an amplification or loss: such a variation perturbs the equilibrium between the dispersion and nonlinearity in such a way that when, for example, the dispersion decreases, the soliton pulse is compressed. It can be seen from Eq. (9.6) that if the value of $|\beta_2|$ decreases along the fiber and to keep the soliton order N ($N = 1$ for the fundamental soliton), the pulse width must indeed decrease as $|\beta_2|^{1/2}$. Hence, the use of fibers with variable dispersion is viewed as a passive and effective method to control optical solitons in soliton communication systems. DDFs, in particular, have been recognized to be very useful for high-quality, stable, polarization-insensitive, adiabatic soliton pulse compression and soliton train generation [16,72,77,78]. These fibers can be made by tapering the core diameter of a single-mode fiber during the drawing process, and hence changing the waveguide contribution to the second-order dispersion. Provided the dispersion variation in the DDF is sufficiently gradual, soliton compression can be an adiabatic process where an input fundamental soliton pulse can be ideally compressed as it propagates, while retaining its soliton character and conserving the energy.

The evolution of the fundamental soliton in a DDF can be described by the following NLSE:

$$i\frac{\partial q}{\partial Z} + \frac{1}{2}p(Z)\frac{\partial^2 q}{\partial T^2} + |q|^2 q = 0 \tag{9.15}$$

where the variable coefficient $p(Z) = |\beta_2(Z)/\beta_2(0)|$ takes into account the variation of dispersion along the fiber. Using the transformations $s = \int_0^Z p(y)dy$ and $u = q/\sqrt{p}$, Eq. (9.15) assumes the form

$$i\frac{\partial u}{\partial s} + \frac{1}{2}\frac{\partial^2 u}{\partial T^2} + |u|^2 u = -i\frac{1}{2p}\frac{dp}{ds}u \tag{9.16}$$

Equation (9.16) clearly shows that the effect of decreasing dispersion is mathematically equivalent to the effect of distributed amplification, adding a gain term to the NLSE. The effective gain coefficient is related to the rate at which GVD decreases along the fiber.

Since decreasing dispersion is equivalent to an effective gain, a DDF can be used in place of a conventional fiber amplifier to generate a train of ultrashort pulses. To achieve such an objective, a continuous-wave (CW) beam with a weak sinusoidal modulation imposed on it is injected in the DDF. The sinusoidal modulation can be imposed, for example, by beating two optical signals with different wavelengths. As a result of the combined effect of GVD, SPM, and decreasing GVD, the nearly CW beam is converted into a high-quality train of ultrashort solitons, whose repetition rate is governed by the frequency of the initial sinusoidal modulation.

If the input to the DDF is a fundamental soliton given by $q(0,T) = A\,\text{sech}(AT)$ and provided that the dispersion variation is sufficiently adiabatic, the pulse after a length L of a lossless DDF will be compressed to a soliton-like pulse of the form [16,77]

$$q(L, T) = \sqrt{G_{\text{eff}}}\, A\text{sech}[G_{\text{eff}}AT]\exp\{iA^2L/2\} \qquad (9.17)$$

where

$$G_{\text{eff}} = \beta_2(0)/\beta_2(L) \qquad (9.18)$$

is commonly called the effective amplification of the fiber. Adiabatic compression means that all the energy of the input pulse remains localized in the compressed pulse. In such a case, the compression factor is given by

$$F_c = \frac{t_{\text{FWHM}}}{t_{\text{comp}}} = \frac{|q(L,0)|^2}{|q(0,0)|^2} = G_{\text{eff}} \qquad (9.19)$$

where energy has been defined as the product of pulse intensity and pulse width. In the case of a real fiber, the loss must be taken into account, which leads to pulse broadening. As a consequence, the final compression factor is smaller than that in the ideal lossless case and becomes

$$F_c = G_{\text{eff}} \exp(-2\Gamma L) \qquad (9.20)$$

where Γ is the normalized fiber attenuation coefficient.

Different approaches have been developed to determine the optimum GVD profile and its dependence on the width and peak power of input pulses. In the case of picosecond soliton pulse compression, direct numerical simulations of the NLSE show that linear, Gaussian, and exponential dispersion profiles may all be used effectively to provide ideal, adiabatic compression, where the input pulse energy is conserved and remains localized within the pulse [77]. Compression factors larger than 50 are possible launching input pulses with peak powers corresponding to the fundamental soliton into a DDF whose length is about one soliton period. This technique takes advantage of soliton-effect compression but requires lower peak powers and produces compressed pulses of better quality than those obtained using higher order solitons.

In the case of subpicosecond soliton pulses, the influence of higher order nonlinear and dispersive effects must be accounted for. Some studies show that, in this case, the linear and Gaussian dispersion profiles are the most suitable to achieve high-quality, pedestal-free, adiabatic compression of fundamental solitons. However, in the presence of higher order effects, the final compression factor is generally lower, and after reaching a maximum at a particular length of DDF, it decreases steadily [16,77]. This can be explained in terms of pulse compression stabilization, which originates from a competition between the rate of dispersion decrease in the DDF and the rate of dispersion increase due to the combined effects of soliton self-frequency shift and third-order dispersion. In contrast to the degradation occurring for a positive

TOD, the combination of a negative TOD and intrapulse Raman scattering can significantly enhance the compression of fundamental solitons [16].

Pulse compression can also be achieved using dissipative solitons, corresponding to pulse solutions of the complex Ginzburg–Landau equation discussed in Section 8.5 [79]. Such a compression effect has been numerically demonstrated for both fixed-amplitude and arbitrary-amplitude solitons. In the first case, the final pulse width can be set by an appropriate choice of the various system parameters, whereas in the second case the filtering strength plays a decisive role [79].

9.3 FIBER BRAGG GRATINGS

A fiber Bragg grating (FBG) is obtained when the refractive index varies periodically along the fiber length. The realization of such index gratings is possible due to the photosensitivity of optical fibers, which change permanently their optical properties when subject to intense ultraviolet radiation. Index changes occur only in the fiber core and are typically of the order of $\sim 10^{-4}$ in the 1550 nm spectral region. However, the amount of index change induced by ultraviolet absorption can be enhanced by more than one order of magnitude in fibers with high Ge concentration [80].

The holographic technique is now routinely used to fabricate such fiber gratings with controllable period [81]. This technique uses two optical beams, obtained from the same laser and making an angle 2θ between them, which are made to interfere at the exposed core of an optical fiber. The grating period, Λ, is determined by that angle and by the wavelength of the ultraviolet radiation, λ_{uv}, according to the relation

$$\Lambda = \frac{\lambda_{uv}}{2\sin\theta} \tag{9.21}$$

Taking into account both the frequency and the intensity dependences, in addition to its periodic variation along the fiber length, the refractive index of a FBG can be written in the form

$$n(\omega, z, I) = \bar{n}(\omega) + n_2 I + \delta n_g(z) \tag{9.22}$$

where n_2 is the Kerr coefficient and $\delta n_g(z)$ describes the periodic variation of the grating refractive index. In the following, we will consider a sinusoidal grating of the form

$$\delta n_g = n_a \cos(2\pi z/\Lambda) \tag{9.23}$$

where n_a is the refractive index modulation depth.

A fiber grating acts as a reflector for a given wavelength of light λ, such that

$$\lambda \equiv \lambda_B = 2\bar{n}\Lambda \tag{9.24}$$

The Bragg wave number and the Bragg frequency are given by $\beta_B = \pi/\Lambda$ and $\omega_B = \pi c/(\bar{n}\Lambda)$, respectively. Equation (9.24) is known as the *Bragg resonance condition* and it ensures that the phase of the reflected light adds up constructively, producing a strong reflection of light wave.

In order to study the wave propagation along the fiber grating, we assume that the nonlinear effects are relatively weak and solve the Helmholtz equation in the frequency domain:

$$\nabla^2 \tilde{E} + \bar{n}^2(\omega, z, I)(\omega/c)^2 \tilde{E} = 0 \qquad (9.25)$$

The field $\tilde{E}$ in Eq. (9.25) includes both forward- and backward-propagating waves and can be written in the form

$$\tilde{E}(r, \omega) = F(r)[\tilde{A}_f(z, \omega)\exp(i\beta_B z) + \tilde{A}_b(z, \omega)\exp(-i\beta_B z)] \qquad (9.26)$$

where $F(r)$ describes the transverse modal distribution in a single-mode fiber and β_B is the Bragg wave number. Using Eqs. (9.22)–(9.26) and using the slowly varying approximation for the amplitudes $\tilde{A}_f$ and $\tilde{A}_b$, the following coupled-mode equations are obtained [82]:

$$\frac{d\tilde{A}_f}{dz} = i[\delta(\omega) + \Delta\beta]\tilde{A}_f + i\kappa_g \tilde{A}_b \qquad (9.27)$$

$$-\frac{d\tilde{A}_b}{dz} = i[\delta(\omega) + \Delta\beta]\tilde{A}_b + i\kappa_g \tilde{A}_f \qquad (9.28)$$

where

$$\delta = (\bar{n}/c)(\omega - \omega_B) \equiv \beta(\omega) - \beta_B \qquad (9.29)$$

is the detuning from the Bragg frequency ω_B and

$$\kappa_g = \frac{\pi n_a}{\lambda_B} \qquad (9.30)$$

is the coupling coefficient. In Eqs. (9.27) and (9.28), the nonlinear effects are included through $\Delta\beta$.

9.3.1   Pulse Compression Using Fiber Gratings

If the input intensity is sufficiently low, the nonlinear effects can be neglected, considering $\Delta\beta = 0$. In these circumstances, we can obtain from Eqs. (9.27) and (9.28) the following result for the reflection coefficient of a fiber Bragg grating of length L_g [81]:

$$r_g = \frac{A_b(0)}{A_f(0)} = \frac{i\kappa_g \sin(QL_g)}{Q\cos(QL_g) - i\delta \sin(QL_g)} \qquad (9.31)$$

where Q satisfies the linear dispersion relation

$$\delta = \pm\sqrt{\kappa_g^2 + Q^2} \tag{9.32}$$

Maximum reflectivity occurs at the center of the stop band and can be obtained from Eqs. (9.31) and (9.32) making $\delta = 0$:

$$R = |r_g|^2 = \tanh^2 \kappa_g L \tag{9.33}$$

We have $R = 0.93$ for $\kappa_g L = 2$ and $R \approx 1$ for $\kappa_g L \geq 3$. The bandwidth of the reflection spectrum, which is defined as the wavelength spacing between the two reflection minima on either side of the central peak, is approximately given by [83]

$$\Delta\lambda = \frac{\lambda_B^2}{\pi n_{\text{eff}} L}(\kappa_g^2 L^2 + \pi^2)^{1/2} \tag{9.34}$$

Figure 9.6 illustrates the relation between Q and δ given by Eq. (9.32), both for a uniform medium where $\kappa_g = 0$ (dashed lines) and for a grating for which $\kappa_g \neq 0$ (solid curves). The most interesting feature in this figure in the case $\kappa_g \neq 0$ is the existence of a photonic band gap for detunings such that $-\kappa_g < \delta < \kappa_g$. Light with frequencies within this band gap cannot propagate and is reflected. Frequencies

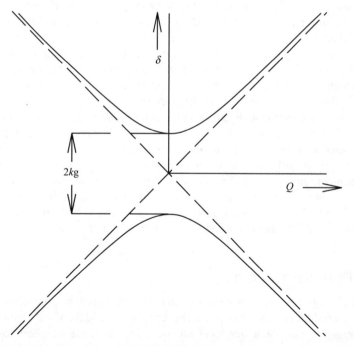

Figure 9.6 Dispersion relation of a fiber grating. The dashed lines correspond to a uniform medium, while the solid curves are for a uniform grating. (After Ref. [94]; © 1997 OSA.)

outside the gap, however, can propagate, but at velocities that can be significantly reduced in comparison to the speed of light in the uniform medium. According to Fig. 9.6, the group velocity vanishes at the band edge and asymptotically approaches $c/\bar{n}$ far from the Bragg resonance.

Outside but close to the stop-band edges, the grating provides large dispersion. Noting that the effective propagation constant of the forward- and backward-propagating waves is $\beta_e = \beta_B \pm Q$, where the choice of sign depends on the sign of δ, we obtain the following result for the group velocity inside the grating:

$$V_G = \left[\frac{dQ}{d\omega}\right]^{-1} = \pm v_g \sqrt{1-x^2} \qquad (9.35)$$

where $x = \kappa_g/\delta$. On the other hand, we can obtain the following results for the dispersion parameters of a fiber grating [84]:

$$\beta_2^g = \frac{d^2 Q}{d\omega^2} = -\frac{1}{v_g^2}\frac{x^2}{\delta(1-x^2)^{3/2}}, \qquad \beta_3^g = \frac{d^3 Q}{d\omega^3} = \frac{3}{v_g^3}\frac{x^2}{\delta^2(1-x^2)^{5/2}} \qquad (9.36)$$

The grating-induced GVD depends on the sign of the detuning δ. The GVD is anomalous ($\beta_2^g < 0$) on the high-frequency side of the stop band and becomes normal ($\beta_2^g > 0$) on the low-frequency side. Near the stop-band edges, the grating exhibits large GVD. Since an FBG acts as a dispersive delay line, it can be used for dispersion compensation in the fiber transmission systems [85]. It can also be used in place of the bulk grating pair in a grating-fiber compressor, providing a compact all-fiber device. In fact, since β_2^g can typically exceed 10^7 ps²/cm, a 1 cm long fiber grating may provide as much dispersion as 10 km of silica fiber or a bulk grating pair with more than 1 m spacing. The main limitation in this case comes from the grating-induced TOD, which can significantly affect the quality of the pulses when the optical frequency falls close to the edges of the stop band.

Both the compression factor and the pulse quality can be significantly enhanced by using an FBG in which the grating period varies linearly with position, the so-called *chirped fiber grating* [84]. In this case, the grating reflects different frequency components of the pulse at different points along its length. Such a device has been used to compensate for dispersion-induced broadening of pulses in fiber transmission systems as well as for pulse compression, where compression factors above 100 have been achieved [86,87]. The only disadvantage of a chirped fiber grating is that the compressed pulse is reflected instead of being transmitted.

9.3.2 Fiber Bragg Solitons

If the frequency of the light wave propagating in the fiber grating approaches the band edge, Fig. 9.6 shows that the group velocity tends to zero and the small group velocity in this region creates a larger local intensity. Due to the Kerr nonlinearity, the refractive index increases in proportion to this intensity, which enables the propagation of a soliton-like pulse within the stop band. Such a soliton was reported for the first time in 1981 [88] and it is often referred to as *Bragg soliton*, or *grating soliton*. It is also called *gap soliton*, because it exists only inside the band gap of the Bragg resonance.

Expanding $\beta(\omega)$ in Eq. (9.29) in a Taylor series, we have

$$\beta(\omega) = \beta_0 + \beta_1(\omega-\omega_0) + \frac{1}{2}\beta_2(\omega-\omega_0)^2 + \frac{1}{6}\beta_3(\omega-\omega_0)^3 + \cdots \qquad (9.37)$$

In most practical situations, the GVD dispersion can be neglected in FBGs. Keeping terms in Eq. (9.37) only to first order and replacing $(\omega-\omega_0)$ with the differential operator $i(\partial/\partial t)$, Eqs. (9.27) and (9.28) can be converted to time domain, assuming the form

$$i\frac{\partial A_f}{\partial z} + i\beta_1\frac{\partial A_f}{\partial t} + \delta A_f + \kappa_g A_b + \gamma(|A_f|^2 + 2|A_b|^2)A_f = 0 \qquad (9.38)$$

$$-i\frac{\partial A_b}{\partial z} + i\beta_1\frac{\partial A_b}{\partial t} + \delta A_b + \kappa_g A_f + \gamma(|A_b|^2 + 2|A_f|^2)A_b = 0 \qquad (9.39)$$

In Eqs. (9.38) and (9.39), $\beta_1 = 1/v_g$, where v_g is the group velocity and $\gamma = n_2\omega_0/(cA_{eff})$ is the nonlinear parameter.

Unlike the nonlinear Schrödinger equation, the system of equations (9.38) and (9.39) is not exactly integrable. Nevertheless, a family of exact soliton solutions of these equations was found following the pattern of the previously known exact solutions in the so-called massive Thirring model [89]. The Bragg soliton solution is given by [90,91]

$$A_f(z,t) = \sqrt{\frac{\kappa_g(1+v)}{\gamma(3-v^2)}}(1-v^2)^{1/4}W(X)\exp(i\psi) \qquad (9.40)$$

$$A_b(z,t) = -\sqrt{\frac{\kappa_g(1-v)}{\gamma(3-v^2)}}(1-v^2)^{1/4}W^*(X)\exp(i\psi) \qquad (9.41)$$

where

$$W(X) = (\sin\theta)\mathrm{sech}(X\sin\theta-i\theta/2), X = \frac{z-V_Gt}{\sqrt{1-v^2}}\kappa_g \qquad (9.42a)$$

$$\psi = vX\cos\theta - \frac{4v}{(3-v^2)}\arctan[|\cot(\theta/2)|\coth(\sin\theta)] \qquad (9.42b)$$

The above solution is a *two-parameter family* of Bragg solitons. The parameter $v = V_G/v_g$ is in the range $-1 < v < 1$ while the parameter θ can be chosen anywhere in the interval $]0,\pi[$. Bragg solitons correspond to specific combinations of counter-propagating waves. Depending on the relative amplitudes of the two waves, the soliton can move forward or backward at the reduced speed V_G. If the counter-propagating waves have equal amplitudes $V_G = 0$, which corresponds to a *stationary gap soliton*.

The first experimental observation of Bragg solitons occurred in 1996, using a 6 cm long fiber grating [93]. Since then, other experiments have been performed that allow a better understanding of these nonlinear pulses. For example, in a 1997 experiment, soliton formation at frequencies slightly above the gap frequency was reported using 80 ps pulses obtained from a Q-switched, mode-locked Nd:YLF laser [94]. In a latter experiment, Bragg solitons were also observed in a 7.5 cm long apodized fiber grating using the same type of input pulses [95]. Figure 9.7 shows the experimental results for three values of the detuning of input frequencies from the band gap [94]. The solid curve corresponds to the input pulse shape, which has a width of 80 ps and a peak intensity of 18 GW/cm². Figure 9.7a shows the transmitted intensity when the pulse was centered on 1053.00 nm, where the linear grating transmission is approximately 90%. As a result of the dispersive and nonlinear effects, the transmitted pulse was compressed to 40 ps, whereas its peak intensity was enhanced by approximately 40%. The compressed pulse also leaves the grating ~40 ps later than the pulse propagating without dispersion, which is due to the smaller group velocity at this frequency, corresponding to 81% of the speed of light in the uniform fiber. Figure 9.7b shows the transmitted intensity when the pulse was tuned closer to the photonic band gap, such that linear transmissivity is 80%. The transmitted pulse now has a width of approximately 25 ps and is delayed by ~45 ps, corresponding to an average velocity of 79% of the speed of light in the uniform medium. The nonlinear losses are responsible for the reduced intensity of the compressed pulse. Finally, Fig. 9.7c shows the transmitted intensity when frequency is further tuned closer to the band gap, where the linear transmissivity is 50%. The transmitted pulse now has a width of approximately 23 ps and is retarded by approximately 55 ps, corresponding to an average velocity of 76% of the speed of light in the uniform medium. These results clearly show the slow down and the compression of light as a result of the reduced group velocity and the soliton effect. However, gap solitons that form within the stop band of a fiber grating have not been observed because of a practical difficulty: the light whose wavelength falls inside the stop band of a Bragg grating is reflected.

It can be shown that, if $\kappa_g L_{\mathrm{LN}} \gg 1$, where $L_{\mathrm{NL}} = (\gamma P_0)^{-1}$ is the nonlinear length, the coupled-mode equations (9.38) and (9.39) reduce to the following effective nonlinear Schrödinger equation [92]:

$$i\frac{\partial U}{\partial z} - \frac{\beta_2^g}{2}\frac{\partial^2 U}{\partial T^2} + \gamma_g |U|^2 U = 0 \tag{9.43}$$

where U corresponds to the envelope associated with Bloch wave formed by a linear combination of A_f and A_b, $z = V_G t$, $T = t - z/V_G$, β_2^g is given by Eq. (9.36), and

$$\gamma_g = \frac{2 - v^2}{2v}\gamma \tag{9.44}$$

Given the above correspondence, fiber Bragg solitons can be used for pulse compression in a way similar to that described in Section 9.2 for higher order solitons in fibers without grating. The relations given there for the compression factor and for the

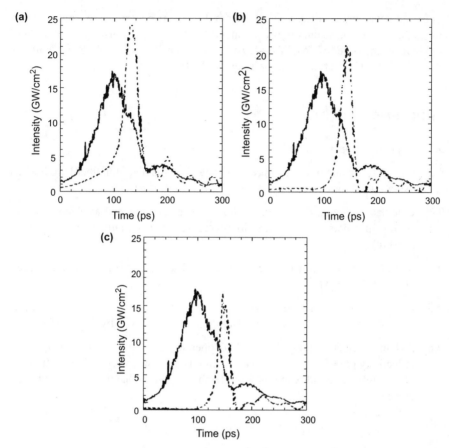

Figure 9.7 Transmitted intensities (dotted lines) and input intensities (solid lines) of light pulse in a fiber grating. Parts (a), (b), and (c) correspond to grating detunings of 15, 13.2, and 11.4 cm^{-1}, respectively. (After Ref. [94]; © 1997 OSA.)

optimum fiber length also apply in this case as long as the GVD parameter β_2 and the nonlinear parameter γ are replaced by the corresponding fiber grating parameters, given by Eqs. (9.36) and (9.44), respectively.

When a beam is launched at one end of the grating, both its intensity and wavelength with respect to the stop band will play an important role concerning the beam transmission along the FBG. At low input powers, if the beam wavelength is located near the Bragg wavelength, the transmissivity will be small. However, increasing the input power and above a given level, most of that power will be transmitted. In fact, due to the power dependence of the refractive index, the effective detuning δ in Eqs. (9.38) and (9.39) also changes. As a consequence, light whose wavelength was inside the stop band can be tuned out of the stop band and get transmitted through the FBG, a behavior known as *SPM-induced switching*. This phenomenon was observed in a 1998 experiment [96], using an 8 cm long grating with its Bragg wavelength centered near 1536 nm. In such an experiment, switching from 3% to 40% of the pulse energy was obtained for nanosecond pulses at internal field strengths of the order of 15 GW/cm^2. At higher intensities, multiple Bragg solitons were obtained with durations in the range 100–500 ps.

PROBLEMS

9.1   Explain the operation of a grating-fiber compressor.

9.2   Calculate the maximum compression factor and the optimum fiber length when using a sixth-order soliton with a soliton period of 8 km.

9.3   Calculate the compression factor for a 3 ps pulse launched as a fundamental soliton in a fiber whose dispersion decreases exponentially from 10 to 1 ps/ (km nm).

9.4   Show that the reflection coefficient of a fiber Bragg grating of length L_g is indeed given by Eq. (9.31).

9.5   Using the dispersion relation $\delta^2 = Q^2 + \kappa_g^2$, show that the second- and third-order dispersion parameters of a Bragg grating are given by Eq. (9.36).

9.6   Assume a fiber grating designed to operate at 1550 nm with a maximum reflectivity of 90% and a length of 15 mm. Determine the coupling coefficient, κ_g, the refractive index modulation depth, n_a, and the bandwidth, $\Delta\lambda$, of such grating.

REFERENCES

1. A. E. Siegman, *Lasers*, University Science Books, Mill Valley, CA, 1986.
2. H. A. Haus, *IEEE J. Sel. Top. Quantum Electron.* **6**, 1173 (2000).
3. J. D. Kafka, T. Baer, and D. W. Hall, *Opt. Lett.* **14**, 1269 (1989).
4. F. X. Kärtner, D. Kopf, and U. Keller, *J. Opt. Soc. Am. B* **12**, 486 (1995).
5. D. J. Richardson, R. I. Laming, D. N. Payne, V. Matsas, and M. W. Philips, *Electron. Lett.* **27**, 542 (1991).
6. I. N. Dulling III, *Opt. Lett.* **16**, 539 (1991).
7. M. Hofer, M. E. Ferman, F. Haberl, M. H. Ober, and A. J. Schmidt, *Opt. Lett.* **16**, 502 (1991).
8. V. J. Matsas, T. P. Newson, D. J. Richardson, and D. N. Payne, *Electron. Lett.* **28**, 1391 (1992).
9. H. Nakatsuka, D. Grishkowsky, and A. C. Balant, *Phys. Rev. Lett.* **47**, 910 (1981).
10. R. L. Fork, C. H. Brito Cruz, P. C. Becker, and C. V. Shank, *Opt. Lett.* **12**, 483 (1987).
11. L. F. Mollenauer, R. H. Stolen, and J. P. Gordon, *Phys. Rev. Lett.* **45**, 1095 (1980).

12. K. A. Ahmed, K. C. Chan, and H. F. Liu, *IEEE J. Sel. Top. Quantum Electron.* **1**, 592 (1995).

13. K. C. Chan and H. F. Liu, *IEEE J. Quantum Electron.* **31**, 2226 (1995).

14. S. V. Chernikov and P. V. Mamyshev, *J. Opt. Soc. Am. B* **8**, 1633 (1991).

15. M. D. Pelusi and H. F. Liu, *IEEE J. Quantum Electron.* **33**, 1430 (1997).

16. K. T. Chan and W.-H. Cao, *Opt. Commun.* **184**, 463 (2000).

17. N. J. Doran and D. Wood, *Opt. Lett.* **13**, 56 (1988).

18. M. E. Fermann, F. Haberl, M. Hofer, and H. Hochreiter, *Opt. Lett.* **15**, 752 (1990).

19. R. H. Stolen, J. Botineau, and A. Ashkin, *Opt. Lett.* **7**, 512 (1982).

20. L. F. Mollenauer and R. H. Stolen, *Opt. Lett.* **9**, 13, (1984).

21. F. M. Mitschke and L. F. Mollenauer, *IEEE J. Quantum Electron.* **22**, 2242 (1986).

22. H. A. Haus and M. N. Islam, *IEEE J. Quantum Electron.* **21**, 1172 (1985).

23. K. J. Blow and D. Wood, *J. Opt. Soc. Am. B* **5**, 629 (1988).

24. K. J. Blow and B. P. Nelson, *Opt. Lett.* **13**, 1026 (1988).

25. E. P. Ippen, H. A. Haus, and L. Y. Liu, *J. Opt. Soc. Am. B* **8**, 2077 (1991).

26. K. Smith, J. R. Armitage, R. Wyatt, and N. J. Doran, *Electron. Lett.* **28**, 1149 (1990).

27. D. J. Richardson, R. I. Laming, D. N. Payne, M. W. Phillips, and V. J. Matsas, *Electron. Lett.* **27**, 730 (1991).

28. N. J. Doran, D. S. Forrester, and B. K. Nayar, *Electron. Lett.* **25**, 267 (1989).

29. K. J. Blow, N. J. Doran, and B. K. Nayar, *Opt. Lett.* **14**, 754 (1989).

30. E. Yamada and M. Nakazawa, *IEEE J. Quantum Electron.* **30**, 1842 (1994).

31. D. J. Richardson, R. I. Laming, D. N. Payne, V. J. Matsas, and M. W. Phillips, *Electron. Lett.* **27**, 1451 (1991).

32. K. Tamura, E. P. Ippen, H. A. Haus, and L. E. Nelson, *Opt. Lett.* **18**, 1080, (1993).

33. H. A. Haus, K. Tamura, L. E. Nelson, and E. P. Ippen, *IEEE J. Quantum Electron.* **31**, 591 (1995).

34. K. Tamura, E. P. Ippen, H. A. Haus, and L. E. Nelson, *Opt. Lett.* **18**, 1080 (1993).

35. T. R. Clark, T. F. Carruthers, P. J. Matthews, and I. N. Duling III, *Electron. Lett.* **35**, 720 (1999).

36. J. P. Gordon, *Opt. Lett.* **8**, 596 (1983).

37. J. P. Gordon, *Opt. Lett.* **11**, 662 (1986).

38. J. P. Gordon and H. A. Haus, *Opt. Lett.* **11**, 665 (1986).

39. J. P. Gordon, *J. Opt. Soc. Am. B* **9**, 91 (1992).

40. L. F. Mollenauer, J. P. Gordon, and S. G. Evangelides, *Opt. Lett.* **17**, 1575 (1992).

41. L. F. Mollenauer, E. Lichtman, M. J. Neubelt, and G. T. Harvey, *Electron. Lett.* **29**, 910 (1993).

42. F. Fontana, L. Bossalini, P. Franco, M. Midrio, M. Romagnoli, and S. Wabnitz, *Electron. Lett.* **30**, 321 (1994).

43. M. Romagnoli, S. Wabnitz, P. Franco, M. Midrio, F. Fontana, and G. E. Town, *J. Opt. Soc. Am. B* **12**, 72 (1995).

44. G. E. Town, K. Sugden, J. A. R. Williams, I. Bennion, and S. B. Poole, *IEEE Photon. Technol. Lett.* **7**, 78 (1995).

45. G. E. Town, J. Chow, A. Robertson, and M. Romagnoli, *CLEO-Europe'96*, Hamburg, September 1996, Postdeadline Paper CPD2.10.

46. P. A. Belanger, *J. Opt. Soc. Am. B* **8**, 2077 (1991).

47. J. D. Moores, *Opt. Commun.* **96**, 65 (1993).

48. H. A. Haus, K. Timura, L. E. Nelson, and E. P. Ippen, *IEEE J. Quantum Electron.* **31**, 591 (1995).

49. N. N. Akhmediev and A. Ankiewicz, *Solitons: Nonlinear Pulses and Beams*, Chapman & Hall, 1997.

50. M. F. Ferreira, M. V. Facão, and S. V. Latas, *Fiber Integr. Opt.* **19**, 31 (2000).

51. N. N. Akhmediev, J. M. Soto-Crespo, and G. Town, *Phys. Rev. E* **63**, 056602 (2001).

52. A. M. Johnson, R. H. Stolen, and W. M. Simpson, *Appl. Phys. Lett.* **44**, 729 (1984).

53. W. J. Tomlinson, R. H. Stolen, and C. V. Shank, *J. Opt. Soc. Am. B* **1**, 139 (1984).

54. K. J. Blow, N. J. Doran, and B. P. Nelson, *Opt. Lett.* **10**, 393 (1985).

55. B. Strickland and G. Mourou, *Opt. Commun.* **55**, 447 (1985).

56. K. Tai and A. Tomita, *Appl. Phys. Lett.* **48**, 309 (1986).

57. B. Valk, Z. Vilhelmsson, and M. M. Salour, *Appl. Phys. Lett.* **50**, 656 (1987).

58. E. M. Dianov, A. Y. Karasik, P. V. Mamyshev, A. M. Prokhorov, and D. G. Fursa, *Sov. J. Quantum Electron.* **17**, 415 (1987).

59. A. M. Weiner, J. P. Heritage, and R. H. Stolen, *J. Opt. Soc. Am. B* **5**, 364 (1988).

60. M. Kuckartz, R. Schulz, and H. Harde, *J. Opt. Soc. Am. B* **5**, 1353 (1988).

61. A. Baltuska, Z. Wei, M. S. Pshenichnikov, and D. A. Wiersma, *Opt. Lett.* **22**, 102 (1997).

62. L. Xu, N. Narasawa, N. Nakagawa, R. Morita, H. Shigekawa, and M. Yamashita, *Opt. Commun.* **162**, 256 (1999).

63. L. F. Mollenauer, R. H. Stolen, J. P. Gordon, and W. J. Tomlinson, *Opt. Lett.* **8**, 289 (1983).

64. E. M. Dianov, Z. S. Nikonova, A. M. Prokhorov, and V. N. Serkin, *Sov. Tech. Phys. Lett.* **12**, 311 (1986).

65. A. S. Gouveia-Neto, A. S. L. Gomes, and J. R. Taylor, *J. Mod. Opt.* **35**, 7 (1988).

66. R. H. Stolen, J. Botineau, and A. Ashkin, *Opt. Lett.* **7**, 512 (1982).

67. B. Nikolaus, D. Grischkowsky, and A. C. Balant, *Opt. Lett.* **8**, 189 (1983).

68. K. R. Tamura and M. Nakazawa, *IEEE Photon. Technol. Lett.* **11**, 230 (1999).

69. M. D. Pelusi, Y. Matsui, and A. Suzuki, *IEEE J. Quantum Electron.* **35**, 867 (1999).

70. K. C. Chan and H. F. Liu, *Opt. Lett.* **19**, 49 (1994).

71. G. P. Agrawal, *Opt. Lett.* **15**, 224 (1990).

72. P. K. A. Wai and W. H. Cao, *J. Opt. Soc. Am. B* **20**, 1346 (2003).

73. Z. Shumin, L. Fuyun, X. Wencheng, Y. Shiping, W. Jian, and D. Xiaoyi, *Opt. Commun.* **237**, 1 (2004).

74. M. Nakazawa, K. Kurokawa, H. Kubota, and E. Yamada, *Phys. Rev. Lett.* **65**, 1881 (1990).

75. R. F. Nabiev, I. V. Melnikov, and A. V. Nazarkin, *Opt. Lett.* **15**, 1348 (1990).

76. M. L. Quiroga-Teixeiro, D. Anderson, P. A. Andrekson, A. Berntson, and M. Lisak, *J. Opt. Soc. Am. B* **13**, 687 (1996).

77. A. Mostofi, H. H. Hanza, and P. L. Chu, *IEEE J. Quantum Electron.* **33**, 620 (1997).

78. S. V. Chernikov, E. M. Dianov, D. J. Richardson, and D. N. Payne, *Opt. Lett.* **18**, 476 (1993).

79. M. F. Ferreira and S. V. Latas, *Opt. Eng.* **41**, 1696 (2002).

80. J. L. Archambault, L. Reekie, and P. S. J. Russel, *Electron. Lett.* **29**, 453 (1993).

81. A. Othonos and K. Kalli, *Fiber Bragg Gratings*, Artech House, Boston, MA, 1999.

82. H. A. Haus, *Waves and Fields in Optoelectronics*, Prentice-Hall, Englewood Cliffs, NJ, 1984.

83. A. K. Ghatak and K. Thyagarajan, *Optical Electronics*, Cambridge University Press, Cambridge, 1989.

84. G. P. Agrawal, *Fiber-Optic Communication Systems*, 2nd ed., Wiley, New York, 1997.

85. N. M. Litchinitser, B. J. Eggleton, and D. B. Patterson, *J. Lightwave Technol.* **15**, 1303 (1977).

86. J. A. R. William, I. Bennion, K. Sugden, and N, Doran, *Electron. Lett.* **30**, 985 (1994).

87. R. Kashyap, S. V. Chernikov, P. F. McKee, and J. R. Taylor, *Electron. Lett.* **30**, 1078 (1994).

88. Yu. I. Voloshchenko, Yu. N. Ryzhov, and V. E. Sotin, *Zh. Tekh. Fiz.* **51**, 902 (1981) (*Sov. Phys. Tech. Phys.* **26**, 541 (1981)).

89. W. E. Thirring, *Ann. Phys. (NY)* **3**, 91 (1958).

90. A. B. Aceves and S. Wabnitz, *Phys. Lett. A* **141**, 37 (1989).

91. D. N. Christodoulides and R. I. Joseph, *Phys. Rev. Lett.* **62**, 1746 (1989).

92. C. M. de Sterke, *J. Opt. Soc. Am. B* **15**, 2660 (1998).

93. B. J. Eggleton, R. R. Slusher, C. M. de Sterke, P. A. Krug, and J. E. Sipe, *Phys. Rev. Lett.* **76**, 1627 (1996).

94. B. J. Eggleton, C. M. de Sterke, and R. E. Slusher, *J. Opt. Soc. Am. B* **14**, 2980 (1997).

95. B. J. Eggleton, C. M. de Sterke, and R. E. Slusher, *J. Opt. Soc. Am. B* **16**, 587 (1999).

96. D. Traverner, N. G. R. Broderick, D. J. Richardson, R. I. Laming, and M. Isben, *Opt. Lett.* **23**, 328 (1998).

10

POLARIZATION EFFECTS

Silica fibers used in long-distance communications are usually called single-mode fibers, but they indeed support two orthogonally polarized modes with the same spatial distribution. The two modes are degenerate in an ideal fiber (maintaining perfect cylindrical symmetry along its entire length) in the sense that their effective refractive indices, n_x and n_y, are identical. In practice, all fibers exhibit some modal birefringence ($n_x \neq n_y$) of unintentional ellipticity in the core geometry and unintentional nonsymmetric stress along the fiber length. Moreover, the magnitude and direction of the asymmetry may change randomly along the fiber, which causes random coupling between the polarization modes. The small random birefringence along the optical fiber is the origin of the polarization mode dispersion (PMD) phenomenon [1–6], which leads to broadening of optical pulses, as seen in Section 3.6. This broadening is in addition to group velocity dispersion (GVD)-induced pulse broadening and cannot be eliminated by the use of dispersion management. For this reason, PMD has become a major source of concern for high-speed long-haul transmission systems, especially for systems with a bit rate of 40 Gb/s per channel and above [7–12].

Different types of birefringent fibers have been proposed and fabricated in order to avoid polarization fluctuation problems. The difference of the propagation constants between the two polarization modes in these fibers is intentionally made large, allowing the propagation of light without changing its state of polarization. Such fibers are called *polarization maintaining fibers* [4,13–17] and their use requires identification of the slow and fast axes before an optical signal can be launched into them.

In a linear system, if the input pulse excites both polarization components it becomes broader at the output of a birefringent fiber because the two components disperse along the fiber as a result of their different group velocities. In soliton-based systems, pulse propagation is described by the Manakov equation [18] and PMD also tends to separate the two principal polarization components. However, the soliton counteracts this separation through the effective potential well produced by the local index change due to the Kerr effect. It is to be expected, therefore, that solitons could withstand PMD, if it is not excessive [19,20].

Nonlinear Effects in Optical Fibers. By Mário F. S. Ferreira.
Copyright © 2011 John Wiley & Sons, Inc. Published 2011 by John Wiley & Sons, Inc.

As seen in Chapter 8, the dispersion management technique has been attracting much interest as an effective way of reducing the timing jitter, which is the most undesirable effect in soliton transmission systems. Since a dispersion-managed soliton (DMS) has an amplitude that is larger than that of a conventional soliton for the same value of the average dispersion, it is expected to have a larger nonlinear trapping effect and to resist better to PMD [21,22].

In this chapter we first introduce the coupled nonlinear Schrödinger equations (CNLSEs) that govern the pulse propagation in linearly birefringent fibers and analyze the nonlinear phase shift between the two polarization components of the field. Afterward, we analyze the propagation of conventional solitons both in fibers with constant birefringence and in fibers with randomly varying birefringence. The last part of this chapter is dedicated to the analysis of PMD-induced pulse broadening, considering both conventional and dispersion-managed solitons.

10.1 COUPLED NONLINEAR SCHRÖDINGER EQUATIONS

As seen in Chapter 4, the lowest order nonlinear effects in optical fibers originate from the third-order susceptibility $\chi^{(3)}$, which is also responsible for the formation of optical solitons. When the nonlinear response of the medium is assumed to be instantaneous, the third-order nonlinear polarization can be written as

$$\mathbf{P}_{NL}(\mathbf{r}, t) = \varepsilon_0 \chi^{(3)} \vdots \mathbf{E}(\mathbf{r}, t)\mathbf{E}(\mathbf{r}, t)\mathbf{E}(\mathbf{r}, t) \tag{10.1}$$

Let us consider the case where the electric field vector has both x and y components:

$$\mathbf{E}(\mathbf{r}, t) = \frac{1}{2}(\hat{i}\bar{E}_x + \hat{j}\bar{E}_y)\exp(-i\omega_0 t) + \text{c.c.} \tag{10.2}$$

where c.c. means the complex conjugate of the previous expression. By substituting (10.2) in (10.1), we obtain the result

$$\mathbf{P}_{NL}(\mathbf{r}, t) = \frac{1}{2}(\hat{i}\bar{P}_{NLx} + \hat{j}\bar{P}_{NLy})\exp(-i\omega_0 t) + \text{c.c.} \tag{10.3}$$

where

$$\bar{P}_{NLx} = \frac{3}{4}\varepsilon_0\chi^{(3)}_{xxxx}\left[\left(|\bar{E}_x|^2 + \frac{2}{3}|\bar{E}_y|^2\right)\bar{E}_x + \frac{1}{3}\bar{E}_x^*\bar{E}_y^2\right] \tag{10.4}$$

$$\bar{P}_{NLy} = \frac{3}{4}\varepsilon_0\chi^{(3)}_{xxxx}\left[\left(|\bar{E}_y|^2 + \frac{2}{3}|\bar{E}_x|^2\right)\bar{E}_y + \frac{1}{3}\bar{E}_y^*\bar{E}_x^2\right] \tag{10.5}$$

In obtaining (10.4) and (10.5), we have used the relation

$$\chi^{(3)}_{xxxx} = \chi^{(3)}_{xxyy} + \chi^{(3)}_{xyxy} + \chi^{(3)}_{xyyx} \tag{10.6}$$

which is valid for an isotropic medium with rotational symmetry [23,24]. Moreover, the three components in Eq. (10.6) were considered to have the same magnitude, as happens in the case of silica fibers.

Using the above expressions for the nonlinear polarization and adopting the procedure of Section 4.4, the slowly varying amplitudes U_x and U_y of the polarization components are found to satisfy the following coupled differential equations [25–28]:

$$\frac{\partial U_x}{\partial z} + \beta_{1x}\frac{\partial U_x}{\partial t} + \frac{i}{2}\beta_{2x}\frac{\partial^2 U_x}{\partial t^2} = i\gamma\left(|U_x|^2 + \frac{2}{3}|U_y|^2\right)U_x$$
$$+ \frac{i\gamma}{3}U_x^*U_y^2\exp\left[-2i(\beta_{0x}-\beta_{0y})z\right] \tag{10.7}$$

$$\frac{\partial U_y}{\partial z} + \beta_{1y}\frac{\partial U_y}{\partial t} + \frac{i}{2}\beta_{2y}\frac{\partial^2 U_y}{\partial t^2} = i\gamma\left(|U_y|^2 + \frac{2}{3}|U_x|^2\right)U_y$$
$$+ \frac{i\gamma}{3}U_y^*U_x^2\exp\left[2i(\beta_{0x}-\beta_{0y})z\right] \tag{10.8}$$

where the amplitudes U_j $(j = x, y)$ are such that their absolute square represents the optical power, γ is the nonlinear coefficient, β_{0j} $(j = x, y)$ represent the propagation constants of the two orthogonal linearly polarized waves, $\beta_{1j} = d\beta_j/d\omega|_{\omega=\omega_0}$, and $\beta_{2j} = d^2\beta_j/d\omega^2|_{\omega=\omega_0}$. Equations (10.7) and (10.8) govern the pulse propagation in linearly birefringent fibers. The first term on the right-hand side of these equations is responsible for self-phase modulation (SPM), whereas the second term is responsible for cross-phase modulation (XPM). This term corresponds to a nonlinear coupling between the two polarization components, through which the nonlinear phase shift acquired by one component depends on the intensity of the other component. Finally, the last term in Eqs. (10.7) and (10.8) results from the phenomenon of four-wave mixing. As seen in Chapter 6, the efficiency of this process depends on the matching of the phases of the involved fields. If the fiber length is much longer than the beat length $L_b = 2\pi/(\beta_{0x}-\beta_{0y})$, these terms often change sign and can be neglected. This is particularly the case in highly birefringent fibers, for which $L_b \sim 1\,\text{cm}$ typically.

10.2 NONLINEAR PHASE SHIFT

Besides neglecting the last terms on the right-hand side of Eqs. (10.7) and (10.8), we will also neglect in this section the terms with time derivatives in these equations. This simplification is acceptable in the case of continuous-wave (CW) radiation or if sufficiently long pulses are used, such that the fiber length L is much shorter than the dispersion length $L_D = t_0^2/|\beta_2|$. In such cases, Eqs. (10.7) and (10.8) reduce to

$$\frac{\partial U_x}{\partial z} = i\gamma\left(|U_x|^2 + \frac{2}{3}|U_y|^2\right)U_x \tag{10.9}$$

$$\frac{\partial U_y}{\partial z} = i\gamma\left(|U_y|^2 + \frac{2}{3}|U_x|^2\right)U_y \qquad (10.10)$$

The above equations can be solved using

$$U_j = \sqrt{P_j}\, e^{i\phi_j}, \quad j = x, y \qquad (10.11)$$

where P_j is the power and ϕ_j is the phase of polarization component j. It can be seen that P_j remains constant, whereas ϕ_j evolves with distance according to the equations

$$\frac{d\phi_x}{dz} = \gamma\left(P_x + \frac{2}{3}P_y\right), \qquad \frac{d\phi_y}{dz} = \gamma\left(P_y + \frac{2}{3}P_x\right) \qquad (10.12)$$

The solutions of these equations after propagation along a distance $z = L$ are given by

$$\phi_x = \gamma\left(P_x + \frac{2}{3}P_y\right)L, \qquad \phi_y = \gamma\left(P_y + \frac{2}{3}P_x\right)L \qquad (10.13)$$

Equation (10.13) shows that the nonlinear phase shift affecting each polarization component is determined by both the SPM and XPM effects. The phase difference between the two polarization components is given by

$$\Delta\phi_{\mathrm{NL}} \equiv \phi_x - \phi_y = \frac{\gamma}{3}(P_x - P_y)L \qquad (10.14)$$

If the light with power P_0 is linearly polarized at an angle θ with respect to the x-axis, we have $P_x = P_0\cos^2\theta$, $P_y = P_0\sin^2\theta$, and the relative phase difference becomes

$$\Delta\phi_{\mathrm{NL}} = \frac{1}{3}\phi_{\max}\cos(2\theta) \qquad (10.15)$$

where

$$\phi_{\max} = \gamma P_0 L \qquad (10.16)$$

is the maximum phase shift induced by SPM (see Eq. (5.4)). If the losses are taken into account, the length L in Eq. (10.16) must be replaced by the effective length, L_{eff}, defined in Section 4.3.

The θ-dependence of the nonlinear phase shift given by Eq. (10.15) can be used in some important applications. An example is given by the optical Kerr shutter, in which an intense, high-power, pump wave is used to control the transmission of a weak probe signal through a nonlinear medium [26]. Both the pump and the probe waves are linearly polarized at the fiber input and their polarizations make an angle of 45° with respect to each other. A polarizer is situated at the fiber output and adjusted perpendicular to the probe input polarization. The pump-induced birefringence causes a slight difference between the refractive indices seen by the probe components parallel and orthogonal to the pump polarization direction. Along the slower axis, which coincides with the pump polarization direction, the refractive index is higher and, therefore, the phase velocity of the wave is lower. Along the faster axis, the behavior is reversed. As a consequence, the polarization state of the probe will be turned and the probe wave will be partially transmitted through the polarizer. Since this transmission depends on the intensity of the pump wave, the probe signal can be modulated. This device is also known as a Kerr modulator.

Another application of the θ-dependence of the nonlinear phase shift $\Delta\phi_{NL}$ consists in the pulse shaping, in which the signal pulse itself is used to modify its own shape. This effect is possible because the transmission through a combination of fiber and polarizer, similar to that of the Kerr shutter described above, depends on the pulse intensity. Such a device can be designed such that it allows the transmission of the central intense part of the pulse, whereas the low-intensity tails are suppressed. This property is especially useful to remove the low-intensity pedestal present in some compressed pulses, as seen in Chapter 9 [29,30]. Due to the relative phase shift $\Delta\phi_{NL}$ introduced between the two polarization components of the pulse, some power is transmitted through the cross polarizer when $\theta \neq 0$. Since this polarizer makes an angle $\theta + \pi/2$ from the x-axis, the transmitted amplitude is given by

$$U_t = \sqrt{P_0}\,\sin\theta\,\cos\theta[1-\exp(i\,\Delta\phi_{NL})] \qquad (10.17)$$

Equations (10.15)–(10.17) show that, for a given angle θ, the transmittivity $T_p(\theta) = |U_t|^2/P_0$ depends on the power P_0. Pulse narrowing can be achieved if the angle θ is chosen such that it maximizes the transmission at the pulse peak and suppresses its low-intensity tails. Figure 10.1 shows the transmittivity $T_p(\theta)$ as a function of the input polarization angle θ for three different peak powers corresponding to $\phi_{max} = 10$, 20, and 30. For $\phi_{max} = 30$, the transmittivity can approach 90% at $\theta = 36.2°$.

10.3 SOLITONS IN FIBERS WITH CONSTANT BIREFRINGENCE

Since the two orthogonally polarized waves have the same frequency, we will consider in the following Eqs. (10.7) and (10.8) with $\beta_{2x} = \beta_{2y} = \beta_2$. Moreover, we use normalized time and distance given by

$$T = \frac{1}{t_0}\left(t - \frac{z}{v_g}\right) \qquad (10.18)$$

$$Z = \frac{z}{L_D} \qquad (10.19)$$

where t_0 is the pulse width, $v_g = 2/(\beta_{1x} + \beta_{1y})$ is the average group velocity of the two polarization modes, and $L_D = t_0^2/|\beta_2|$ is the dispersion distance. Introducing also the normalized field amplitudes

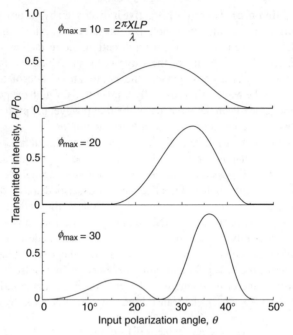

Figure 10.1 Transmittivity $T_p(\theta)$ as a function of θ for $\phi_{max} = 10, 20$, and 30. (After Ref. [29]; © 1982 OSA.)

$$q_j = \sqrt{\gamma L_D} U_j, \quad j = x, y \tag{10.20}$$

we obtain the following normalized coupled nonlinear Schrödinger equations:

$$\frac{\partial q_x}{\partial Z} + \delta_g \frac{\partial q_x}{\partial T} - \frac{i}{2} d(Z) \frac{\partial^2 q_x}{\partial T^2} = i(|q_x|^2 + \hat{\varepsilon}|q_y|^2) q_x + i(1-\hat{\varepsilon}) q_x^* q_y^2 \exp(-2i\Delta\beta Z) \tag{10.21}$$

$$\frac{\partial q_y}{\partial Z} - \delta_g \frac{\partial q_y}{\partial T} - \frac{i}{2} d(Z) \frac{\partial^2 q_y}{\partial T^2} = i(|q_y|^2 + \hat{\varepsilon}|q_x|^2) q_y + i(1-\hat{\varepsilon}) q_y^* q_x^2 \exp(2i\Delta\beta Z) \tag{10.22}$$

where $d(Z)$ is the normalized dispersion, $\hat{\varepsilon} = 2/3$, and

$$\Delta\beta = L_D(\beta_{0x} - \beta_{0y}) \tag{10.23}$$

is the differential wave number of the orthogonal components, whereas

$$\delta_g = \frac{L_D}{2t_0}(\beta_{1x} - \beta_{1y}) \tag{10.24}$$

represents the differential group velocity.

As referred above, the last term on the right-hand side of Eqs. (10.21) and (10.22) can be neglected if the fiber length is much longer than the beat length L_b. In such a case, the coupled nonlinear Schrödinger equations can be restated as a variational problem [31] in terms of the Lagrangian density L_d:

$$\delta \iint L_d \, dT \, dZ = 0 \tag{10.25}$$

where

$$L_d = L_x + L_y - L_{xy} \tag{10.26}$$

$$L_x = \frac{i}{2}\left(q_x \frac{\partial q_x^*}{\partial Z} - q_x^* \frac{\partial q_x}{\partial Z} \right) + \frac{i}{2}\delta_g\left(q_x \frac{\partial q_x^*}{\partial T} - q_x^* \frac{\partial q_x}{\partial T} \right) - \frac{1}{2}|q_x|^4 + \frac{1}{2}\left|\frac{\partial q_x}{\partial T}\right|^2 \tag{10.27}$$

$$L_y = L_x \quad (x \to y) \tag{10.28}$$

$$L_{xy} = \hat{\varepsilon}|q_x|^2|q_y|^2 \tag{10.29}$$

When the initial conditions for q_x and q_y are given by

$$q_x(0, T) = q_y(0, T) = \frac{A}{\sqrt{2}}\operatorname{sech}(T) \tag{10.30}$$

the asymptotic soliton solution of Eqs. (10.21) and (10.22) with $d(Z) = 1, \delta_g = 0$, and rapidly oscillating terms neglected becomes

$$q_x = q_y = \frac{1}{\sqrt{1+\hat{\varepsilon}}}\eta \operatorname{sech}(\eta T) \exp\left(\frac{i}{2}\eta^2 Z \right) \tag{10.31}$$

with

$$\eta = A\sqrt{2(1+\hat{\varepsilon})} - 1 \tag{10.32}$$

We treat the birefringence as a perturbation and find approximate solutions of Eqs. (10.21) and (10.22) in the form

$$\begin{bmatrix} q_x \\ q_y \end{bmatrix} = \frac{\eta_j}{\sqrt{1+\hat{\varepsilon}}}\operatorname{sech}\left[\eta_j(T-T_{0j})\right]\exp\left[-i\kappa_j(T-T_{0j}) + i\theta_j\right] \tag{10.33}$$

where $j = x, y$. Evolution of the parameters η_j, κ_j, T_{0j}, and θ_j may be found from the Euler equation

$$\frac{\partial L}{\partial v} = \frac{d}{dZ}\left[\frac{\partial L}{\partial v_Z} \right] \tag{10.34}$$

where v stands for the above-mentioned soliton parameters, $v_Z = \partial v/\partial Z$, and L is a Lagrangian, obtained from the Lagrangian density L_d as

$$L = \int\limits_{-\infty}^{+\infty} L_d \, dT \qquad (10.35)$$

Substituting Eq. (10.33) into Eqs. (10.26)–(10.29), with Eqs. (10.34) and (10.35), the following equations for the soliton parameters are obtained [31]:

$$\frac{d\eta_j}{dZ} = 0 \qquad (10.36)$$

$$\frac{dT_{0j}}{dZ} = (-1)^{j+1}\delta_g - \kappa_j \qquad (10.37)$$

$$\frac{d\kappa_j}{dZ} = -\frac{1}{2\eta_j}\frac{\partial\langle L_{xy}\rangle}{\partial T_{0j}} \qquad (10.38)$$

$$\frac{d\theta_j}{dZ} = \frac{1}{2}\kappa_j^2 + \frac{1}{2}\left(\frac{1-\hat{\varepsilon}}{1+\hat{\varepsilon}}\right)\eta_j^2 + \frac{1}{2}(1+\hat{\varepsilon})\frac{\partial\langle L_{xy}\rangle}{\partial\eta_j} \qquad (10.39)$$

where

$$\langle L_{xy}\rangle = \frac{\hat{\varepsilon}\eta_x^2\eta_y^2}{(1+\hat{\varepsilon})}\int\limits_{-\infty}^{+\infty} \text{sech}^2[\eta_x(T-T_{0x})]\text{sech}^2[\eta_y(T-T_{0y})]dT \qquad (10.40)$$

Since the birefringence parameter appears in Eq. (10.37) in a symmetric way, we may consider the case corresponding to the perturbation-induced dynamics of a symmetric solution, assuming

$$\eta_x = \eta_y = \eta, \qquad \kappa_x = -\kappa_y = \Delta\kappa/2, \qquad T_{0x} = -T_{0y} = \Delta T/2 \qquad (10.41)$$

Then, Eqs. (10.37) and (10.38) are reduced to

$$\frac{d\Delta T}{dZ} = 2\delta_g - \Delta\kappa \qquad (10.42)$$

$$\frac{d\Delta\kappa}{dZ} = \frac{d}{d(\Delta T)}U(\Delta T) \qquad (10.43)$$

where

$$U(\Delta T) = \frac{8\hat{\varepsilon}\eta^2}{1+\hat{\varepsilon}}\frac{\sinh(\eta\Delta T)-\eta\Delta T\cosh(\eta\Delta T)}{\sinh^3(\eta\Delta T)} \qquad (10.44)$$

From Eqs. (10.42) and (10.43), it follows that

$$\frac{1}{2}\left(\frac{d\Delta T}{dZ}\right)^2 + U(\Delta T) = \text{constant} \qquad (10.45)$$

This result indicates that ΔT behaves as the position of a particle moving in the potential $U(\Delta T)$. The particle will be trapped in the potential well if the initial kinetic energy is smaller than the potential barrier:

$$\frac{1}{2}\left(\frac{d\Delta T}{dZ}\bigg|_{Z=0}\right)^2 < U(\Delta T \to \infty) - U(\Delta T = 0) \qquad (10.46)$$

This condition determines a threshold value for the soliton amplitude for a capture of separate polarizations into a bound intermode state. Assuming that the initial frequency difference between these polarizations is zero, Eq. (10.46) becomes

$$\eta^2 > \frac{15}{8}\delta_g^2 \qquad (10.47)$$

Taking into account the relation (10.32), we obtain from Eq. (10.47) the threshold amplitude of the input pulse, A_{thr}, as a function of the birefringence, δ_g:

$$A_{thr} = \frac{1}{\sqrt{2(1+\hat{\varepsilon})}} + \frac{1}{2}\sqrt{\frac{3}{2\hat{\varepsilon}}}\delta_g \qquad (10.48)$$

When $\hat{\varepsilon} = 2/3$, Eq. (10.48) takes the approximate form

$$A_{thr} \approx 0.55 + 0.75\delta_g \qquad (10.49)$$

Figure 10.2 illustrates the temporal separation between the two polarization components, ΔT, against the propagation distance for the initial condition given by Eq. (10.30), considering two different values of A and $\delta_g = 0.5$. The solid curves correspond to the numerical solution of (10.42) and (10.43), whereas the dashed curves were obtained from Eqs. (10.21) and (10.22) with an initial soliton amplitude given by Eq. (10.32) with $\hat{\varepsilon} = 2/3$. A qualitative agreement is observed between the solid and the dashed curves in Fig. 10.2. In the case $A = 0.8$, which is smaller than the threshold value $A_{th} = 0.925$ given by Eq. (10.49), the separation between the two polarization components increases monotonously. However, for $A = 1.1$ this separation becomes limited and oscillates around zero during propagation, which confirms the mutual trapping of the two polarization components by the nonlinear birefringent fiber. In this case, the two polarization components propagate at a common group velocity due to the XPM-induced nonlinear coupling between them. In order to realize such temporal synchronization, the two components shift their carrier frequencies in the opposite directions, such that the one along the fast axis slows down while the other along the slow axis speeds up.

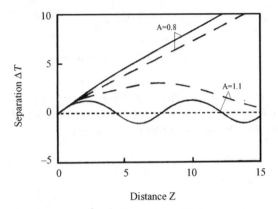

Figure 10.2 Pulse separation between the two polarization components, ΔT, against the propagation distance for the initial condition (10.30), considering two different values of A and $\delta_{\mathrm{g}} = 0.5$. The solid curves are the numerical solutions of (10.42) and (10.43), while the dashed curves were obtained from Eqs. (10.21) and (10.22) with $\hat{\varepsilon} = 2/3$.

10.4 SOLITONS IN FIBERS WITH RANDOMLY VARYING BIREFRINGENCE

The propagation of solitons in fibers with randomly varying birefringence can be described assuming that the fiber has linear birefringence of fixed strength with randomly varying direction of axes, as proposed by Wai and Menyuk [18]. In this model, the coupled nonlinear Schrödinger equations for the lossless fiber whose birefringence direction is rotated by α from the fixed coordinates x and y can be written in the form [18]

$$i\frac{\partial \mathbf{Q}}{\partial Z} + \frac{\Delta\beta}{2}\sum \mathbf{Q} + i\delta_{\mathrm{g}}\sum\frac{\partial \mathbf{Q}}{\partial T} + \frac{1}{2}\frac{\partial^2 \mathbf{Q}}{\partial T^2} + \frac{5}{6}|\mathbf{Q}|^2\mathbf{Q} + \frac{1}{6}(\mathbf{Q}^{\mathrm{t}}\sigma_3\mathbf{Q})\sigma_3\mathbf{Q} + \frac{1}{3}\mathbf{R} = 0$$

$$(10.50)$$

where $\mathbf{Q} = [q_x\ q_y]^{\mathrm{t}}$, $\mathbf{Q}^{\mathrm{t}} = [q_x^*\ q_y^*]$, $\mathbf{R} = [q_x^* q_y^2\ q_x^2 q_y^*]^{\mathrm{t}}$, and

$$\Sigma = \sigma_3 \cos(2\alpha) + \sigma_1 \sin(2\alpha)$$

$$(10.51)$$

is defined in terms of the Pauli's matrices

$$I = \begin{bmatrix} 1 & 0 \\ 0 & 1 \end{bmatrix}, \quad \sigma_1 = \begin{bmatrix} 0 & 1 \\ 1 & 0 \end{bmatrix}$$

$$\sigma_2 = \begin{bmatrix} 0 & -i \\ i & 0 \end{bmatrix}, \quad \sigma_3 = \begin{bmatrix} 1 & 0 \\ 0 & -1 \end{bmatrix}$$

$$(10.52)$$

The field $\mathbf{Q}$ is then transformed to the field $\Psi = \begin{bmatrix} U & V \end{bmatrix}^t$, obtained when one uses the polarization states of the linear and continuous waves propagating in the fiber as a base, according to the relations

$$
\mathbf{Q} = \begin{bmatrix} q_x \\ q_y \end{bmatrix} = \begin{bmatrix} \cos\alpha & -\sin\alpha \\ \sin\alpha & \cos\alpha \end{bmatrix} \begin{bmatrix} U' \\ V' \end{bmatrix} \quad \text{and} \quad \begin{bmatrix} U' \\ V' \end{bmatrix} = \begin{bmatrix} u_1 & u_2 \\ -u_2^* & u_1^* \end{bmatrix} \begin{bmatrix} U \\ V \end{bmatrix} \tag{10.53}
$$

with $|u_1|^2 + |u_2|^2 = 1$, where $\begin{bmatrix} U' & V' \end{bmatrix}^t$ is the electric field expressed in local axes of birefringence. The electric field Ψ satisfies the equation

$$
i\frac{\partial\Psi}{\partial Z} + i\delta_g\bar{\sigma}\frac{\partial\Psi}{\partial T} + \frac{1}{2}\frac{\partial^2\Psi}{\partial T^2} + \frac{5}{6}|\Psi|^2\Psi + \frac{1}{6}(\Psi^t\sigma_3\Psi)\sigma_3\Psi + \frac{1}{3}N = 0 \tag{10.54}
$$

where

$$
\bar{\sigma} = \begin{bmatrix} a_1 & a_4^* \\ a_4 & -a_1 \end{bmatrix} \tag{10.55}
$$

and $\mathbf{N} = [N_1 \quad N_2]^t$, with

$$
\mathbf{N}_1 = a_3^2(2|V|^2 - |U|^2)U - a_3 a_6^*(2|U|^2 - |V|^2)V - a_3 a_6 U^2 V^* - a_6^{*2}V^2 U^* \tag{10.56a}
$$

$$
\mathbf{N}_2 = a_3^2(2|U|^2 - |V|^2)V + a_3 a_6(2|V|^2 - |U|^2)U + a_3 a_6^* V^2 U^* - a_6^2 U^2 V^* \tag{10.56b}
$$

The coefficients a_i, $i = 1, \ldots, 6$, are defined in terms of u_1 and u_2 as

$$
\begin{aligned}
a_1 &= |u_1|^2 - |u_2|^2, & a_2 &= -(u_1 u_2 + u_1^* u_2^*) \\
a_3 &= i(u_1 u_2 - u_1^* u_2^*), & a_4 &= 2u_1 u_2^* \\
a_5 &= u_1^2 - u_2^{*2}, & a_6 &= -i(u_1^2 + u_2^{*2})
\end{aligned} \tag{10.57}
$$

Since $[u_1 \quad -u_2^*]^t$ is rapidly varying as it represents the evolution of the polarization state of a continuous wave in the fiber in response to birefringence, $\bar{\sigma}$ and $\mathbf{N}$ are also rapidly varying. The long-term average of these quantities is given by

$$
\langle\bar{\sigma}\rangle = 0 \tag{10.58}
$$

$$
\langle\mathbf{N}\rangle = \begin{bmatrix} -\dfrac{1}{3}(|U|^2 - 2|V|^2)U \\[2mm] \dfrac{1}{3}(2|U|^2 - |V|^2)V \end{bmatrix} \tag{10.59}
$$

Moving the rapid variation about the long-term average to the right-hand side of Eq. (10.54), we obtain the so-called Manakov-PMD equation [18,32]

$$i\frac{\partial \Psi}{\partial Z} + \frac{1}{2}\frac{\partial^2 \Psi}{\partial T^2} + \frac{8}{9}|\Psi|^2 \Psi = -i\delta_{\mathrm{g}}\bar{\sigma}\frac{\partial \Psi}{\partial T} - \frac{1}{3}(\mathbf{N}-\langle\mathbf{N}\rangle) \tag{10.60}$$

Regarding the left-hand side of (10.60), the second term includes the effect of chromatic dispersion, while the third term includes the effect of Kerr nonlinearity averaged over the Poincaré sphere with the well-known 8/9 factor [27,33,34]. Concerning the right-hand side of Eq. (10.60), the first term leads to the linear PMD, while the second term corresponds to the nonlinear PMD, which is usually negligible [32]. When the evolution of the polarization state is sufficiently rapid, the terms on the right-hand side of Eq. (10.60) can be neglected and the Manakov equation is obtained [35]. Since this equation has soliton solutions, it is expected that solitons may survive even in the presence of PMD.

10.5   PMD-INDUCED SOLITON PULSE BROADENING

In the following, the fiber will be modeled as a cascade of many short segments having randomly varying birefringence. Such random birefringence scatters a part of the energy of the solitons to dispersive radiation in both polarization states parallel and orthogonal to that of the soliton. Because of the generation of these dispersive radiations, the degree of polarization of the soliton is degraded and its energy decays during propagation over long distances.

Let us consider one of the fiber segments constituting the transmission link. We can denote the orthogonal principal states of polarization (PSPs) of this piece of fiber by the unit vectors $\hat{i}$ and $\hat{j}$. At the entrance of the segment, the electric field of a Manakov soliton can be expressed as

$$\mathbf{q}(Z,T) = (r\hat{i}+s\hat{j})\sqrt{\frac{9}{8}}\eta\,\mathrm{sech}(\eta T) \tag{10.61}$$

where $|r|^2 + |s|^2 = 1$ and the factor $\sqrt{9/8}$ accounts for the energy enhancement required for a soliton be formed in the randomly birefringent fiber. At the output of the fiber segment of normalized length $Z_{\mathrm{f}} = z_{\mathrm{f}}/L_{\mathrm{D}}$, the soliton has the form

$$\mathbf{q}(Z+Z_{\mathrm{f}},T) = \sqrt{\frac{9}{8}}e^{i\theta}\eta\{r\,e^{i\phi}\hat{i}\,\mathrm{sech}[\eta(T-\Delta T/2)] + s\,e^{-i\phi}\hat{j}\,\mathrm{sech}[\eta(T+\Delta T/2)]\} \tag{10.62}$$

where ϕ arises from the wave number birefringence and ΔT is the differential group delay between the two polarization components of the soliton after traversing Z_{f}. Assuming that Δ is much smaller than the pulse width, we can expand (10.62) and rewrite $\mathbf{q}(Z+Z_{\mathrm{f}},T)$ in the form [19]

$$\mathbf{q}(Z+Z_f,T)$$

$$= \sqrt{\frac{9}{8}}e^{i\theta}\eta\left\{(r\,e^{i\phi}\hat{i}+s\,e^{-i\phi}\hat{j})\mathrm{sech}(\eta T) - \frac{1}{2}\eta\Delta T(r\,e^{i\phi}\hat{i}-s\,e^{-i\phi}\hat{j})\mathrm{sech}(\eta T)\tanh(\eta T)\right\} \tag{10.63}$$

The first term in Eq. (10.63) corresponds to the majority of the soliton, whose polarization state is $\mathbf{u} = r\,e^{i\phi}\hat{i} + s\,e^{-i\phi}\hat{j}$. Projecting the second term on the polarization state $\mathbf{u}$ and its orthogonal polarization state $\mathbf{v} = s\,e^{i\phi}\hat{i} - r\,e^{-i\phi}\hat{j}$, we can write the field (10.63) as

$$\mathbf{q}(Z + Z_f, T) = A(T)\mathbf{u} + B(T)\mathbf{v} \tag{10.64}$$

where

$$A(T) = \sqrt{\frac{9}{8}}e^{i\theta}\eta \operatorname{sech}\left[\eta T - \frac{1}{2}\eta\Delta T(|r|^2 - |s|^2)\right] \tag{10.65}$$

$$B(T) = \sqrt{\frac{9}{8}}e^{i\theta}\eta^2\Delta Trs\,\operatorname{sech}(\eta T)\tanh(\eta T) \tag{10.66}$$

Equation (10.65) shows that $A(T)$ is the soliton displaced by the effect of birefringence. On the other hand, $B(T)$ can be interpreted as the radiation shed by the soliton over the length Z_f. The energy of this radiation is given by

$$\delta E = \int_{-\infty}^{\infty} |B(T)|^2\,dT = \frac{3}{4}\eta^3|rs|^2\Delta T^2 \tag{10.67}$$

When the energy δE is averaged over all possible soliton polarization states (which gives $\langle |rs|^2 \rangle = 1/6$), we obtain

$$\langle \delta E \rangle = \frac{1}{8}\eta^3\langle \Delta T^2 \rangle = \frac{8}{729}E^3\langle \Delta T^2 \rangle \tag{10.68}$$

where the relation between the soliton energy and amplitude $E = 9\eta/4$ has been used in the last equation.

The average of the square of the normalized differential group delay for the fiber of normalized length Z_f is given by

$$\langle \Delta T^2 \rangle = \frac{3\pi}{8}\langle |\Delta T| \rangle^2 = \frac{3\pi}{8t_0^2}D_p^2 z_f \tag{10.69}$$

where D_p is the so-called PMD parameter and $z_f = Z_f L_D$ is the real fiber length. Using Eq. (10.69) in Eq. (10.68), we can write the following equation for the average energy decay:

$$\frac{dE}{dZ} = -\frac{\pi L_D D_p^2}{243 t_0^2}E^3 \tag{10.70}$$

which has a solution

$$E(z) = \frac{E(0)}{\sqrt{1 + 2\pi L_D D_p^2 E^2(0) Z / (243 t_0^2)}} \tag{10.71}$$

The above solution indicates that the soliton pulse is gradually broadened by a factor

$$b_{\text{sol}} = \sqrt{1 + \frac{\pi D_p^2 z}{24 t_0^2}} \tag{10.72}$$

where we considered that $E(0) = 9/4$. Using the relation given by Eq. (10.69) with $\Delta T = \Delta \tau / t_0$ and the rms initial pulse width τ_0, this broadening factor can be written in the form

$$b_{\text{sol}} = \sqrt{1 + \frac{\pi^2}{108} \frac{\langle \Delta \tau^2 \rangle}{\tau_0^2}} \tag{10.73}$$

The broadening factor for the soliton pulse given by Eq. (10.72) or (10.73) has the same form as the broadening factor for a linear pulse but has a slower broadening rate, showing better robustness of solitons to PMD.

Figure 10.3 shows the simulation results for the PMD-induced pulse broadening as a function of transmission distance for different GVD values [36]. In the simulations, an initial soliton $A(0, t) = \sqrt{9/8} \operatorname{sech}(t/t_0)$ was used. The initial pulse width

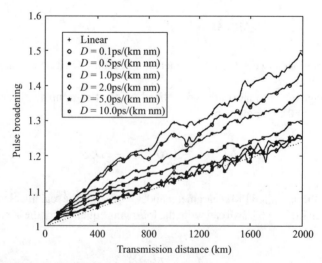

Figure 10.3 PMD-induced pulse broadening as a function of transmission distance for different GVD values. (After Ref. [36]; © 2000 IEEE.)

$t_{FWHM0} = 20\,\mathrm{ps}$, coupling length $L_c = 500\,\mathrm{m}$, and the PMD coefficient $D_p = 0.5\,\mathrm{ps/km}^{1/2}$. The dotted line corresponds to the analytical result given by Eq. (10.72). The figure shows that the soliton pulse broadening decreases with the increase of GVD, and when GVD is large enough, the pulse broadening approaches the analytical result. The reason is that the analytical result was obtained assuming that the Manakov soliton is weakly perturbed. This is only the case if the birefringence terms are much smaller than the GVD terms in the coupled nonlinear Schrödinger equations.

Figure 10.4 illustrates the broadening of soliton pulses (curve b) and linear Gaussian pulses with (curve c) and without (curve a) first-order PMD compensation at the carrier frequency [37–40]. Comparing the curves a and b in Fig. 10.4, we verify that the solitons broaden at a significantly lower rate than linear pulses without PMD compensation. Moreover, comparing the curves b and c, we observe that the solitons broaden at a lower rate than linear pulses with first-order PMD compensation for $\langle \Delta \tau^2 \rangle^{1/2}/\tau_0 > 2.5$, demonstrating that soliton robustness to PMD would become definitively better at high values of the DGD.

The first experimental observation of soliton robustness to PMD in a real transmission system was reported in 1999 [41]. Such robustness was confirmed later both experimentally and numerically [22,42]. Because of self-trapping effects of the two polarization components, the soliton can adjust itself and maintain its shape after radiating some dispersive waves. As a result, the net broadening will be at a lower rate than linear pulses. However, significant degradations in soliton systems are caused by such dispersive waves. Besides the pulse broadening determined by the loss of energy, the dispersive waves interact with soliton pulses and cause timing jitter and pulse distortion [22]. It has been shown that some of the soliton control methods discussed in Chapter 8 can significantly improve the soliton's robustness to PMD, particularly for long-distance systems [34,43].

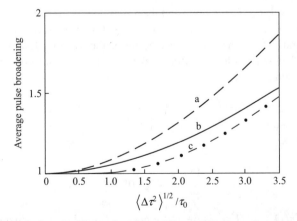

Figure 10.4 Broadening of soliton pulses (b) and linear Gaussian pulses with (c) and without (a) first-order PMD compensation at the carrier frequency.

10.6 DISPERSION-MANAGED SOLITONS AND PMD

As discussed in Chapter 8, dispersion-managed (DM) solitons offer significant advantages with respect to conventional solitons in high-capacity transmission systems [44–46]. The main idea is to combine a high local group velocity dispersion with low path-average dispersion. The former feature results in the reduction of the four-wave mixing while the latter one reduces the Gordon–Haus timing jitter effect [47,48]. Moreover, the DM solitons show an enhanced pulse energy [44], which not only leads to improved SNR, but also contributes to reduce the Gordon–Haus timing jitter and enhance the nonlinear trapping effect. As a consequence, a dispersion-managed soliton is believed to be more effective in reducing PMD-induced pulse broadening than a conventional soliton. The enhancement factor of the pulse energy is proportional to a parameter S called the map strength, defined as

$$S = \frac{\lambda^2}{2\pi c} \frac{D_1 z_1 - D_2 z_2}{t_0^2} \tag{10.74}$$

where D_1 and D_2 are the group velocity dispersions of fiber with lengths z_1 and z_2, respectively, whereas t_0 is the input pulse width.

As seen in Section 8.6, we can derive the ordinary differential equations that describe the evolution of the DM pulse parameters by using the variational approach with a proper ansatz of the propagating pulse. We consider that such ansatz corresponds to a Gaussian pulse shape, given by

$$q_j(Z,T) = Q_j \exp\left\{ -\left(\frac{1}{2\tau_j^2} + \frac{iC_j}{2} \right)(T - T_{0j})^2 - i\kappa_j(T - T_{0j}) + i\psi_j \right\}, \quad j = x, y \tag{10.75}$$

where Q_j, τ_j, C_j, κ_j, T_{0j}, and ψ_j represent the amplitude, pulse width, frequency chirp, frequency, time position, and phase, respectively. Using the variational approach to Eqs. (10.21) and (10.22), in which the last terms on the right-hand sides are neglected, we obtain the following ordinary differential equations for the difference of frequency and time position:

$$\frac{d\Delta\kappa}{dZ} = \frac{2\hat{e}E\,\Delta T}{\sqrt{\pi}(\tau_x^2 + \tau_y^2)^{3/2}} \exp\left\{ -\frac{\Delta T^2}{\tau_x^2 + \tau_y^2} \right\} \tag{10.76}$$

$$\frac{d\Delta T}{dZ} = 2\delta_g - d(Z)\Delta\kappa \tag{10.77}$$

where $E = \sqrt{\pi}(Q_x^2 \tau_x + Q_y^2 \tau_y)$ is the pulse energy. From Eqs. (10.76) and (10.77) we can see that, in the presence of nonlinearity, $\Delta\kappa$ and ΔT behave periodically along Z, whereas in the linear case $\Delta\kappa = 0$ and ΔT increases linearly. The period of these trajectories becomes shorter for strong dispersion management due to large cross-phase modulation induced by higher nonlinearity. Hence, nonlinear trap becomes effective in controlling the polarization mode dispersion in fibers with constant birefringence.

In order to analyze the effect of PMD in randomly varying birefringent fibers, let us expand Eqs. (10.76) and (10.77) around $\Delta T = 0$. Taking the first order in this expansion and assuming that δ_g is a white Gaussian process, such that $\langle\delta_g\rangle = 0$ and $\langle\delta_g(Z)\delta_g(Z')\rangle = \sigma^2\delta(Z-Z')$, we obtain [21]

$$\sqrt{\langle\Delta T^2\rangle} = \left[2\sigma^2 Z\left\{1 + \frac{\sin(2b|d_0|Z)}{2b|d_0|Z}\right\}\right]^{1/2} \tag{10.78}$$

where $b = 2\hat{\varepsilon}E/\sqrt{\pi(\tau_x^2 + \tau_y^2)^3}$. In the linear case ($b=0$), we have

$$\sqrt{\langle\Delta T^2\rangle} = 2\sqrt{\sigma^2 Z} \tag{10.79}$$

In obtaining Eq. (10.78), the local dispersion d was replaced with the average dispersion d_0, which is reasonable for weak dispersion management. Equation (10.79) is consistent with the result obtained above for the case of an ideal soliton and it shows that the effect of PMD in dispersion-managed soliton transmission systems is reduced with the help of nonlinear trap by $1/\sqrt{2}$ compared to linear systems. However, for strong dispersion management, this analysis does not apply. In such a case, the orthogonally polarized modes are completely trapped and the pulse broadening increases almost as $\ln(Z)$ [21,49].

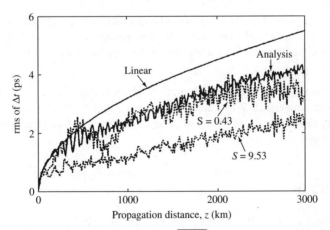

Figure 10.5 Comparison of the evolution of $\sqrt{\langle\Delta T^2\rangle}$ in linear and nonlinear systems with the initial pulse width 2 ps. Solid lines represent the results of linear and nonlinear systems given by Eqs. (10.78) and (10.79), respectively, and dashed lines are the results of numerical simulations for $S = 0.43$ and 9.53, obtained by averaging over 50 trials. Dispersion management period $z_a = 10$ km and PMD coefficient $D_p = 0.1$ ps/km$^{1/2}$. (After Ref. [21]; © 2000 IEEE.)

The robustness of dispersion-managed solitons to PMD has been confirmed both experimentally [50] and numerically [21,51,52]. In general, DM solitons are more robust to PMD than conventional solitons. Such robustness increases with the map strength and average GVD, since this determines an increased power of DM solitons.

Figure 10.5 shows a comparison of the evolution of $\sqrt{\langle \Delta T^2 \rangle}$ in linear and dispersion-managed nonlinear systems obtained by direct numerical simulations of Eqs. (10.21) and (10.22) and the approximate results provided by Eqs. (10.78) and (10.79) [21]. The PMD-induced pulse broadening is suppressed considerably by increasing the map strength by the help of enhanced nonlinear trap. The analytical result of the Langevin equation is in good agreement with the numerical results when $S = 0.43$. Indeed, Eq. (10.78) is valid especially for small values of the map strength S, when the pulse dynamics is dominated by the average dispersion.

PROBLEMS

10.1 Derive expressions for the nonlinear contributions to the refractive index when an optical beam propagates inside a birefringent fiber.

10.2 Derive the result given by Eq. (10.15) for the relative phase difference between the two polarization components of a CW field propagating inside a linearly birefringent fiber.

10.3 Explain how fiber birefringence can be used to remove the low-intensity pedestal associated with an optical pulse.

10.4 Starting from Eqs. (10.37) and (10.38), derive Eqs. (10.45) and (10.48). Provide a physical explanation for the existence of a threshold value for the soliton amplitude for a capture of the separate polarizations into a bound state.

10.5 Explain the origin of PMD in optical fibers and of the PMD-induced pulse broadening effect. Why can solitons resist better to PMD than linear pulses? Why DM solitons are expected to be more robust to PMD than conventional solitons?

REFERENCES

1. C. D. Poole and R. E. Wagner, *Electron. Lett.* **22**, 1029 (1986).
2. C. D. Poole and J. Nagel, *Polarization effects in lightwave systems*, in I. P. Kaminow and T. L. Koch (Eds.), *Optical Fiber Telecommunications*, Vol. IIIA, Academic Press, San Diego, CA, 1997.
3. H. Bullow, *IEEE Photon. Technol. Lett.* **10**, 696 (1998).
4. J. N. Damask, *Polarization Optics in Telecommunications*, Springer, New York, 2005.

5. A. Galtarrossa and C. R. Menyuk (Eds.), *Polarization Mode Dispersion*, Springer, New York, 2005.

6. N. J. Muga, A. N. Pinto, M. F. Ferreira, and J. F. Rocha, *J. Lightwave Technol.* **24**, 3932 (2006).

7. F. Matera, M. Settembre, M. Tamburrini, F. Favre, D. Le Guen, T. Georges, M. Henry, G. Michaud, P. Franco, A. Shiffini, M. Romagnoli, M. Guglielmucci, and S. Casceli, *J. Lightwave Technol.* **17**, 2225 (1999).

8. E. Kolltveit, P. A. Andrekson, J. Brentel, B. E. Olsson, B. Bakhshi, J. Hansryd, P. O. Hedekvist, M. Karlsson, H. Sunnerud, and J. Li, *Electron. Lett.* **35**, 75 (1999).

9. H. Sunnerud, M. Karlsson, C. Xie, and P. A. Andrekson, *J. Lightwave Technol.* **20**, 2204 (2002).

10. A. Galtarrossa, P. Griggio, L. Palmieri, and A. Pizzinat, *J. Lightwave Technol.* **22**, 1127 (2004).

11. V. Chernyak, M. Chertkov, I. Gabitov, I. Kolokolov, and V. Lebedev, *J. Lightwave Technol.* **22**, 1155 (2004).

12. G. Ning, S. Aditya, P. Shum, Y. D. Gong, H. Dong, and M. Tang, *Opt. Commun.* **260**, 560 (2006).

13. D. N. Payne, A. J. Barlow, and J. J. R. Hansen, *IEEE J. Quantum Electron.* **18**, 477 (1982).

14. S. C. Rashleigh, *J. Lightwave Technol.* **1**, 312 (1983).

15. J. Noda, K. Okamoto, and Y. Sasaki, *J. Lightwave Technol.* **4**, 1071 (1986).

16. K. Tajima, M. Ohashi, and Y. Sasaki, *J. Lightwave Technol.* **7**, 1499 (1989).

17. R. B. Dyott, *Elliptical Fiber Waveguides*, Artech House, Boston, MA, 1995.

18. P. K. A. Wai and C. R. Menyuk, *J. Lightwave Technol.* **14**, 148 (1996).

19. L. F. Mollenauer, J. P. Gordon, and P. V. Mamyshev, in I. P. Kaminow and T. L. Koch (Eds.), *Optical Fiber Telecommunications*, Vol. IIIA, Academic Press, San Diego, CA, 1997, Chapter 12.

20. F. Matera and M. Settembre, *Opt. Lett.* **20**, 28 (1995).

21. I. Nishioka, T. Hirooka, and A. Hasegawa, *IEEE Photon. Technol. Lett.* **12**, 1480 (2000).

22. C. Xie, M. Karlsson, P. Andrekson, and H. Sunnerud, *IEEE J. Sel. Top. Quantum Electron.* **8**, 575 (2002).

23. D. A. Kleinman, *Phys. Rev.* **126**, 1977 (1962).

24. R. W. Boyd, *Nonlinear Optics*, 2nd ed., Academic Press, San Diego, CA, 2003.

25. C. R. Menyuk, *IEEE J. Quantum Electron.* **23**, 176 (1987).

26. G. P. Agrawal, *Nonlinear Fiber Optics*, 3rd ed., Academic Press , San Diego, CA, 2001.

27. A. Hasegawa, *Physica D* **188**, 241 (2004).

28. M. F. Ferreira, *Fiber Int. Opt.* **27**, 113 (2008).

29. R. H. Stolen, J. Botineau, and A. Ashkin, *Opt. Lett.* **7**, 512 (1982).

30. B. Nikolaus, D. Grischkowsky, and A. C. Balant, *Opt. Lett.* **8**, 189 (1983).

31. Y. S. Kivshar, *J. Opt. Soc. Am. B* **7**, 2204 (1990).

32. D. Marcuse, C. R. Menyuk, and P. K. Wai, *J. Lightwave Technol.* **15**, 1735 (1997).

33. P. K. Wai, C. R. Menyuk, and H. H. Chen, *Opt. Lett.* **16**, 1231 (1991).

34. M. Matsumoto, Y. Akagi, and A. Hasegawa, *J. Lightwave Technol.* **15**, 584 (1997).

35. C. V. Manakov, *Sov. Phys. JETP* **38**, 248 (1974).

36. C. Xie, M. Karlsson, and P. A. Andrekson, *IEEE Photon. Technol. Lett.* **12**, 801 (2000).

37. F. Heismann, D. Fishman, and D. L. Wilson, *Proceedings of ECOC'98* Madrid, Spain, 1998, pp. 529–530.

38. M. Karlsson, *Opt. Lett.* **23**, 688 (1998).

39. H. Sunnerud, M. Karlsson, and P. A. Andrekson, *IEEE Photon. Technol. Lett.* **12**, 50 (2000).

40. M. F. Ferreira, S. V. Latas, M. H. Sousa, A. N. Pinto, J. F. Rocha, P. S. André, N. J. Muga, R. N. Nogueira, and J. E. Machado, *Fiber Int. Opt.* **24**, 261 (2005).

41. B. Bakhshi, J. Hansryd, P. A. Andrekson, J. Brentel, E. Kolltveit, B. E. Olsson, and M. Karlsson, *Electron. Lett.* **35**, 65 (1999).

42. H. Sunnerud, J. Lie, C. Xie, and P. Andrekson, *J. Lightwave Technol.* **19**, 1453 (2001).

43. C. Xie, M. Karlsson, and H. Sunnerud, *Opt. Lett.* **26**, 672 (2001).

44. N. J. Smith, F. M. Knox, N. J. Doran, K. J. Blow, and I. Bennion, *Electron. Lett.* **32**, 54 (1996).

45. A. Hasegawa (Ed.), New Trends in Optical Soliton Transmission Systems, Kluwer, Dordrecht, The Netherlands, 1998.

46. M. F. Ferreira, M. V. Facão, S. V. Latas, and M. H. Sousa, *Fiber Integr. Opt.* **24**, 287 (2005).

47. M. Suzuki, I. Morita, N. Edagawa, S. Yamamoto, H. Taga, and S. Akiba, *Electron. Lett.* **31**, 2027 (1995).

48. G. M. Carter, J. M. Jacob, C. R. Menyuk, E. A. Golovchenko, and A. N. Pilipetskii, *Opt. Lett.* **22**, 513 (1997).

49. Y. Chen and H. Haus, *Opt. Lett.* **25**, 29 (2000).

50. H. Sunnerud, J. Li, P. A. Andrekson, and C. Xie, *IEEE Photon. Technol. Lett.* **13**, 118 (2000).

51. X. Zhang, M. Karlsson, P. Andrekson, and K. Brtilsson, *Electron. Lett.* **34**, 1122 (1998).

52. C. Xie, M. Karlsson, P. Andrekson, and H. Sunnerud, *IEEE Photon. Technol. Lett.* **13**, 221 (2000).

11

STIMULATED RAMAN SCATTERING

Spontaneous Raman scattering is named after the Indian physicist Sir Chandrasekhara Raman, who discovered this phenomenon in 1928 [1] and received the Nobel Prize in Physics in 1930 for this work. During the investigations of light scattering in different media, Raman observed the existence of a frequency-shifted field in addition to the incident field. The frequency shift of the scattered light is determined by the vibrational oscillations that occur between constituent atoms within the molecules of the material. A slight excitation of the molecular resonances due to the presence of the input field, referred to as the pump wave, results in spontaneous Raman scattering. As the scattered light increases in intensity, the scattering process eventually becomes stimulated. In this regime, the scattered field interacts coherently with the pump field to further excite the resonances, which significantly enhances the transfer of power between the two optical waves. Raman scattering can occur in all materials, but the dominant Raman lines in silica glass are due to the bending motion of the Si−O−Si bond. The theoretical description of stimulated Raman scattering (SRS) has been presented in several publications [2–5].

Both downshifted (Stokes) and upshifted (anti-Stokes) light can, in principle, be generated through the Raman scattering process. However, in the case of optical fibers, the Stokes radiation is generally much stronger than the anti-Stokes radiation due to two main reasons. On the one hand, the anti-Stokes process is phase mismatched, whereas the Stokes process is phase matched, for collinear propagation. On the other hand, the anti-Stokes process involves the interaction of the pump light with previously excited molecular resonances, which is not a condition for the Stokes process.

The first experimental demonstration of stimulated Raman scattering in fibers was realized by Erich Ippen in early 1970, who constructed a CS_2-core continuous-wave (CW) fiber Raman laser [6]. Soon after, Stolen et al. observed for the first time the same effect in single-mode silica fibers and demonstrated its use in a Raman oscillator [7]. This work was followed by an amplifier experiment to directly measure the Raman gain in silica fibers [8].

SRS is a major problem in wavelength division multiplexed (WDM) systems, as it can induce crosstalk between different channels. In such a case, channels with a higher carrier frequency deliver part of their power to channels with a lower frequency. However, the same effect may also find useful applications, namely, for efficient amplification of injected signals and for the generation of new frequencies, as will be discussed in the last part of this chapter.

11.1 RAMAN SCATTERING IN THE HARMONIC OSCILLATOR MODEL

The Raman scattering can be described using the harmonic oscillator model [9], which was presented in Section 2.6. In this model, the vibratory molecules of the medium correspond to the various oscillators, which are assumed to be identical, with a mass m, a damping coefficient γ_m, and a resonant frequency ω_r. The equation of motion of each of the oscillators is then given by:

$$\frac{\partial^2 X}{\partial t^2} + \gamma_m \frac{\partial X}{\partial t} + \omega_r^2 X = \frac{F(z,t)}{m} \tag{11.1}$$

where $X(z,t)$ is the displacement of the mass from its equilibrium position and $F(z,t)$ is a forcing function.

The model assumes that each oscillator already oscillates with the resonant frequency ω_r in the absence of the external force. Therefore, both the displacement $X(z,t)$ and the dipole moment are periodic functions of time. According to the discussion of Section 2.6, this determines a temporal modulation of the refractive index. If a monochromatic wave with frequency ω_p irradiates the medium, the refractive index modulation forms new waves with frequencies $\omega_p \pm \omega_r$. The wave with frequency $\omega_S = \omega_p - \omega_r$ corresponds to the Stokes wave, whereas the wave with frequency $\omega_A = \omega_p + \omega_r$ is the anti-Stokes wave. As mentioned above, in optical fibers, the Stokes wave is generally much more intense than the anti-Stokes wave. Due to this fact, only the Stokes wave will be considered in the following.

Considering the quantum mechanical picture, in the SRS process one has simultaneously the absorption of a photon from the pump beam at frequency ω_p and the emission of a photon at the Stokes frequency ω_S (Fig. 11.1). The difference in energy is taken up by a high-energy phonon (molecular vibration) at frequency ω_r. Thus, SRS provides energy gain at the Stokes frequency at the expense of the pump. This process is considered nonresonant because the upper state is a short-lived virtual state.

The optical field resulting from the interference between the original and the Stokes waves has an intensity that is temporally modulated with a frequency $\omega_p - \omega_S$. Such modulated intensity causes a periodic force acting on the oscillators. If the frequency of the modulated intensity is similar to the natural resonant frequency of

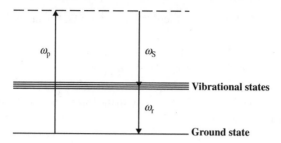

Figure 11.1   Energy level illustration of the stimulated Raman scattering process.

the oscillators, stronger oscillations of the dipoles are induced. This produces a stronger Stokes wave that leads to stronger oscillations of the dipoles and so on.

As pointed out above, the displacement $X(z,t)$ and the forcing function $F(z,t)$ in Eq. (11.1) are expected to be proportional to the difference frequency term in the product of Stokes and pump waves. These waves are assumed to have the same polarization and are given by

$$E_j(z,t) = \frac{1}{2}E_{0j}\exp\left[i(\beta_j z - \omega_j t)\right] + \text{c.c.}, \quad j = S, p \tag{11.2}$$

As a result of the displacement $X(z,t)$ of some of its atoms, the molecular polarizability, α_p, becomes time dependent and can be approximated as

$$\alpha_p = \alpha_{p0} + \alpha_{p1}X(z,t) \tag{11.3}$$

where $\alpha_{p1} = \partial\alpha_p/\partial X|_{X=0}$ is the change of polarizability with displacement. The permittivity of the medium depends on the molecular polarizability as

$$\varepsilon = \varepsilon_0[1 + N\alpha_p(z,t)] \tag{11.4}$$

where N is the number density of molecules. The nonlinear polarization is related to the variation of the medium permittivity arising from the change in polarizability with displacement and can be written in the form

$$P_{NL}(z,t) = \varepsilon_0 N\alpha_{p1}X(z,t)E(z,t) \tag{11.5}$$

where $E = E_S + E_p$ is the total electric field, given by the sum of the Stokes and pump fields.

Calculating the gradient of the energy density, $u_e = \varepsilon E^2/2$, and using Eqs. (11.3) and (11.4), one can derive the following result for the forcing function $F(z,t)$:

$$F(z,t) = \frac{\varepsilon_0\alpha_{p1}}{4}E_{0p}E_{0S}^*\exp\left\{i\left[(\beta_p - \beta_S)z - (\omega_p - \omega_S)t\right]\right\} + \text{c.c.} \tag{11.6}$$

Using this result in Eq. (11.1), the displacement $X(z,t)$ can be given as

$$X(z,t) = \frac{\varepsilon_0\alpha_{p1}}{4m} \frac{E_{0p}E_{0S}^*\exp\{i[(\beta_p-\beta_S)z-(\omega_p-\omega_S)t]\}}{\omega_r^2-(\omega_p-\omega_S)^2-i\gamma_m(\omega_p-\omega_S)} + \text{c.c.} \qquad (11.7)$$

Substituting Eq. (11.7) into (11.5), we obtain the following results for the nonlinear polarizations at frequency ω_j (j = p, S):

$$P_{NL}^{\omega_p}(z,t) = \frac{1}{2}\varepsilon_0\chi_R^{\omega_p}|E_{0S}^2|E_{0p}\exp\left[i(\beta_p z-\omega_p t)\right] + \text{c.c.} \qquad (11.8)$$

$$P_{NL}^{\omega_S}(z,t) = \frac{1}{2}\varepsilon_0\chi_R^{\omega_S}|E_{0p}^2|E_{0S}\exp[i(\beta_S z-\omega_S t)] + \text{c.c.} \qquad (11.9)$$

where $\chi_R^{\omega_j}$ represent the Raman susceptibilities, given by

$$\chi_R^{\omega_S} = \chi_R^{\omega_p *} = -\frac{\varepsilon_0 N\alpha_{p1}^2}{4m\omega_r\gamma_m(\Delta_r+i)} \qquad (11.10)$$

with

$$\Delta_r = \frac{2}{\gamma_m}\left[(\omega_p-\omega_S)-\omega_r\right] \qquad (11.11)$$

In obtaining Eq. (11.10), the near-resonance operation was assumed, such that $\omega_p-\omega_S \approx \omega_r$. Figure 11.2 illustrates the real and imaginary parts of the susceptibilities given by Eq. (11.10).

The inclusion of the above results for the nonlinear polarizations in the wave equation (4.13) and the use of the slowly varying envelope approximation (SVEA)

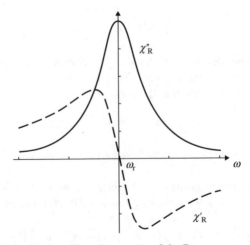

Figure 11.2   Real and imaginary parts of the Raman susceptibilities.

provide the following coupled equations for the amplitude of the Stokes and pump fields:

$$\frac{dE_{0S}}{dz} = \frac{i\omega_S}{2nc}\chi_R|E_{0p}|^2 E_{0S} \qquad (11.12)$$

$$\frac{dE_{0p}}{dz} = \frac{i\omega_p}{2nc}\chi_R^*|E_{0S}|^2 E_{0p} \qquad (11.13)$$

where n is the refractive index and $\chi_R \equiv \chi_R^{\omega_S}$.

The wave intensity is related to the amplitude as $I_j = (1/2)\varepsilon_0 cn|E_{0j}|^2$. Using this relation, Eqs. (11.12) and (11.13) can be written in terms of the wave intensities as

$$\frac{dI_S}{dz} = g_R I_S I_p - \alpha_S I_S \qquad (11.14)$$

$$\frac{dI_p}{dz} = -\frac{\omega_p}{\omega_S}g_R I_S I_p - \alpha_p I_p \qquad (11.15)$$

where the last term in each equation was added to take into account the losses in the medium. The parameter g_R in Eqs. (11.14) and (11.15) is the Raman gain coefficient, given by

$$g_R = \frac{N\omega_S\alpha_{p1}^2}{2m\gamma_m\omega_r c^2 n^2}f(\Delta_r) \qquad (11.16)$$

where $f(\Delta_r) = (1 + \Delta_r^2)^{-1}$ is the lineshape function. However, in the case of amorphous materials such as fused silica, this lineshape can assume a more complicated profile. From Eq. (11.16), we can observe the dependence of the Raman gain coefficient on various parameters, namely, its proportionality to the frequency ω_S.

Using the relation $P_j = I_j A_{\text{eff}}$ ($j = p, S$), where P_j is the optical power and A_{eff} is the effective core area, Eqs. (11.14) and (11.15) can be rewritten in terms of the optical power as follows:

$$\frac{dP_S}{dz} = -\alpha_S P_S + \frac{g_R}{A_{\text{eff}}}P_p P_S \qquad (11.17)$$

$$\frac{dP_p}{dz} = -\alpha_p P_p - \frac{\omega_p}{\omega_S}\frac{g_R}{A_{\text{eff}}}P_S P_p \qquad (11.18)$$

In Eqs. (11.17) and (11.18), the terms containing a product of the pump and signal powers describe their coupling via SRS. The strength of this coupling is determined by the ratio between the Raman gain coefficient and the effective core area, $C_R = g_R/A_{\text{eff}}$, which is known as the Raman gain efficiency.

11.2   RAMAN GAIN

The Raman scattering reflects the resonances of a given material. In the case of optical fibers, they are basically made of fused silica, which is an amorphous material. Several vibrational modes occur in the structure of amorphous silica, with resonance frequencies that overlap with each other and form broad frequency bands. As a consequence, the Raman scattering in optical fibers occurs over a relatively large frequency range.

Figure 11.3 shows the normalized Raman gain spectra for bulk fused silica, measured when the pump light and signal light were either copolarized (solid curve) or orthogonally polarized (dotted curve) [10]. The most significant feature of the Raman gain in silica fibers is that g_R extends over a large frequency range (up to 40 THz). Optical signals whose bandwidths are of this order or less can be amplified using the Raman effect if a pump wave with the right wavelength is available. The Raman shift, corresponding to the location of the main peak in Fig. 11.3, is close to 13 THz for silica fibers. The Raman gain depends on the relative state of polarization of the pump and signal fields. The copolarized gain is almost an order of magnitude larger than the orthogonally polarized gain near the peak of the gain curve. The peak value of the Raman gain decreases with increasing pump wavelength, as can be inferred from Eq. (11.16), and it is about 6×10^{-14} m/W in the wavelength region around 1.5 μm.

Oxide glasses used as dopants of pure silica in fiber manufacture exhibit Raman gain coefficients that can be significantly higher than that of silica. Figure 11.4 shows the Raman spectra of some of these glasses along with that of silica. The most important dopant in communication fiber is GeO_2, whose Raman gain coefficient is about 8.2 times that for pure SiO_2 glass.

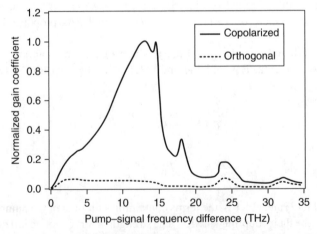

Figure 11.3   Normalized Raman gain coefficient for copolarized (solid curve) and orthogonally polarized (dotted curve) pump and signal beams. (After Ref. [10]; © 2004 IEEE.)

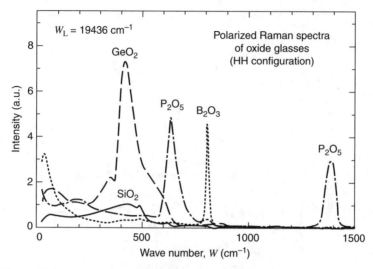

Figure 11.4   Raman spectra of several glasses. (After Ref. [11]; © 1978 AIP.)

Doping the core of a silica fiber with GeO_2 provides two main beneficial effects. First, the relative index difference, Δ, between the core and the cladding is raised, which increases the effective waveguiding and in turn reduces the effective cross-sectional area A_{eff} [12]. As a consequence, the rate of stimulated Raman scattering is increased. Second, the Raman gain coefficient increases above that of pure SiO_2 in proportion to the concentration GeO_2 [13]. In general, the contribution of germania to the Raman gain increases in a manner that depends on its fractional concentration within the fiber core and on the spatial overlap between pump and Stokes transverse modes [14]. In addition to germania, fluorine is also used as the dopant material in some actual fibers [17].

Figure 11.5 shows the relationship between relative index difference Δ and the Raman gain coefficient at a 12.9 THz frequency shift from the pump wavelength of 1450 nm [17]. In this figure, the filed and open circles show the measured results for germanium- and fluorine-doped optical fibers, respectively. The solid and dashed lines show the best fit with the measured results. It can be seen that the Raman gain coefficient in the Ge-doped fibers increases with increasing Ge concentration, whereas it decreases as the F concentration in the core increases. The relationship between the relative index difference and the Raman gain coefficient in both cases can be approximated as follows [17]:

$$\text{Ge-doped fiber:} \quad g_R \approx 2.75 \times 10^{-14} + 2.16 \times 10^{-14} |\Delta| \qquad (11.19)$$

$$\text{F-doped fiber:} \quad g_R \approx 2.75 \times 10^{-14} - 0.32 \times 10^{-14} |\Delta| \qquad (11.20)$$

In spite of increasing the Raman gain coefficient, doping with GeO_2 also increases the fiber losses, due to the concomitant rise in Rayleigh scattering and in the number of

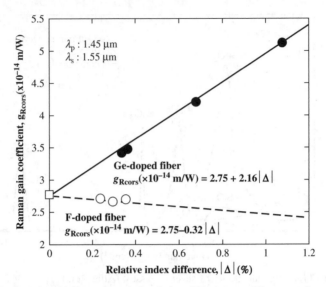

Figure 11.5 Relationship between relative index difference and Raman gain coefficient in Ge- and F-doped optical fibers. (After Ref. [17]; © 2004 OSA.)

dopant-dependent loss centers. Thus, greatly increasing the GeO_2 concentration will not be worthwhile if the increased fiber losses outweigh the improvement in gain.

Significantly higher values of the Raman gain parameter can be achieved in some kinds of nonsilica fibers. For example, Raman gain coefficients with peak values up to 30 times that of the fused silica and more than twice its spectral coverage were measured for tellurite glasses [18]. A Raman gain parameter that was several hundred times greater than that of silica case was also recently reported for chalcogenide fibers based on As–S–Se glasses [19,20].

11.3 RAMAN THRESHOLD

If only the pump wave is launched into the fiber, the generation of the Stokes wave begins with spontaneous Raman scattering. As this scattered light increases its intensity, the process becomes stimulated, which can lead to Stokes power levels similar to that of the pump wave.

Assuming that $\alpha_S = \alpha_p = \alpha$ (which is a reasonable approximation around the 1.5 µm wavelength region in a low-loss fiber), the following analytical solutions of Eqs. (11.17) and (11.18) can be obtained:

$$P_S(z) = \frac{\omega_S}{\omega_p} \frac{F(z)}{1 + F(z)} P_0 \exp(-\alpha z) \qquad (11.21)$$

$$P_p(z) = \frac{P_0 \exp(-\alpha z)}{1 + F(z)} \qquad (11.22)$$

where

$$F(z) = \frac{\omega_p}{\omega_S} \frac{P_{S0}}{P_{p0}} \exp\left\{ g_R \frac{P_0}{A_{eff}} \frac{1-\exp(-\alpha z)}{\alpha} \right\} \qquad (11.23)$$

In Eqs. (11.21)–(11.23), we have $P_0 = P_{p0} + (\omega_S/\omega_p)P_{S0}$, whereas P_{S0} and P_{p0} represent the input power of the Stokes and pump waves, respectively.

When the input power of the Stokes wave is weak, such that $P_{S0} \ll P_{p0}$, Eqs. (11.21) and (11.22) become

$$P_S(z) \approx P_{S0} \exp\left\{ g_R \frac{P_{p0}}{A_{eff}} \frac{1-\exp(-\alpha z)}{\alpha} - \alpha z \right\} \qquad (11.24)$$

$$P_p(z) \approx P_{p0} \exp(-\alpha z) \qquad (11.25)$$

In the absence of an input signal P_{S0}, the Stokes wave arises from spontaneous Raman scattering along the fiber. It was shown that this process is equivalent to injecting one fictitious photon per mode at the input end of the fiber [21]. Thus, considering a single-mode fiber, we can calculate the output Stokes power using Eq. (11.24) and integrating over the Raman gain spectrum

$$P_S(L) = \int_{-\infty}^{\infty} \hbar\omega \exp\left\{ g_R(\omega) \frac{P_{p0}L_{eff}}{A_{eff}} - \alpha L \right\} d\omega \qquad (11.26)$$

where

$$L_{eff} = \frac{1-\exp(-\alpha L)}{\alpha} \qquad (11.27)$$

is the effective fiber length. Using the method of steepest descent and making $\omega = \omega_S$, we obtain

$$P_S(L) = \hbar\omega_S B_{eff} \exp\left\{ g_R(\omega_S) \frac{P_{p0}L_{eff}}{A_{eff}} - \alpha L \right\} \qquad (11.28)$$

where

$$B_{eff} = \left(\frac{2\pi A_{eff}}{P_{p0}L_{eff}} \right)^{1/2} \left| \frac{\partial^2 g_R}{\partial \omega^2} \right|_{\omega=\omega_S}^{-1/2} \qquad (11.29)$$

is the effective bandwidth of the Stokes radiation.

The threshold for stimulated Raman scattering is defined as the input pump power at which the output power for the pump and Stokes waves become equal [21]:

$$P_S(L) = P_p(L) = P_{p0} \exp(-\alpha L) \tag{11.30}$$

Using Eq. (11.28) and assuming a Lorentzian shape for the Raman gain spectrum, the threshold pump power is given approximately by [21]

$$P_{p0}^{th} = 16 \frac{A_{eff}}{g_R L_{eff}} \tag{11.31}$$

Equation (11.31) was derived assuming that the polarization of the pump and Stokes waves is maintained along the fiber. However, in the case of standard single-mode fibers, due to the arbitrary distribution of the polarization states of both waves, the Raman threshold is increased by a factor of 2. For example, the threshold for the stimulated Raman scattering is $P_{p0}^{th} \approx 600$ mW at $\lambda_P = 1.55\,\mu$m in long polarization maintaining fibers, such that $L_{eff} \approx 22$ km, considering an effective core area of $A_{eff} = 50\,\mu\text{m}^2$. However, in standard single-mode fibers with similar characteristics, the threshold would be $P_{p0}^{th} \approx 1.2$ W.

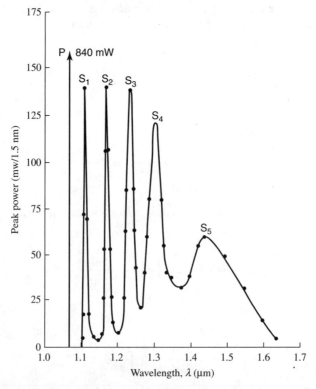

Figure 11.6   Output spectrum from a single-mode fiber Raman laser, showing multiple Stokes orders. The pump wavelength is $\lambda = 1.06\,\mu$m. (After Ref. [22]; © 1978 IEEE.)

If the first-order Stokes wave is sufficiently strong, it can act as a pump source itself, resulting in the generation of a higher order Stokes waves. Figure 11.6 shows an example of a fiber output spectrum at a pump power of about 1 kW in which multiple Stokes orders occur [22]. The spectral broadening that can be observed in the higher orders results from several competing nonlinear processes.

11.4 IMPACT OF RAMAN SCATTERING ON COMMUNICATION SYSTEMS

The impact of SRS on the performance of lightwave systems has been studied extensively [21,23–33]. Because SRS has a relatively high threshold, it is not of concern for single-channel systems. However, in WDM systems, SRS can cause crosstalk between channel signals whose wavelength separation falls within the Raman gain curve. Specifically, the long-wavelength signals are amplified by the short-wavelength signals, leading to power penalties for the latter signals. The shortest wavelength signal is the most depleted, since it acts as a pump for all other channels. The Raman-induced power transfer between two channels depends on the bit pattern, which leads to power fluctuations and determines additional receiver noise. The magnitude of these deleterious effects depends on several parameters, such as the number of channels, their frequency spacing, and the power in each of them. Thus, it is important to provide some guidelines concerning these parameters in order to minimize the Raman crosstalk between different channels.

Figure 11.7 illustrates the influence of SRS for a simple two-channel case. It is assumed that both channels are launched with the same input power and that the wavelength of channel 1 is shorter than the wavelength of channel 2. As can be observed from Fig. 11.7b, if both channels transmit a logical "one," it is amplified in channel 2, whereas it is depleted in channel 1. However, if one of the channels transmit a logical "zero," no interaction occurs between the channels.

In practice, this result can be somewhat affected by the fiber dispersion. Due to the walk-off between the initially overlapping pulses of the two channels, the interaction length is reduced and, consequently, both the amplification and depletion effects are attenuated. However, it is possible to observe the overlapping and, consequently, the nonlinear interaction between two pulses, even if they do not overlap at the fiber input.

In WDM systems with a large number of channels, the picture is similar to the two-channel case. The channels with a higher carrier wavelength are amplified, whereas the channels with a lower wavelength are depleted. If dispersion is neglected and considering the worst case of "1" bits being simultaneously transmitted on all N channels of a WDM system, spaced by Δf_{ch} and each of them carrying a power P_{ch}, it can be shown that the product of total power (NP_{ch}) and total bandwidth ($(N-1)\Delta f_{ch}$) must be smaller than 500 GHz W to guarantee a penalty for the shortest wavelength channel lower than 1 dB [28].

In WDM systems that contain no optical amplifiers, the SRS leads to a power reduction of the short-wavelength channels and, therefore, a degradation of the SNR.

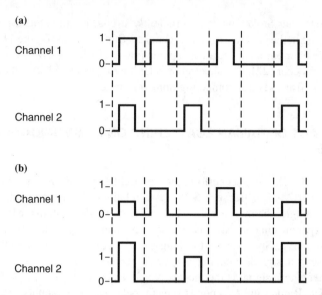

Figure 11.7   Schematic of the influence of SRS in a two-channel WDM system at (a) fiber input and (b) fiber output for $\lambda_1 < \lambda_2$. (After Ref. [28]; © 1990 IEEE.)

However, in long-haul transmission systems, a number of optical amplifiers are generally used. Besides providing the desired amplification of the signal, such optical amplifiers also add noise. Since noise is added periodically over the entire length of a system, it experiences less Raman loss than the signal. For small degradations, the fractional depletion of the noise is half the fractional depletion of the signal. Therefore, the SRS reduces the SNR and the capacity in amplified systems.

If the degradation of the SNR in the shortest wavelength channel of a WDM system is smaller than 0.5 dB, the product of total power, total bandwidth, and total effective length of the system should not exceed 10 THz mW Mm [24]. This condition is illustrated in Fig. 11.8, which shows the number of channels against the system length and the corresponding total bit rate for four different amplifier spacings: 25, 50, 100, and 150 km. Ideal amplifiers (3 dB noise figure) were assumed, as well as a fiber loss of 0.2 dB/km, a transmission rate of 2.5 Gb/s per channel, a 0.5 nm channel spacing, a 10 GHz receiver optical bandwidth, and an optical signal-to-noise ratio (SNR) = 9 (for average power), corresponding to a bit error rate (BER) = 10^{-14}.

A simple model to establish a relationship between the foregoing parameters considers a link having N channels with equal frequency spacing Δf_{ch} and equal powers P_{ch}, as well as the depletion of the shortest wavelength channel in the worst case, in which all channels transmit "1" bits simultaneously [24]. The amplification factor for the nth channel corresponding to a propagation distance $z = L$ is given from Eq. (11.24) by

$$G_n = \exp\left\{ g_{Rn} \frac{P_{ch}L_{eff}}{A_{eff}} \right\} \qquad (11.32)$$

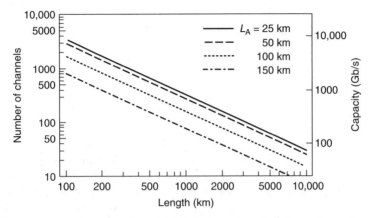

Figure 11.8 The maximum number of 2.5 Gb/s channels and total capacity against system length for four different amplifier spacings in the worst-case assumption of marks interacting on all channels. (After Ref. [29]; © 1993 IEEE.)

where L_{eff} is given by Eq. (11.27) and g_{Rn} is the Raman gain coefficient at the nth channel frequency provided by the shortest wavelength channel, which acts as a pump source. Within the weak amplification condition, the fraction of power lost by the shortest wavelength channel is given by

$$D_n = g_{Rn} \frac{P_{ch} L_{eff}}{A_{eff}} \qquad (11.33)$$

The fractional power lost by the shortest wavelength channel to the $N - 1$ channels of longer wavelengths is

$$D_R = \sum_{n=1}^{N-1} g_{Rn} \frac{P_{ch} L_{eff}}{2 A_{eff}} \qquad (11.34)$$

where the Raman gain for each channel was reduced by a factor of 2 to account for polarization scrambling.

The summation in Eq. (11.34) can be carried out analytically considering that the Raman gain spectrum is approximately triangular, such that g_R is zero for $\Delta f = 0$, increases linearly up to a maximum value $g_R = g_{Rp}$ for $\Delta f = 1.5 \times 10^4$ GHz, and then drops to zero. Thus, for $\Delta f < 1.5 \times 10^4$ GHz, the dependence of the Raman gain coefficient on the frequency is given by

$$g_{Rn} = \frac{\Delta f}{1.5 \times 10^4} g_{Rp} \qquad (11.35)$$

Using Eq. (11.35) and assuming that all channels fall within the Raman gain bandwidth, the fractional power loss for the shortest wavelength channel becomes

$$D_R = \frac{N(N-1)}{2} K \qquad (11.36)$$

where

$$K = \frac{\Delta f_{ch} P_{ch} g_{Rp} L_{eff}}{(3 \times 10^4) A_{eff}} \qquad (11.37)$$

is the fractional power loss due to a neighboring channel with Δf_{ch} in GHz. Equations (11.36) and (11.37) show that increasing the channel spacing or the number of channels requires a corresponding decrease in the power per channel so that the total power depletion (measured by D_R) does not exceed a given value.

The above description of the SRS impact in multichannel systems is only approximate. In fact, if random modulation is considered in intensity modulation-direct detection systems, the power change due to SRS is reduced because fiber nonlinearities do not occur unless at least two channels are at level "one" simultaneously. The effect of the random modulation on the crosstalk among different WDM channels can be conveniently described through statistical analyses, taking into account simultaneously the power depletion and the Stokes amplification of each channel [32,33].

Raman crosstalk can be suppressed by reducing the channel power, but such approach may not be practical in some circumstances. Another possibility is to use the technique of mid-span spectral inversion [29]. This technique leads to an inversion of the whole WDM spectrum in the middle of the transmission link. Hence, channels with higher wavelengths would become short-wavelength channels and vice versa. As a result, the direction of Raman-induced power transfer will be reversed in the second half of the fiber span and a balance of the channel powers will be achieved at the end of the fiber link. Spectral inversion can be realized inside a fiber through phase conjugation provided by the FWM effect.

11.5 RAMAN AMPLIFICATION

Raman amplification in optical fibers was demonstrated in the early 1970s by Stolen and Ippen [8]. The benefits of Raman amplification were elucidated by many research papers in the mid-1980s [34–40]. However, due principally to the poor pumping efficiency and the scarcity of high-power pumps at appropriate wavelengths, much of that work was overtaken by erbium-doped fiber amplifiers (EDFAs) by the late 1980s. Then, in the mid-1990s, the development of suitable high-power pumps and the availability of higher Raman gain fibers motivated a renewed interest in Raman amplification. As a result, almost every long-haul transmission system developed during the 2000s uses Raman amplification.

Raman amplifiers offer some important advantages compared to other types of optical amplifiers. First, Raman gain can be provided by every fiber. Second, Raman gain can be achieved at any wavelength, provided that an adequate pump source is available. Third, Raman amplification offers a relatively broad bandwidth (>5 THz), which allows the simultaneous amplification of several channels in a WDM system or the amplification of short optical pulses. Fourth, the Raman gain spectrum can be shaped by combining multiple pump wavelengths. This aspect is particularly important and it has motivated many studies searching for optimization approaches that give the flattest gain with the fewest number of pumps [41]. In this way, amplifiers with gain bandwidths greater than 100 nm were demonstrated [42].

Figure 11.9 shows a numerical example of a broadband Raman gain obtained using a broad pump spectrum to pump a nonzero dispersion fiber (NZDF). The short-wavelength pumps amplify the longer wavelengths, and so more power is typically needed at the shortest wavelengths; this is indicated by the height of the bars in Fig. 11.9. When using a broad pump spectrum, an important issue is the interaction between the pumps, which affects the noise properties of the amplifier. Problems arise particularly due to FWM between the pumps, since it can create light at new frequencies within the signal band. This new light can interfere with the signal channels, producing beat noise [10].

Raman amplifiers can be realized considering two main options. One is the distributed Raman amplifier (DRA) that utilizes the transmission fiber itself as the Raman gain medium to obtain amplification. This option has the merit of reducing the overall excursion experienced by the signal power. Consequently, nonlinear effects are reduced at higher signal levels, whereas the SNR remains relatively high at lower signal levels. The other option is the lumped Raman amplifier (LRA), in which all the pump power is confined to a relatively short fiber element that is inserted into

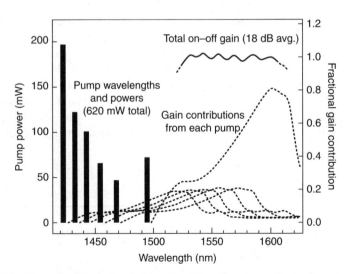

Figure 11.9 Numerical example of a broadband Raman gain obtained using a broad pump spectrum to pump a NZDF. Bars show the counterpump wavelengths and input powers. Solid line shows the total small signal on–off gain. Dashed lines show the fractional gain contribution from each pump wavelength. (After Ref. [10]; © 2004 IEEE.)

the transmission line to provide gain. The primary use of the LRA is to open new wavelength bands between about 1280 and 1530 nm, a wavelength range that is inaccessible by EDFAs.

The Raman amplifier can be pumped using three different configurations: forward pumping, backward pumping, and bidirectional pumping. In fact, the SRS can occur in both directions, forward and backward. In the first case, the signal amplification occurs near the fiber input, where the signal power is relatively high. This configuration for the Raman amplification does not generally affect the transmission quality and can compensate for the transmission loss by approximately the value of the Raman gain. For example, 10 dB Raman gain provides an increase in transmission distance of 50 km when the fiber attenuation is 0.2 dB/km. In the case of backward pumping, the signal amplification occurs near the fiber output end before photodetection, where the signal power may be very weak, comparable to the Raman noise power. This scheme for the Raman amplification can considerably reduce the minimum detectable signal power, which results in a corresponding increase in the transmission distance. However, the transmission quality is affected due to backward noise light. The bidirectional pumping configuration uses both forward and backward pumping simultaneously and its performance can be estimated from those for the respective unidirectional schemes considered before.

Figure 11.10 shows schematically a fiber Raman amplifier. The pump and the signal beams are injected into the fiber through a WDM fiber coupler. The case illustrated in Fig. 11.10 shows the two beams copropagating inside the fiber. It was confirmed experimentally that the Raman gain is almost the same for the copropagating and for the counterpropagating cases [43].

Figure 11.11 shows the output signal power $P_S(L)$ versus the input signal power $P_S(0)$ obtained numerically from Eqs. (11.17) and (11.18), considering a Raman amplifier with length $L = 8$ km and several values of the input pump power. Typical values were assumed for the calculation: $g_R = 6.7 \times 10^{-14}$ m/W, $\alpha = 0.2$ dB/km, $A_{eff} = 30 \times 10^{-12}$ m^2, $\lambda_p = 1.46$ μm, and $\lambda_S = 1.55$ μm. The transfer characteristics are linear for $P_p(0) = 0.1, 0.2$, and 0.4 W, but the effects of nonlinear pump depletion become visible for input signal levels $P_S(0) > -15$ dBm when $P_p(0) = 0.6$ W. For a pump power $P_p(0) = 0.8$ W, the linear behavior of the Raman amplifier is lost even for signal power levels as low as -35 dBm.

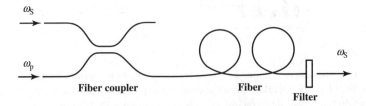

Figure 11.10 Schematic of a fiber Raman amplifier.

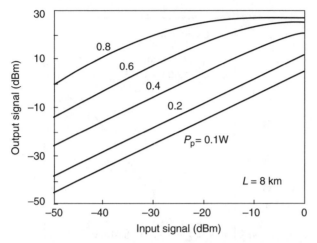

Figure 11.11   Transfer characteristics of a Raman amplifier with length $L = 8$ km for several values of the input pump power. The amplifier parameters are $g_R = 6.7 \times 10^{-14}$ m/W, $\alpha = 0.2$ dB/km, $A_{\text{eff}} = 30 \times 10^{-12}$ m^2, $\lambda_p = 1.46$ μm, and $\lambda_S = 1.55$ μm.

In the absence of Raman amplification, the signal power at the amplifier output would be $P_S(L) = P_S(0)\exp(-\alpha L)$. Hence, the amplifier gain is given by [44]

$$G_R = \frac{P_S(L)}{P_S(0)\exp(-\alpha L)} = \frac{\exp(g_R P_p(0)L_{\text{eff}}/A_{\text{eff}})}{1 + F(L)} \tag{11.38}$$

where L_{eff} is given by Eq. (11.27) and $F(L)$ is obtained from Eq. (11.23) considering $z = L$. The amplifier gain given by Eq. (11.38) is seen to be a function of the input signal power (i.e., a saturation nonlinearity) through the term $F(L)$. When $F(L) \gg 1$, Eq. (11.21) gives

$$P_S(L) = \frac{\omega_S}{\omega_p} P_p(0)\exp(-\alpha L) \tag{11.39}$$

Then, the amplifier output reaches the pump level irrespective of the input signal level. This implies that any spontaneous Raman scatter in the fiber will be amplified up to power levels comparable to that of the pump, which is clearly undesirable in an amplifier application. On the other hand, when $F(L) \ll 1$, we have from Eq. (11.21) that

$$P_S(L) = P_S(0)\exp(g_R L_{\text{eff}} P_p(0)/A_{\text{eff}} - \alpha L) \tag{11.40}$$

In this case, the amplifier gain is

$$G_R = \frac{P_s(L)}{P_s(0)\exp(-\alpha L)} = \exp(g_R L_{\text{eff}} P_p(0)/A_{\text{eff}}) \tag{11.41}$$

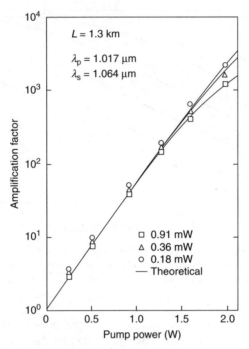

Figure 11.12 Experimentally observed variation of amplifier gain with the pump power operating at 1.064 μm by using a 1.017 μm pump for several values of the input signal power for a 1.3 km FRA. (After Ref. [34]; © 1981 Elsevier.)

From Eq. (11.41), the Raman gain in decibels is expected to increase linearly with the pump input power. This fact was confirmed experimentally, as illustrated in Fig. 11.12 [34]. The beginning of saturation of the amplifier gain can be observed in Fig. 11.12 for high values of the pump and signal powers.

The variation of the amplifier gain with the pump power $P_p(0)$ is illustrated in Fig. 11.13 for a fixed input signal power $P_S(0) = 0$ dBm and amplifier lengths $L = 2, 4$, and 8 km. The amplifier gain, in dB, shows a linear dependence on the pump power for $L = 2$ km, but it is clearly limited by the pump depletion for longer amplifier lengths. The black dot on these curves indicates the value of the pump power at which the depletion parameter F becomes equal to unity. Only operating at pump powers below such a value ensures a linear operation of the amplifier.

Short fiber Raman amplifiers can be realized using some of the highly nonlinear fibers developed during the recent years. The required pump power is also lower in such cases. In particular, conventional fibers based on tellurite or arsenic selenide (As_2Se_3) glass compositions have been drawn to obtain the Raman amplification and lasing characteristics [19,45–48]. The Raman gain efficiency is further increased using microstructured fibers. A peak gain of 10 dB was recently reported for a 1.1 m long As_2Se_3 photonic crystal fiber pumped at 1500 nm with an input power of 500 mW [48].

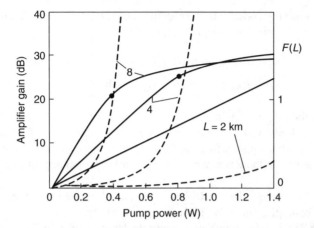

Figure 11.13 Raman amplifier gain (solid curves, left scale) and the saturation parameter $F(L)$ (dashed curves, right scale) against the pump power $P_p(0)$ for amplifier lengths $L = 2, 4$, and 8 km. The other parameters are equal to those used in Fig. 11.11.

The dominant source of noise in fiber Raman amplifiers is amplified spontaneous emission (ASE), which is generated by spontaneous Raman scattering. In fact, a part of the pump energy is spontaneously converted into Stokes radiation extending over the entire bandwidth of the Raman gain spectrum and is amplified together with the signal. The output thus consists not only of the desired signal but also of background noise extending over a wide frequency range (~ 10 THz or more).

The generation and amplification of ASE can be described by the following equation [10]:

$$\frac{dP_A}{dz} = -\alpha P_A + g_R \left[P_A + h\nu_A B_{eff}(N+1) \right] \frac{P_p}{A_{eff}} \qquad (11.42)$$

In Eq. (11.42), P_A is the ASE power in a bandwidth B_{eff} and N a phonon occupancy factor, given by

$$N = \frac{1}{\exp(h\Delta\nu/k_B T) - 1} \qquad (11.43)$$

where h is Plank's constant, k_B is Boltzmann's constant, T is the absolute temperature, and $\Delta\nu$ is the frequency separation between pump and signal. At $25°C$, $N \approx 0.14$ at the Raman gain peak.

It has been shown that for both forward and backward pumping, the noise light power is equivalent to a hypothetical injection of a single photon per unit frequency at the fiber input end for forward pumping, and at some distance away from the fiber output for backward pumping. Assuming a Raman gain such that $G_R \gg 1$ and

$\alpha L \gg 1$, the noise light powers for forward pumping, $P_{Af}(L)$, and backward pumping, $P_{Ab}(L)$, are approximately given by [44,49,50]

$$P_{Af}(L) \approx h\nu_s B_{eff}(G_R-1)\exp(-\alpha L) \qquad (11.44)$$

$$P_{Ab}(L) \approx h\nu_s B_{eff}(G_R-1)/\ln G_R \qquad (11.45)$$

where G_R is the amplifier gain given by Eq. (11.41) and ν_s is the signal frequency. Equation (11.44) shows that the noise light power in the case of forward pumping decreases as the fiber length is increased. However, in the case of backward pumping, it depends mainly on the Raman gain, being nearly independent of fiber length and loss [49,50].

In a fiber Raman amplifier, SRS has to compete frequently with other nonlinear effects. For example, since the intrinsic gain coefficient for stimulated Brillouin scattering (SBS) is two orders of magnitude larger than that for SRS, in some cases the SBS may occur at lower pump powers and may significantly affect the SRS process [51–53]. In particular, the Raman gain may not only be reduced but also become unstable [40,54]. However, as will be seen in Chapter 12, the SBS can be suppressed taking into account its narrow linewidth, which is typically less than 100 MHz. Therefore, if a pump laser with linewidth broader than the Brillouin bandwidth is employed, then the SBS effect would not be significant. When a semiconductor laser is used as pump source, its spectral linewidth can be artificially increased by direct frequency modulation, resulting in effective SBS suppression.

11.6 RAMAN FIBER LASERS

One of the first works on stimulated Raman scattering also demonstrated a Raman oscillator using mirrors to provide feedback in a 190 cm fiber [7]. The second harmonic radiation (532 nm) of a pulsed Nd:YAG laser was used for pumping the Raman oscillator. The threshold for oscillation was high, about 500 W of power in the fiber. After this pioneer work, a great interest was shown in the development of Raman fiber oscillators. As a result of such wide researches, low-threshold tunable Raman oscillators for both visible [55,56] and near-IR [57,58] regions were created in the second half of the 1970s. The fabrication of high-quality low-loss fibers helped to reduce the oscillation threshold below 1 W [56].

At high pump powers, higher order Stokes wavelengths are generated inside the fiber cavity. The use of an intracavity prism allows the spatial dispersion of these wavelengths. The Raman fiber laser (RFL) can operate at several wavelengths simultaneously if different mirrors are used for each Stokes wave. Tuning of each wavelength can be performed by turning the corresponding mirror. The first tunable RFL was demonstrated in a 1977 experiment [56], in which a tuning range of 8 nm was achieved. However, the use of bulk optics (prisms and mirrors) inside the laser cavity is not convenient, since it introduces both additional losses and mechanical instabilities.

The integration of cavity mirrors within the fiber in order to achieve a compact structure became one main goal in the research about RFLs. In an early approach, an all-single-mode RFL using a ring optical cavity and a fiber coupler was proposed [59]. With such a structure, a low threshold of 740 mW was achieved. Afterward, the discovery of the photosensitivity of optical fibers and the creation of fiber Bragg gratings (FBGs) [60,61] represented an important step in the development of Raman lasers. High-power CW radiation can be efficiently converted to several higher order Stokes beams by utilizing highly reflective FBGs as mirrors in RFLs [62,63]. Using this approach, a cascaded Raman laser can be developed, such that each intermediate Stokes radiation is efficiently converted to the next higher order Stokes until the desired order is achieved. Thus, the use of FBGs made it possible to transform a bulky setup of the multiresonant RFL into an elegant all-fiber structure.

Cascaded RFLs has been developed in order to satisfy the great demand for pumping optical amplifiers (Er doped and Raman), as well as for sources in optical fiber communication systems, especially in WDM ones. In spite of several advantages offered by germanosilicate (GeO_2/SiO_2) fibers, they present a major drawback in this context, which is related to their relatively small Stokes shift ($\sim$440 cm^{-1}). Using such fibers, six Stokes orders are necessary to convert the CW radiation of high-power double-clad fiber lasers (DCFLs) operating around 1 µm into the 1450–1480 nm spectral region, where RFLs are especially demanded. This limitation can be alleviated using phosphosilicate (P_2O_5/SiO_2) fibers, which show a Stokes shift of about 1330 cm^{-1} [64]. Almost any wavelength of the telecom range can be generated using P-doped RFLs pumped by high-power DCFLs at around 1.06 µm [63,65,66]. It is also possible to generate multiwavelength lasers simultaneously using cascaded or composite cavities [63,67,68].

An example of a two-cascaded all-fiber RFL structure is shown in Fig. 11.14 [62]. The RFL is constituted by a laser diode array pump module, a Yb-doped double-clad fiber laser, and a 1.48 µm cascaded Raman laser. The cascaded RFL cavity was formed by two pairs of FBGs with a 1 km long phosphosilicate fiber between them. The reflectivity of the FBGs was >99% except for the 1.48 µm output coupler, which was 15%. The output power of the laser is 1 W, which corresponds to 34% of the pump power of the laser diodes. Figure 11.15 shows the emission spectrum measured at the output of the cascaded RFL. The suppression of the radiation at 1.24 µm, corresponding to the first phosphorous Stokes order, is 20 dB. On the other hand, the expected silica Stokes peaks at 1.12 and 1.31 µm are absent, avoiding the use of any rejection filters.

Multiwavelength Raman fiber lasers have been investigated in view of the demand of new high-power laser sources in the near-infrared range and, in particular, for Raman amplification of high-speed WDM communication systems with a flat gain profile. Such multiwavelength RFLs have been realized in a variety of configurations, including a single Ge-doped fiber [69], or a combination of P-doped fibers together with Ge-doped fibers [63,68]. In Ref. 63, a three-wavelength RFL was demonstrated based on a chain of Raman cavities. The structure of such a laser is illustrated in Fig. 11.16 and consists of three major parts: a high-brightness optical pump source based on an Yb double-clad fiber laser, a one-stage Raman laser based on a P-doped

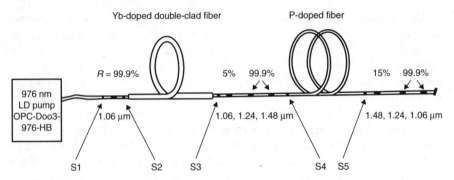

Figure 11.14 Experimental setup of 1.48 μm two-cascaded Raman laser pumped by Yb-doped double-clad fiber laser. (After Ref. [62]; © 2000 IEEE.)

fiber, and a two-stage Raman laser based on a Ge-doped fiber. The selection of 1425, 1454, and 1463 nm wavelengths was made in order to provide a flat Raman gain profile for the C-band telecommunications window. The first two parts serve to generate 1278 nm light that will be used for pumping the third part (the two-stage Raman laser). High reflector HR 1278 is used for double-pass pumping. The first stage of the Raman laser generates a relatively weak wave at 1463 nm and a strong wave at 1353 nm. The power distribution between these two waves is controlled by a variable output coupler (VOC). The 1353 nm wave serves as a double-pass pump in the second (1425/1354) stage, whereas the 1463 nm wave makes a single pass through the cavity. The power distribution between the 1425 and 1454 nm waves is defined by VOC 1454. Careful selection of pump and output wavelengths combined with the proper cavity parameters allows one to generate practically any set of 14xx output wavelengths using the structure represented in Fig. 11.16.

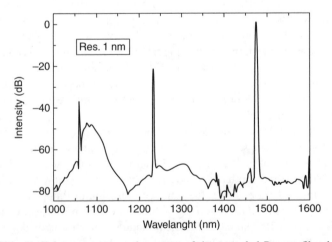

Figure 11.15 Emission spectrum at the output of the cascaded Raman fiber laser. (After Ref. [62]; © 2000 IEEE.)

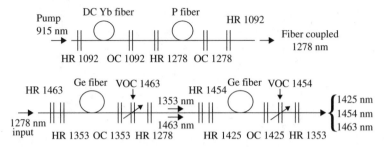

Figure 11.16   Scheme of a three-wavelength FRL. DC, double clad; OC, output coupler; HR, high reflector. (After Ref. [63]; © 2003 OSA.)

The tunability of the emitted light is an important property of any laser source. In the case of Raman fiber lasers, it is difficult to have simultaneously a high output power, high efficiency, and wide tuning range. In a 2006 experiment, an RFL with a linear cavity configuration was demonstrated, showing an output power that varied from 7.4 to 8.7 W, whereas the lasing threshold went from 1.2 to 2.2 W over the whole tuning range. However, such tuning range was limited to 15 nm, going from 1134.9 to 1150.5 nm. The narrow spectral coverage was due to the beam bending technique used in such setup [71,72].

In a recent experiment, a continuous tuning over 60 nm, from 1075 to 1135 nm, was reported with an all-fiber Raman laser also using a linear cavity configuration and a purely axial compression of the FBG [73,74]. The RFL provided up to 5 W of Stokes output power for 6.5 W of launched pump power. The RFL configuration in shown in Fig. 11.17. The Fabry–Perot cavity was composed of a 125 m long fiber and two

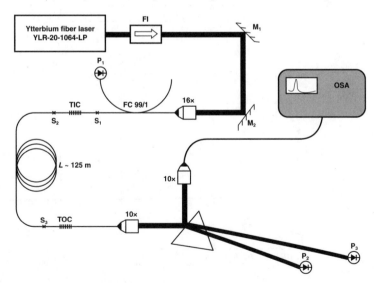

Figure 11.17   Experimental scheme of the RFL: FI (Faraday isolator), TIC (tunable input coupler), TOC (tunable output coupler), P_1, P_2, P_3 (photodiodes), S_1, S_2, S_3 (splices). (After Ref. [74]; © 2008 IEEE.)

(a)

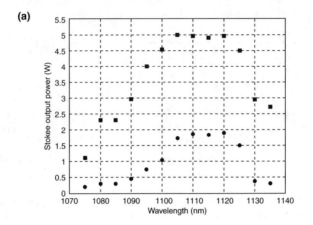

(b)

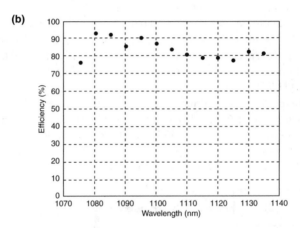

(c)

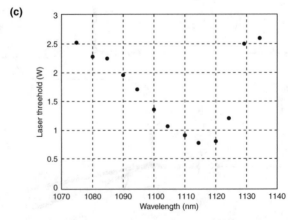

Figure 11.18 (a) Stokes output power for 3 W (circles) and 6.5 W (squares) of launched pump power, (b) laser efficiency, and (c) laser threshold as a function of the emission wavelength. (After Ref. [74]; © 2008 IEEE.)

different pairs of tunable fiber Bragg gratings (TFBGs) were used to cover the whole gain bandwidth. The first pair of TFBGs was used to tune the RFL from 1075 to 1105 nm, while the other one covers the range from 1110 to 1135 nm. The pump laser was a CW 20 W ytterbium-doped fiber laser operating at 1064 nm. At the output end, a prism spatially separated the residual pump power and the Stokes power.

Using the configuration of Fig. 11.17, a maximum Stokes output power of several watts over 60 nm for 6.5W of launched pump power was achieved, as shown by Fig. 11.18a. Variations of laser efficiency from 76.1% to 93.1% were observed along the tuning range (Fig. 11.18b). Since the length of the fiber and the reflectivity of the gratings were constant over the whole tuning range, a minimum laser threshold was expected to occur near the peak of the Raman gain spectrum, located at 1115 nm. A minimum laser threshold of 0.78 W was indeed observed at this wavelength, as shown in Fig. 11.18c.

PROBLEMS

11.1 Explain the origin of Raman scattering and distinguish between spontaneous and stimulated Raman scattering.

11.2 Verify that Eqs. (11.21)–(11.23) are indeed solutions of Eqs. (11.17) and (11.18), when $\alpha_S = \alpha_p = \alpha$.

11.3 Starting from Eqs. (11.21)–(11.23) and assuming a sufficiently low loss coefficient such that $\alpha z \ll 1$ for all the considered distances, find an expression for the distance at which the intensities of the pump and Stokes waves become equal.

11.4 Estimate the threshold power for stimulated Raman scattering in a single-mode fiber with a length of 50 km and an effective core area $A_{eff} = 50\,\mu m^2$ at (i) 1.55 μm and (ii) 1.3 μm. Assume a Raman gain coefficient $g_R = 5 \times 10^{-14}$ m/W and that the fiber loss is in the case (i) $\alpha = 0.2$ dB/km and in case (ii) $\alpha = 0.6$ dB/km.

11.5 Derive Eq. (11.36) for the fractional power loss of the shortest wavelength channel in a WDM system, assuming that the Raman gain spectrum has a triangular profile.

11.6 Consider two WDM channels at wavelengths $\lambda_1 = 1535$ nm and $\lambda_2 = 1550$ nm transmitted in a standard single-mode fiber that has an attenuation constant $\alpha = 0.3$ dB/km, an effective area $A_{eff} = 50\,\mu m^2$, and a Raman gain coefficient $g_R = 4 \times 10^{-14}$ m/W. Above which power level does one of the channels experience a growth during propagation?

11.7 Describe the main advantages offered by Raman fiber amplifiers compared to erbium-doped fiber amplifiers.

11.8 Consider a Raman amplifier constituted by a 4 km long fiber with an effective area $A_{eff} = 30 \times 10^{-12}$ m^2, a loss coefficient $\alpha = 0.2$ dB/km, and a Raman coefficient $g_R = 6.7 \times 10^{-14}$ m/W. The pump and signal wavelengths are $\lambda_p = 1.46$ μm and $\lambda_S = 1.55$ μm, respectively. Assuming an input pump power $P_p(0) = 800$ mW, calculate the signal power at which the depletion parameter F given by Eq. (11.23) becomes unity. Calculate the amplifier gain in such a case.

REFERENCES

1. C. V. Raman and K. S. Krishnan, *Nature* **121**, 501 (1928).

2. R. W. Hellwarth, *Phys. Rev.* **130**, 1850 (1963).

3. Y. R. Chen and N. Bloembergen, *Am. J. Phys.* **35**, 989 (1967).

4. A. Penzkofer, A. Laubereau, and W. Kaiser, *Prog. Quantum Electron.* **6**, 55 (1982).

5. R. H. Stolen, J. P. Gordon, W. J. Tomlinson, and H. A. Haus, *J. Opt. Soc. Am. B* **6**, 1159 (1989).

6. E. P. Ippen, *Appl. Phys. Lett.* **16**, 303 (1970).

7. R. H. Stolen, E. P. Ippen, and A. R, Tynes, *Appl. Phys. Lett.* **20**, 62 (1972).

8. R. H. Stolen and E. P. Ippen, *Appl. Phys. Lett.* **22**, 276 (1973).

9. A. Yariv, *Quantum Electronics*, 3rd ed., Wiley, New York, 1989.

10. J. Bromage, *J. Lightwave Technol.* **22**, 79 (2004).

11. F. L. Galeener, J. C. Mikkelsen, R. H. Geils, and W. J. Mosby, *Appl. Phys. Lett.* **32**, 34 (1978).

12. N. Shibata, M. Horigudhi, and T. Edahiro, *J. Non-Cryst. Solids*, **45**, 115 (1981).

13. T. Nakashima, S. Seikai, and M. Nakazawa, *Opt. Lett.* **10**, 420 (1985).

14. S. K. Sharma, D. W. Matson, J. A. Philpotts, and T. L. Roush, *J. Electrochem. Soc.* **133**, 431 (1984).

15. S. T. Davey D. L. Williams, B. J. Ainslie, W. J. M. Rothwell, and B. Wakefield, *Proc. Inst. Elect. Eng.* **136**, 301 (1989).

16. J. Bromage, K. Rottwitt, and M. E Lines, *IEEE Photon. Technol. Lett.* **14**, 24 (2002).

17. C. Fukai, K. Nakajima, J. Zhou, K. Tajima, K. Kurokawa, and I. Sankawa, *Opt. Lett.* **29**, 545 (2004).

18. R. Stegeman, L. Jankovic, H. Kim, C. Rivero, G. Stegeman, K. Richardson, P. Delfyett, Y. Guo, A. Schulte, and T. Cardinal, *Opt. Lett.* **28**, 1126 (2003).

19. L. B. Shaw, P. C. Pureza, V. Q. Nghuyen, J. S. Sanghera, and I. D. Aggarwal, *Opt. Lett.* **28**, 1406 (2003).

20. R. E. Slusher, G. Lenz, J. Hodelin, J. S. Sanghera, L. B. Shaw, and I. D. Aggarwal, *J. Opt. Soc. Am. B* **21**, 1146 (2004).

21. R. G. Smith, *Appl. Opt.* **11**, 2489 (1972).

22. L. G. Cohen and C. Lin, *IEEE J. Quantum Electron.* **14**, 855 (1978).

23. A. Tomita, *Opt. Lett.* **8**, 412 (1983).

24. A. R. Chraplyvy, *Electron. Lett.* **20**, 58 (1984).

25. D. Cotter and A. M. Hill, *Electron. Lett.* **20**, 185 (1984).

26. M. S. Kao and J. Wu, *J. Lightwave Technol.* **7**, 1290 (1989).

27. S. Chi and S. C. Wang, *Electron. Lett.* **26**, 1509 (1990).

28. A. R. Chraplyvy, *J. Lightwave Technol.* **8**, 1548 (1990).

29. A. R. Chraplyvy and R. W. Tkach, *IEEE Photon. Technol. Lett.* **5**, 666 (1993).

30. S. Tariq and J. C. Palais, *J. Lightwave Technol.* **11**, 1914 (1993).

31. M. E. Marhic, F. S. Yang, and L. G. Kazpvsky, *J. Opt. Soc. Am. B* **15**, 957 (1998).

32. K.-P. Ho, *J. Lightwave Technol.* **18**, 915 (2000).

33. T. Yamamoto and S. Norimatsu, *Opt. Commun.* **225**, 101 (2003).

34. M. Ikeda, *Opt. Commun.* **39**, 148 (1981).

35. A. R. Chraplyvy, J. Stone, and C. A. Burrus, *Opt. Lett.* **8**, 415 (1983).

36. M. Nakazawa, *Appl. Phys. Lett.* **46**, 628 (1985).

37. M. Nakazawa, T. Nakashima, and S. Seikai, *J. Opt. Soc. Am. B* **2**, 215 (1985).

38. M. L. Dakss and P. Melman, *J. Lightwave Technol.* **3**, 806 (1985).

39. N. A. Olsson and J. Hegarty, *J. Lightwave Technol.* **4**, 391 (1986).

40. Y. Aoki, S. Kishida, and K. Washio, *Appl. Opt.* **25**, 1056 (1986).

41. Y. Emori, K. Tanaka, and S. Namiki, *Electron. Lett.* **35**, 1355 (1999).

42. C. Fukai, K. Nakajima, J. Zhou, K. Tajima, K. Kurokawa, and I. Sankawa, *Opt. Lett.* **29**, 545 (2004).

43. Y. Aoki, S. Kishida, H. Honmou, K. Washio, and M. Sugimoto, *Electron. Lett.* **19**, 620 (1983).

44. M. F. Ferreira, in B. Guenther, A. Miller, L. Bayvel, and J. Midwinter (Eds.), *Encyclopedia of Modern Optics*, Academic Press, 2004.

45. A. Mori, H. Masuda, K. Shikano, and M. Shimizu, *J. Lightwave Technol.* **21**, 1300 (2003).

46. H. Masuda, A. Mori, K. Shikano, and M. Shimizu, *J. Lightwave Technol.* **24**, 504 (2006).

47. P. A. Thielen, L. B. Shaw, J. Sanghers, and I. Aggarwal, *Opt. Express* **11**, 3248 (2003).

48. S. K. Varshney, K. Saito, K. Lizawa, Y. Tsuchida, M. Koshiba, and R. K. Sinha, *Opt. Lett.* **33**, 2431 (2008).

49. Y. Aoki, *Opt. Quantum Electron.* **21**, S89 (1989).

50. Y. Aoki, *J. Lightwave Technol.* **6**, 1225 (1988).

51. B. Foley, M. L. Dakss, R. W. Davies, and P. Melman, *J. Lightwave Technol.* **7**, 2024 (1989).

52. M. F. Ferreira, J. F. Rocha, and J. L. Pinto, *Electron. Lett.* **27**, 1576 (1991).

53. S. Hamidi, D. Simeonidou, A. S. Siddiqui, and T. Chaleon, *Electron. Lett.* **28**, 1768 (1992).

54. G. A. Koepf, D. M. Kalen, and K. H. Greene, *Electron. Lett.* **18**, 942 (1982).

55. K. O. Hill, B. S. Kawasaki, and D. C. Johnson, *Appl. Phys. Lett.* **29**, 181 (1977).

56. R. K. Jain, C. Lin, R. H. Stolen, W. Pleibel, and P. Kaiser, *Appl. Phys. Lett.* **30**, 162 (1977).

57. C. Lin, L. G. Cohen, R. H. Stolen, G. H. Tasker, and W. G. French, *Opt. Commun.* **20**, 426 (1977).

58. C. Lin, R. H. Stolen, and L. G. Cohen, *Appl. Phys. Lett.* **31**, 97 (1977).

59. E. Desurvire, A. Imamoglu, and H. J. Shaw, *J. Lightwave Technol.* **5**, 89 (1987).

60. K. O. Hill, Y. Fujii, D. C. Johnson, and B. S. Kawasaki, *Appl. Phys. Lett.* **32**, 647 (1978).

61. G. Meltz, W. W. Morey, and W. H. Glenn, *Opt. Lett.* **14**, 823 (1989).

62. E. M. Dianov and A. M. Prokhorov, *IEEE J. Sel. Top. Quantum Electron.* **6**, 1022 (2000).

63. A. A. Dernidov, A. N. Starodumov, X. Li, A. Martinez-Rios, and H. Po, *Opt. Lett.* **28**, 1540 (2003).

64. V. V. Grigoryants, B. L. Davydov, M. E. Zhabotinski, V. F. Zolin, G. A. Ivanov, V. I. Sminorv, and Y. K. Chamorovski, *Opt. Quantum Electron.* **9**, 351 (1977).

65. Z. Xiong, N. Moore, Z. G. Li, and G. C. Lim, *J. Lightwave Technol.* **21**, 2377 (2003).

66. Z. Xiong, N. Moore, Z. G. Li, G. C. Lim, D. M. Liu, and D. X. Huang, *Opt. Commun.* **239**, 137 (2004).

67. C. S. Kim, R. M. Sova, and J. U. Kang, *Opt. Commun.* **218**, 291 (2003).

68. Z. Xiong and T. Chen, *Opt. Fiber Technol.* **13**, 81 (2007).

69. M. D. Mermelstein, C. Headley, J. C. Bouteiller, P. Steinvurzel, C. Horn, K. Feder, and B. J. Eggleton, *IEEE Photon. Technol. Lett.* **13**, 1286 (2001).

70. E. Bélanger, D. Faucher, M. Bernier, and R. Vallée, *Proceedings of FIO 2006*, 2006, Paper JWD86.

71. C. S. Gogh, M. R. Mokhtar, S. A. Bitler, S. Y. Set, K. Kikuchi, and M. Ibsen, *IEEE Photon. Technol. Lett.* **15**, 557 (2003).

72. E. Bélanger, S. Gagnon, M. Bernier, J. P. Bérubé, and R. Vallée, *Appl. Opt.* **46**, 3089 (2007).

73. E. Bélanger, S. Gagnon, M. Bernier, J. P. Bérubé, D. Cotê, and R. Vallée, *Appl. Opt.* **47**, 652 (2008).

74. E. Bélanger, M. Bernier, D. Faucher, D. Côté, and R. Vallée, *J. Lightwave Technol.* **26**, 1696 (2008).

12

STIMULATED BRILLOUIN SCATTERING

Brillouin scattering, a phenomenon named after the French physicist Leon Brillouin, was used to investigate the scattering of light at acoustic waves during the 1920s. For low incident intensities, the scattered part of the field remains very weak. However, the process becomes stimulated and strong scattered fields are generated at high input intensities, which are readily available with lasers. Stimulated Brillouin scattering (SBS) in optical fibers was observed for the first time in 1972 [1], using a pulsed narrowband xenon laser operating at 535.3 nm.

SBS is similar to stimulated Raman scattering, in that a Stokes wave that is downshift in frequency from that of a strong pump wave is also generated in this case. However, instead of the material resonances, which play a fundamental role in the Raman scattering, the interaction between the pump and Stokes waves in the SBS process occurs by way of an acoustic wave. In the classical picture, the pump wave creates a pressure wave in the medium through electrostriction. A material density wave is established that propagates at the velocity of sound in the medium in the direction of the pump. The periodic changes in material density appear as a moving refractive index grating. Thus, the incident wave pumps the acoustic wave that scatters it, and the scattering creates a Stokes wave. A significant portion of the pump optical power may be converted into the Stokes wave that travels in the backward direction. Due to the Doppler shift that occurs as the pump diffracts from the moving index grating, the Stokes wave is shifted to a lower frequency with respect to the pump. The response of the material to the interference of the pump and Stokes fields tends to increase the amplitude of the acoustic wave. On the other hand, the beating of the pump wave with the acoustic wave tends to reinforce the Stokes wave. This explains the appearance of the stimulated Brillouin scattering process.

In spite of the apparent similarity between SBS and stimulated Raman scattering (SRS), these processes differ in three important aspects: (1) Brillouin amplification occurs only when the pump and signal beams counterpropagate inside the fiber; (2) the

Nonlinear Effects in Optical Fibers. By Mário F. S. Ferreira.
Copyright © 2011 John Wiley & Sons, Inc. Published 2011 by John Wiley & Sons, Inc.

Stokes shift for SBS is smaller by three orders of magnitude compared to that of SRS; and (3) the Brillouin gain spectrum is extremely narrow, with a bandwidth <100 MHz. On the other hand, the peak of the Brillouin gain coefficient is over 100 times greater than the Raman gain peak, which makes SBS the dominant nonlinear process in silica fibers under some circumstances [2–5]. This is particularly the case in fiber systems that use narrow-linewidth lasers. SBS can be detrimental to such systems in a number of ways: by introducing a severe signal attenuation, by causing multiple frequency shifts, and by introducing a high-intensity backward coupling into the transmission optics. However, Brillouin gain can also find some useful applications, namely, for optical amplification [6,7], lasing [8], channel selection in closely spaced wavelength-multiplexed network [9], optical phase conjugation [10], temperature and strain sensing [11,12], all-optical slow-light control [13,14], optical storage [15], and so on.

12.1   LIGHT SCATTERING AT ACOUSTIC WAVES

As in the case of Raman scattering, the SBS process can be described as a classical three-wave interaction involving the incident (pump) wave of frequency ω_p, the Stokes wave of frequency ω_S, and an acoustic wave of frequency ω_a. The pump creates a pressure wave in the medium owing to electrostriction, which in turn causes a periodic modulation of the refractive index. Such index grating scatters some pump light, giving rise to the Stokes wave. This process is illustrated schematically in Fig. 12.1. Physically, each pump photon in the SBS process gives up its energy to create simultaneously a Stokes photon and an acoustic phonon.

Since the acoustic wave fronts are moving away from the incident pump wave, the scattered light is shifted downward in frequency to the Stokes frequency

$$\omega_S = \omega_p - \omega_a \tag{12.1}$$

On the other hand, the momentum conservation is given by the condition

$$\boldsymbol{\beta}_S = \boldsymbol{\beta}_p - \boldsymbol{\beta}_a \tag{12.2}$$

where $\boldsymbol{\beta}_a$ is the wave vector of the acoustic wave, whereas $\boldsymbol{\beta}_p$ and $\boldsymbol{\beta}_S$ are the wave vectors of the pump and Stokes waves, respectively. Assuming the approximation

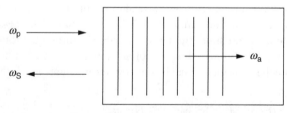

Figure 12.1   Schematic illustration of the stimulated Brillouin scattering process.

$\beta_p \approx \beta_S$, the frequency ω_a and the wave vector $\boldsymbol{\beta}_a$ of the acoustic wave can be related through the following dispersion relation:

$$\omega_a = |\boldsymbol{\beta}_a| v_a = 2v_a |\boldsymbol{\beta}_p| \sin(\theta) \tag{12.3}$$

where θ is the half angle between the pump and Stokes waves and v_a is the acoustic velocity. The acoustic velocity in turn depends on Young's modulus, T_e, and on the mass density, ρ, being given by

$$v_a = \sqrt{\frac{T_e}{\rho}} \tag{12.4}$$

Equation (12.3) shows that ω_a vanishes in the forward direction ($\theta = 0$), whereas it is maximum in the backward direction ($\theta = \pi/2$). In optical fibers, only the forward and backward directions are possible. The Brillouin frequency shift of the back-scattered wave is given by

$$v_B = \frac{\omega_a}{2\pi} = \frac{2nv_a}{\lambda_p} \tag{12.5}$$

where Eq. (12.3) was used with $\beta_p = 2\pi n/\lambda_p$, where λ_p is the pump wavelength and n is the fiber refractive index. Considering $n = 1.45$ and $v_a = 5.96$ km/s as typical values for silica glass, we obtain a Brillouin frequency shift $v_B = 11.1$ GHz at $\lambda_p = 1.55$ μm. However, the amount of the Brillouin shift depends on the dopant concentration in the fiber [16–18]. Increasing the dopant concentration increases the refractive index and reduces the acoustic velocity, but the last effect is dominant. As a consequence, the frequency shift decreases when the dopant concentration is increased. The dependence of the Brillouin frequency shift on fluorine and germania concentrations of a codoped silica fiber at 1550 nm can be described by the following empirical relation [18]:

$$v_B = 11.045 - (0.277[F] + 0.045[GeO_2]) \quad (\text{GHz}) \tag{12.6}$$

where both dopant concentrations are in wt%. Equation (12.6) shows that the Brillouin shift is especially affected by a change of the F concentration.

The Brillouin spectra in optical fibers can differ significantly from that of bulk silica not only due to the presence of dopants in the fiber core but also because of the guided nature of light [19–23]. Besides the longitudinal acoustic mode, which is responsible for SBS, new vibrational modes can also exist in the core and cladding depending on their respective densities and on the structure of the interface between them [24]. These modes can also scatter light, giving rise to multiple resonances in the Brillouin spectra [19].

Figure 12.2a shows the Brillouin spectra for various optical fibers obtained by using a DFB laser at 1552 nm [19]. The fibers have different refractive index profiles, which are represented in Fig. 12.2b. The amount of frequency shift, the number of

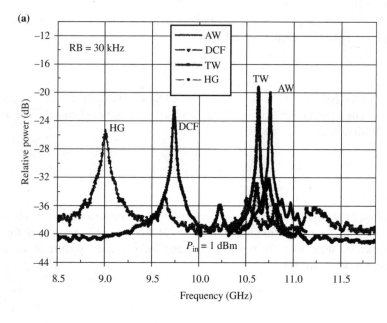

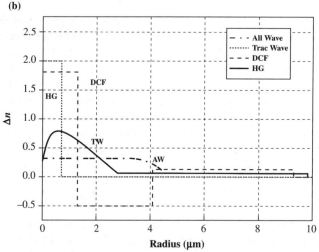

Figure 12.2 (a) Spontaneous Brilouin gain spectra and (b) refractive index profiles of AW, TW, DCF, and HG fibers. (After Ref. [19]; © 2002 IEEE.)

modes, and the separation between them are different for each fiber. The Brillouin shift is inversely proportional to the GeO$_2$ concentrations, which were calculated to be of 21%, 17.7%, 5.4%, and 4.6%, respectively, for highly germanium-doped (HG), dispersion-compensating fibers (DCFs), TrueWave (TW) fibers, and AllWave (AW) fibers. Besides the main Brillouin peak, multiple resonance modes are observed in Fig. 12.2a for each fiber. However, it was observed that, above the threshold of stimulated scattering, only the main peak increases with input power, whereas the other resonance peaks decrease and eventually vanish [19].

12.2 THE COUPLED EQUATIONS FOR STIMULATED BRILLOUIN SCATTERING

The displacement from equilibrium, $X(z,t)$, of a small volume in the medium as a result of the applied electrostrictive force is governed by the following equation of motion [25]:

$$\frac{\partial^2 X}{\partial t^2} + \Gamma_B \frac{\partial X}{\partial t} - v_a^2 \frac{\partial^2 X}{\partial z^2} = \frac{\gamma_e}{2\rho} \frac{\partial E^2}{\partial z} \tag{12.7}$$

where Γ_B is a phenomenological damping coefficient, γ_e is the electrostrictive coefficient, and $E = E_S + E_p$ is the total field, given by the sum of the Stokes and pump fields. The forcing function on the right-hand side of Eq. (12.7) is obtained considering that the force per unit volume in the material is given by the gradient of the stored energy in the electric fields, in analogy with the treatment of Section 11.1.

The electric fields of the pump and Stokes waves, as well as the acoustic wave, are assumed in the form

$$E_p(z,t) = \frac{1}{2}E_{0p}(z)\exp\left[i(\beta_p z - \omega_p t)\right] + \text{c.c.} \tag{12.8}$$

$$E_S(z,t) = \frac{1}{2}E_{0S}(z)\exp[i(-\beta_S z - \omega_S t)] + \text{c.c.} \tag{12.9}$$

$$X(z,t) = \frac{1}{2}X_0(z)\exp[i(\beta_a z - \omega_a t)] + \text{c.c.} \tag{12.10}$$

Considering only the terms that oscillate at $\omega_a = \omega_p - \omega_S$ in E^2 and using the slowly varying envelope approximation, we find from (12.7) the following equation for X_0:

$$\frac{dX_0}{dz} = -\frac{\gamma_e}{4\rho v_a^2} E_{0p} E_{0S}^* \exp(i\Delta\beta z) - \frac{\alpha_a}{2} X_0 \tag{12.11}$$

where $\alpha_a = \Gamma_B/v_a$ is the attenuation coefficient and $\Delta\beta = \beta_p + \beta_S - \beta_a$ is the phase mismatch per unit distance. Integrating Eq. (12.11) and assuming that the rate of change of the two electric fields is much lower than the acoustic wave damping rate, the following result can be obtained for $X_0(z)$:

$$X_0(z) = -\frac{\gamma_e}{4\rho v_a^2} \frac{\exp(i\Delta\beta z) - \exp(-\alpha_a z/2)}{(\alpha_a/2 + i\Delta\beta)} E_{0p}(z) E_{0S}^*(z) \tag{12.12}$$

In writing Eq. (12.12), it was assumed that $X(0) = 0$.

The nonlinear polarization is related to the variation of the medium permittivity, $\Delta\varepsilon$, determined by the pressure wave induced strain, and is given by [25]

$$P_{NL} = \Delta\varepsilon E = -\gamma_e \frac{\partial X}{\partial z} E \qquad (12.13)$$

Substituting Eqs. (12.8), (12.9), and (12.13) into the nonlinear wave equation (4.13), an equation at ω_S and another at ω_p can be obtained. Using the slowly varying envelope approximation, the following equation at ω_S is derived:

$$\frac{dE_{0S}}{dz} = \frac{\gamma_e \beta_a \beta_S}{4\varepsilon_0 n^2} E_{0p} X_0^* \exp(i\Delta\beta z) + \frac{\alpha}{2} E_{0S} \qquad (12.14)$$

where the last term was added phenomenologically to account for linear losses in the medium; the exponential power attenuation coefficient α will be assumed to be the same at the Stokes and pump wavelengths.

Inserting the result (12.12) for X_0 in Eq. (12.14) provides the following equation describing the growth of the Stokes field:

$$\frac{dE_{0S}}{dz} = \left[-\frac{K|E_{0p}|^2}{(\alpha_a/2 - i\Delta\beta)} + \frac{\alpha}{2} \right] E_{0S} \qquad (12.15)$$

where

$$K = \frac{\gamma_e^2 \beta_a \beta_S}{16\varepsilon_0 n^2 \rho v_a^2} \qquad (12.16)$$

and the condition $\exp(-\alpha_a z/2) \ll 1$ was assumed. Following a similar procedure, the following equation can also be obtained for the evolution of the pump field:

$$\frac{dE_{0p}}{dz} = \left[-\frac{K|E_{0S}|^2}{(\alpha_a/2 + i\Delta\beta)} - \frac{\alpha}{2} \right] E_{0p} \qquad (12.17)$$

The phase mismatch $\Delta\beta$ appearing in the above equations can be written in terms of the deviation $\Delta\omega_S = \omega_S - \omega_{S0}$ of the Stokes frequency relatively to its phase-matched value $\omega_{S0} = \omega_p - \omega_a$ as

$$\Delta\beta \approx \frac{\Delta\omega_S}{v_a} \qquad (12.18)$$

12.3 BRILLOUIN GAIN AND BANDWIDTH

The wave intensity is related to its amplitude through

$$I_j = \frac{1}{2}\varepsilon_0 cn|E_{0j}|^2, \quad j = p, S \qquad (12.19)$$

Using Eq. (12.19) and the relation $P_j = I_j A_{eff}$, where P_j is the optical power and A_{eff} is the effective core area, Eqs. (12.15) and (12.17) can be written in terms of optical power as follows:

$$\frac{dP_S}{dz} = -\frac{g_B(\Delta\omega_S)}{A_{eff}} P_p P_S + \alpha P_S \tag{12.20}$$

$$\frac{dP_p}{dz} = -\frac{g_B(\Delta\omega_S)}{A_{eff}} P_S P_p - \alpha P_p \tag{12.21}$$

where $g_B(\Delta\omega_S)$ is the Brillouin gain coefficient, given by

$$g_B(\Delta\omega_S) = \frac{(\Gamma_B/2)^2}{(\Delta\omega_S)^2 + (\Gamma_B/2)^2} g_{B0} \tag{12.22}$$

As shown by Eq. (12.22), the spectrum of the Brillouin gain is Lorentzian with a FWHM $\Delta v_B = \Gamma_B/2\pi$, where

$$\Gamma_B = \alpha_a v_a \tag{12.23}$$

is determined by the acoustic phonon lifetime. For bulk silica, the width is expected to be about 17 MHz at 1.5 μm. However, several experiments have shown much larger bandwidths for silica-based fibers, which in some cases can exceed 100 MHz [18,20]. Such larger bandwidths are generally attributed both to the guided nature of acoustic modes and to inhomogeneities in fiber core cross section along the fiber length.

The peak value of the Brillouin gain coefficient, g_{B0}, is given by

$$g_{B0} = \frac{\gamma_e^2 \beta_a \beta_S}{2\varepsilon_0^2 c n^3 \rho v_a \Gamma_B} \tag{12.24}$$

Considering the parameter values typical of fused silica, the Brillouin coefficient g_{B0} is estimated to be about 2.5×10^{-11} m/W, which is between two and three orders of magnitude larger than the Raman gain coefficient in silica fibers at $\lambda_p = 1.55$ μm. Larger values of the Brillouin gain coefficient can yet be obtained in some non-silica-based fibers. For example, a peak value of 1.6989×10^{-10} m/W was recently reported for a tellurite fiber [26], whereas a value of $\sim 6.08 \times 10^{-9}$ m/W, which is more than 200 times larger than that of silica, has been measured in a single-mode As_2Se_3 chalcogenide fiber [27].

The expression for the Brillouin gain peak in Eq. (12.24) assumes that the pump and Stokes waves have the same polarization. However, this is not verified in a standard single-mode fiber, in which case the Brillouin gain coefficient is reduced to half. On the other hand, Eq. (12.24) is valid only when the spectral width of the pump beam (Δv_p) is much narrower than the Brillouin linewidth (Δv_B). In general, however, to take into account the finite spectral width of the pump beam, the convolution of the power spectrum of the pump field with the intrinsic Brillouin frequency spectrum must be calculated [3,28]. Assuming that the pump spectrum has a Lorentzian profile, such convolution is a Lorentzian of width $\Delta v_B + \Delta v_p$ and thus the Brillouin gain coefficient is reduced according to the following relation [3,28]:

$$\tilde{g}_{B0} = \frac{\Delta v_B}{\Delta v_B + \Delta v_p} g_{B0} \qquad (12.25)$$

The result in Eq. (12.25) applies also to the case of a laser output comprising a very large number of longitudinal modes with completely random phase relationships and contained in a spectral envelope of width Δv_p greater than the Brillouin linewidth Δv_B.

12.4 THRESHOLD OF STIMULATED BRILLOUIN SCATTERING

In order to estimate the threshold in the Brillouin scattering process, we can consider the case of a nondepleted pump wave, where the solutions of Eqs. (12.20) and (12.21) are given by

$$P_p(z) = P_{p0} \exp(-\alpha z) \qquad (12.26)$$

$$P_S(z) = P_{S0} \exp(\alpha z) \exp\left[-\frac{g_B P_{p0} (1 - \exp(-\alpha z))}{A_{eff} \alpha} \right] \qquad (12.27)$$

where $P_{p0} = P_p(z = 0)$ and $P_{S0} = P_S(z = 0)$ are the input pump power and the output Stokes power, respectively. Actually, a Stokes input signal at $z = L$ grows to produce an output signal at $z = 0$, given by

$$P_{S0} = P_S(L) \exp(-\alpha L) \exp\left[\frac{g_B P_{p0} L_{eff}}{A_{eff}} \right] \qquad (12.28)$$

where $L_{eff} = [1 - \exp(-\alpha L)]/\alpha$ is the effective fiber length.

In the absence of an input signal, the Stokes wave builds up from spontaneous scattering. In a treatment analogous to that of Raman scattering, the noise power provided by spontaneous Brillouin scattering is equivalent to injecting a fictitious photon per mode at a distance where the gain is equal to the fiber loss. The threshold pump power is defined as the input pump power that is equal to the backward output Stokes power, which gives the following result [29]:

$$P_{p0}^{th} = 21 \frac{K_B A_{eff}}{g_{B0} L_{eff}} \qquad (12.29)$$

where K_B is a factor varying between 1 and 2 that takes into account the polarization dependence of the pump and Stokes waves. If both waves propagate in a polarization maintaining fiber and have the same polarization, we have $K_B = 1$. In the case of a nonpolarization maintaining fiber, it is $K_B = 1.5$ [30]. For contemporary low-loss transmission fibers, the factor 21 in Eq. (11.29) should be replaced with a smaller number between 17 and 18 [31,32]. Considering an effective area $A_{eff} = 50\,\mu m^2$, an attenuation constant $\alpha = 0.2\,dB/km$, a gain coefficient $g_{B0} = 2.5 \times 10^{-11}\,m/W$, and $K_B = 1$, we obtain from Eq. (12.29) a threshold of $P_{p0}^{th} \approx 2\,mW$. This value is around three orders of magnitude smaller than the threshold required for Raman scattering, which makes stimulated Brillouin scattering the dominant nonlinear effect in some circumstances.

In practice, the SBS threshold shows very strong differences when considering different types of fibers. A threshold of only $\sim$3 mW was measured in a TrueWave fiber with a core diameter of 7.2 μm and an effective length of 16.1 km, whereas it was about 80 mW for a highly Ge-doped fiber with a core diameter of 2.5 μm and an effective length of 0.3 km [19]. Figure 12.3 shows the transmitted (right axis) and reflected SBS (left axis) powers as a function of the injected power for a 13 km long dispersion-shifted fiber [33]. It can be observed that no more than 3 mW could be transmitted through the fiber in this experiment after the onset of SBS.

Launch powers well above 10 mW are needed in some applications, namely, in repeaterless transmission systems. In order to make possible the transmission of such power levels, the SBS threshold is increased by using several schemes. In general, these schemes are based on the increase of either the spectral width of the optical carrier or the Brillouin bandwidth of the fiber. Modulating the phase of the optical carrier at a frequency Δv_m increases the SBS threshold by a factor $(1 + \Delta v_m/\Delta v_B)$. This factor can be greater than 10 for typical modulation frequencies of some hundreds of MHz. On the other hand, broadening of the Brillouin gain bandwidth can be achieved through the concatenation of fibers with different Brillouin shifts. Such Brillouin shifts can be changed by making the core radius nonuniform or by varying the dopant concentration along the fiber length. The fiber's Brillouin

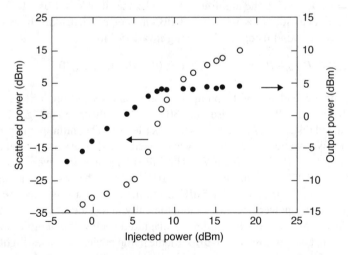

Figure 12.3 Transmitted power (solid circles, right axis) and reflected SBS power (open circles, left axis) against the injected power into a 13 km long fiber. (After Ref. [33]; © 1992 IEEE.)

linewidth can also be broadened by strain applied in cabling, or varying the draw conditions when the fiber is manufactured [34].

12.5 SBS IN ACTIVE FIBERS

The SBS process was discussed in previous sections considering mainly the case of conventional lossy fibers. However, if the fiber is used as an amplifying medium (e.g., as a distributed Raman or erbium-doped fiber amplifier), in which case it is called an *active* fiber, some new features are expected [35–39]. In particular, considering the long interaction length provided by such a fiber, we can anticipate that even a relatively weak narrowband signal (pump) will lead to the ready appearance of higher order Stokes waves.

The equations describing the evolution of the signal (pump) power, P_p, first-order Stokes power, P_{S1}, and second-order Stokes power, P_{S2}, along an optical fiber with distributed amplification can be written in the following form [37,38]:

$$\frac{dP_p}{dz} = -\frac{g_B}{A_{eff}} P_p P_{S1} + \delta P_p \qquad (12.30)$$

$$\frac{dP_{S1}}{dz} = -\frac{g_B}{A_{eff}} P_p P_{S1} + \frac{g_B}{A_{eff}} P_{S2} P_{S1} - \delta P_{S1} \qquad (12.31)$$

$$\frac{dP_{S2}}{dz} = \frac{g_B}{A_{eff}} P_{S2} P_{S1} + \delta P_{S2} \qquad (12.32)$$

where $\delta = g - \alpha$, g being the amplifier gain and α the fiber loss. Eqs. (12.30–12.32) have no analytical solutions when $\delta \neq 0\,dB/km$. For $\delta = 0\,dB/km$, however, such solutions can be found using the following invariant [40]:

$$P = P_p(z) - P_{S1}(z) + P_{S2}(z) = P_p(0) - P_{S1}(0) + P_{S2}(0) \qquad (12.33)$$

The spatial dependence of the pump, first-order, and second-order Stokes powers, obtained by numerically solving Eqs. (12.30–12.32), is illustrated in Fig. 12.4 against the normalized distance z/L, where L is the fiber length. The initial pump power was $P_p(0) = 1\,mW$, whereas the Stokes waves were seeded assuming initial noise powers $P_{S1}(L) = P_{S2}(0) = 0.5\,nW$. The fiber parameters were $L = 20\,km$, $A_{eff} = 28.26 \times 10^{-12}\,m^2$, and $g_B = 2.2 \times 10^{-11}\,m/W$, whereas the net gain/loss parameter was $\delta = -0.5, 0$, and $1.5\,dB/km$. In the case of a lossy fiber ($\delta = -0.5\,dB/km$), the pump power decreases monotonically along the fiber, while the first-order Stokes power remains five orders of magnitude below the pump power. For a fiber with no net gain or loss ($\delta = 0\,dB/km$), the pump power decreases slightly in the initial stage of the fiber and remains practically constant thereafter, whereas the first-order Stokes power increases to $0.19\,mW$ and the second-order Stokes power remains negligible. Finally, when the fiber has a positive distributed gain ($\delta = 1.5\,dB/km$), we

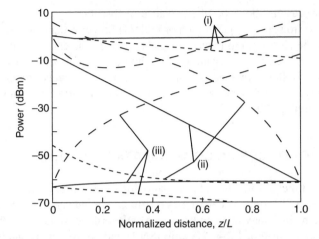

Figure 12.4 Pump (i), first-order (ii), and second-order (iii) Stokes powers against normalized fiber length for $\delta = -0.5$ (short-dashed curves), 0 (solid curves), and 1.5 dB/km (long-dashed curves).

observe that the pump power presents a nonmonotonic evolution and increases after passing through a minimum value. The first-order Stokes power becomes similar to the pump power, whereas the output second-order Stokes power achieves a value of 0.13 mW.

Figure 12.5 shows that, despite the fact that the output pump power increases monotonically with the net gain δ, this growth is slower for $\delta > 0$ dB/km. This is due to the nonlinear pump depletion effect determined by the first-order Stokes wave, which becomes of the same order of the pump for $\delta > 0$ dB/km. In this range, the second-order Stokes output power increases almost exponentially with the fiber gain by seven orders of magnitude.

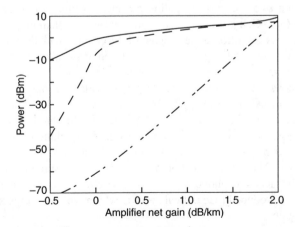

Figure 12.5 Pump (solid curve), first-order (dashed curve), and second-order (dash-dotted curve) powers against fiber net gain for $L = 20$ km.

The above results show that the presence of distribute gain in an optical fiber generally leads to the ready generation of multiple Stokes orders and to a significant variation of the signal pump power along the fiber. In the case of Raman-pumped fibers, these characteristics depend on the particular pump conditions. Numerical calculations have shown that the SBS threshold power is inversely proportional to the path-average integral of Raman gain and that it is lower for forward pumping than for backward pumping [39].

12.6   IMPACT OF SBS ON COMMUNICATION SYSTEMS

SBS can affect the performance of a transmission system in several ways. First, the threshold of the SBS process determines the maximum power that can be launched into the system. Such maximum power can be of the order of some few mW. This fact limits the maximum signal-to-noise ratio (SNR) and the transmission distance that can be reached without amplification. Once the SBS threshold is surpassed, as a consequence of the power transfer to the Stokes wave, the pump signal is depleted, which again determines a degradation of the SNR and leads to an increase of the bit error rate (BER). Moreover, the backward propagating Stokes wave can destabilize and even destroy the signal transmitter if no optical isolator is appropriately inserted in the system.

In actual transmission systems, optical amplifiers are periodically inserted to compensate for the fiber losses. Each amplifier generally includes an optical isolator, which avoids the passage and successive growth of the backward propagating Stokes wave. In spite of this action, SBS between consecutive amplifiers can still degrade the system performance if the signal power is above the threshold.

Another main detrimental effect of SBS is related to the interchannel crosstalk in wavelength division multiplexed (WDM) systems. Such crosstalk occurs only if the fiber link supports the propagation of channels in opposite direction and if the channel spacing between two counterpropagating channels is approximately equal to the Brillouin shift ($\sim$11 GHz). If both these conditions are fulfilled, the channel with the Stokes frequency is amplified at the expense of the channel with the pump frequency. In fact, impairments resulting from SBS-induced crosstalk can be observed in bidirectional transmission systems at power levels far below the SBS threshold [40,41]. However, this kind of crosstalk can be easily suppressed with a slight change of the channel spacing.

Much attention has been paid to estimating the SBS limitations in practical fiber transmission systems. SBS is very sensitive to signal modulation because the origin of SBS involves a process that is not instantaneous on the timescale of the information rate. The narrow Brillouin linewidth is a consequence of the long lifetimes of the acoustic phonons involved in light scattering. In general, high modulation rates produce broad optical spectra, which will determine a reduction of the Brillouin gain.

Concerning the coherent transmission systems, the SBS threshold depends on whether the amplitude, phase, or frequency of the optical carrier is modulated for information coding. Assuming a fixed bit pattern and that the fundamental modulation frequency for amplitude-shift keying (ASK) and phase-shift keying (PSK), as well as that the difference between the two frequencies of the frequency-shift keying (FSK), is much higher than the bandwidth of the Brillouin gain, it can be shown that the powers of the distinct spectral components of pump and Stokes waves satisfy a pair of coupled equations similar to Eqs. (12.20) and (12.21) [42]. In these circumstances, the different frequency components of the modulated wave will not influence each other. For WDM systems, SBS will not occur if each frequency in each individual channel remains below threshold. Under the same conditions, it was shown that the threshold for ASK, PSK, and FSK systems is 2, 2.5, and 4 times, respectively, that of a CW wave [42].

The evaluation of the SBS threshold becomes more complicated when considering a pseudorandom modulation of the pump wave. The results then depend on the particular encoded scheme and on the ratio $R/\Delta v_B$, where R is the bit rate. For NRZ-ASK, for example, the launched field amplitude can be described by [43]

$$E(t) = E_0(1-[1-m(t)][1-(1-k_a)^{1/2}]) \tag{12.34}$$

where k_a is the depth of intensity modulation $(0 < k_a < 1)$ and the function $m(t)$ represents the binary data stream. In this case, assuming that the pump is not depleted by the SBS process, the Brillouin gain becomes [44]

$$g = g_B\left[\left(1-\frac{a}{2}\right)^2 + \frac{a^2}{4}\left(1-\frac{R}{\Delta v_B}(1-e^{-\Delta v_B/R})\right)\right] \tag{12.35}$$

where $a = 1-(1-k_a)^{1/2}$. According to Eq. (12.35), the SBS gain for ASK is minimized when $k_a = 1$, which corresponds to a 100% modulation depth. When $R \ll \Delta v_B$, the Brillouin gain becomes $g \approx g_B/2$, whereas for $R \gg \Delta v_B$, we have $g \approx g_B/4$. The dependence of g on $R/\Delta v_B$ is illustrated in Fig. 12.6.

For PSK modulation, the information is encoded by modulating the phase of the carrier, as given by [43]

$$E(t) = E_0\, e^{ik_p m(t)} \tag{12.36}$$

where k_p is the keyed phase shift. The SBS gain in this case is given by [44]

$$g = g_B\left[\frac{1}{2}(1 + \cos k_p) + \frac{1}{2}(1-\cos k_p)\left[1-\frac{R}{\Delta v_B}(1-e^{-\Delta v_B/R})\right]\right] \tag{12.37}$$

According to Eq. (12.37), the smallest gain for PSK is achieved when the modulation index is $k_p = \pi(2n+1)$. For high bit rates, the Brillouin gain decreases linearly with $R/\Delta v_B$ (see Fig. 12.6).

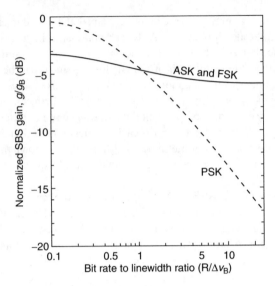

Figure 12.6 Normalized SBS gain as a function of the ratio of bit rate to Brillouin linewidth. (After Ref. [44]; © 1989 IEEE.)

If the modulation index of the FSK, $k_f = \omega_1 - \omega_2$, is much higher than the bit rate R, the FSK spectrum is just the sum of two ASK spectra with $k_a = 1$ centered about ω_1 and ω_2. Therefore, the Brillouin gain for FSK is [43,44]

$$g = g_B \left[\frac{1}{2} - \frac{R}{4\Delta v_B} \left(1 - e^{-\Delta v_B/R} \right) \right] \qquad (12.38)$$

The dependence of g on $R/\Delta v_B$ for FSK is the same as ASK, as illustrated in Fig. 12.6.

In summary, the above results show that pseudorandom modulation of the optical wave determines a reduction of the SBS effects, which becomes more significant with increasing bit rates. In ASK and FSK systems, the maximum reduction of the Brillouin gain is a factor of 4, whereas in PSK systems the SBS gain decreases linearly with the bit rate.

12.7 FIBER BRILLOUIN AMPLIFIERS

SBS in optical fibers is a highly efficient nonlinear amplification mechanism with which large gains can be achieved using pump powers of only a few milliwatts. Such a nonlinear process can be used to construct a fiber Brillouin amplifier (FBA), which is basically an optical fiber in which the pump and signal waves propagate in opposite directions. The only required condition is that the spacing between the pump and signal frequencies must correspond to the Brillouin shift in that fiber.

12.7.1 Amplifier Gain

Considering that $\alpha \ll 1$, the following approximate solutions of Eqs. (12.20) and (12.21) can be derived [7]:

$$P_S(z) = P_S(0)D(z)e^{\alpha z} \tag{12.39}$$

$$P_p(z) = P_p(0)D(z)H(z)e^{-\alpha z} \tag{12.40}$$

where

$$D(z) = \frac{[P_p(0) - P_S(0)]}{[P_p(0)H(z) - P_S(0)]} \tag{12.41}$$

and

$$H(z) = \exp\{g_B[P_p(0) - P_S(0)][1 - \exp(-\alpha z)]/(A_{\text{eff}}\alpha)\} \tag{12.42}$$

The Brillouin amplifier gain is given from Eq. (12.39) as

$$G_B = \frac{P_S(0)}{P_S(L)e^{-\alpha L}} = \frac{1}{D(L)} \tag{12.43}$$

where $D(L)$ is given by Eq. (12.41) with $z = L$.

Figure 12.7 shows the transfer characteristics for a Brillouin amplifier with length $L = 8$ km, considering different input pump powers and typical parameter values: $g_B(0) = 2.5 \times 10^{-11}$ m/W, $\alpha = 0.2$ dB/km, $A_{\text{eff}} = 30 \times 10^{-12}$ m^2. The transfer characteristics are linear for $P_p(0) = 0.2$ mW, but the effects of pump depletion become

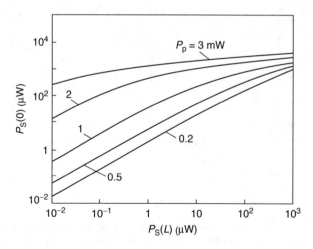

Figure 12.7 Transfer characteristics for a Brillouin amplifier with a length $L = 8$ km and several values of the input pump power.

clearly observable for higher pump levels. For a pump power $P_p(0) = 2\,\text{mW}$, a linear behavior of the Brillouin amplifier cannot be observed even for signal powers as low as $\sim 0.1\,\mu\text{W}$.

The dependence of the Brillouin amplifier gain on the input pump power is illustrated in Fig. 12.8 for several amounts of the signal detuning from the gain peak and for two values of the signal power. A fiber length $L = 5\,\text{km}$, a gain bandwidth $\Delta \nu_B = 30\,\text{MHz}$, and parameter values equal to those in Fig. 12.7 were assumed. Figure 12.8 shows that the amplifier gain depends on the signal magnitude in the saturation regime. An input signal of 1 nW, which is comparable to the original value of spontaneous emission, can be amplified by about 60 dB for a pump power $P_p(0) \approx 5\,\text{mW}$. However, the amplifier gain can be considerably reduced when the signal is slightly detuned from the gain peak. For example, for an input signal power $P_s(L) = 1\,\text{nW}$ and a pump power $P_p(0) \approx 3\,\text{mW}$ a detuning of 15 MHz, which corresponds to a reduction of the gain coefficient to half of its maximum value, determines a reduction of about 25 dB on the amplifier gain.

The narrow Brillouin linewidth strictly limits the bandwidth of data signals that can be amplified in an FBA. However, as discussed previously, the intrinsic Brillouin linewidth is generally enhanced by compositional inhomogeneities in the fiber and can be intentionally extended by more than one order of magnitude by applying frequency modulation to the pump laser. Of course, the bandwidth enhancement is accompanied by a reduction of the peak gain. Consequently, a higher pump power will be necessary to achieve the same gain.

In a 1986 experiment, 5 dB of net gain was obtained in a 37.5 km long FBA by using a pump power of 3.5 mW [45]. The gain occurred over a 150 MHz bandwidth,

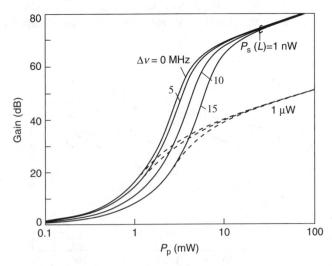

Figure 12.8 Brillouin amplifier gain against the input pump power for an amplifier length $L = 5\,\text{km}$ and different amounts of detuning from the gain center. The solid curves and the dashed curves correspond to input signal powers $P_s(L) = 1\,\text{nW}$ and $P_S(L) = 1\,\mu\text{W}$, respectively.

due to fiber nonuniformity. Later on, the technique of broadening the gain bandwidth by pump laser frequency sweeping was demonstrated and applied to the amplification of data-carrying signals at 10 and 90 Mb/s [46]. The receiver sensitivities showed the full improvement of the gain achieved in these measurements [47], since the signals were attenuated before reaching the receiver.

Brillouin amplifiers made with conventional silica fibers generally use fiber lengths of some kilometers. However, shorter Brillouin amplifiers can be realized using fibers made of highly nonlinear glasses [27,48–50]. Tellurite fibers offer the advantage of having a relatively low background loss of 0.02 dB/m at 1550 nm [51,52], which means a larger effective length and consequently a greater amplifier gain. A Brillouin gain of 29 dB and a linewidth of 20.98 MHz were recently reported using a tellurite fiber with a length of only 100 m and a pump power of 10 mW at 1550 nm [26].

12.7.2 Amplifier Noise

The Brillouin gain discussed above amplifies not only the signal light but also the Stokes light generated by scattering from thermally generated acoustic waves. This is the source of the amplified spontaneous emission (ASE) noise that affects the performance of the FBA. To describe the ASE noise, we must add a spontaneous emission term to the equations describing the evolution of the Stokes and pump waves. The Stokes power per unit frequency satisfies the following equation [6,7]:

$$\frac{dP_S(v,z)}{dz} = \alpha P_S(v,z) - g_B[P_S(v,z) + hv(N+1)]\frac{P_p}{A_{eff}} \qquad (12.44)$$

where h is Planck's constant and N is the thermal equilibrium number of acoustic phonons, given by [6]

$$N = \frac{1}{\exp(hv_B/kT)-1} \approx kT/hv_B \approx 500 \qquad (12.45)$$

where k is the Boltzmann constant and T is the absolute temperature. Assuming that the pump power distribution is given by Eq. (12.40), the solution of Eq. (12.44) is given by [7]

$$P_{asp}(v,z) = \left(\frac{hv}{A_{eff}}\right)g_B(N+1)G(v,z)\int_z^L P_p(z')[G(v,z')]^{-1}\,dz' \qquad (12.46)$$

where

$$G(v,z) = \exp\left[\int_z^L\left(\frac{P_p(z)g_B}{A_{eff}} - \alpha\right)dz\right] \qquad (12.47)$$

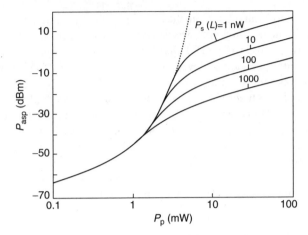

Figure 12.9 Total amplified spontaneous power $P_{asp}(0)$ against the input pump power for four values of the input signal power tuned at the line center. The dashed line is obtained by neglecting the nonlinear pump depletion. The amplifier length is $L = 5$ km.

is the gain function. The total ASE power, P_{asp}, is given by integration of $p_{asp}(v, z)$ over the gain profile.

Figure 12.9 shows the total ASE power against the pump power for four values of the signal tuned at the line center, assuming the same parameters used for Figs 12.7 and 12.8. Similarly to the behavior of the gain (Fig. 12.8), the noise power depends significantly on the signal magnitude in the saturation regime and becomes particularly high for low signal powers. This high noise level, which is about 20 dB above that of an ideal amplifier, imposes severe limitations on the use of FBA as a receiver preamplifier.

From a practical standpoint, the quantity of interest is the so-called on/off ratio, defined as the power ratio of the amplified signal to the amplified ASE. It has been shown that a significant degradation of the on/off ratio occurs when the signal is detuned from the gain peak [7]. In practice, not only the gain but also the on/off ratio imposes a very stringent requirement on the frequency alignment of the pump and signal.

12.7.3 Other Applications of the SBS Gain

Thee narrowband Brillouin gain can be used to advantage in coherent lightwave systems to selectively amplify the carrier of a phase- or amplitude-modulated wave, thus allowing homodyne detection, using the amplified carrier as local oscillator [53]. With sufficient carrier gain, it should be possible to achieve quantum noise-limited detection. This offers advantages of increased sensitivity similar to that of coherent optical homodyne detection, but without the engineering problems of phase locking the local oscillator laser in a coherent homodyne receiver. The scheme should work well at bit rates >100 Mb/s because the modulation sidebands then fall outside the amplifier bandwidth, and the optical carrier can be amplified selectively.

Substituting Eq. (12.40) for $P_p(z)$ in Eq. (12.15), the following result is obtained for the Stokes field at the amplifier output:

$$E_{0S}(0) = E_{0S}(L)\sqrt{G_B}\exp\left\{-\frac{\alpha L}{2}-j\phi\right\} \qquad (12.48)$$

where

$$\phi = \frac{\Delta\omega_S}{\Gamma_B}\ln G_B \qquad (12.49)$$

represents a nonlinear phase change, which depends on the SBS gain, input signal power, and frequency detuning $\Delta\omega_S$. This nonlinear phase shift can lead to undesirable amplitude modulation of a frequency-modulated signal [54]. Moreover, it imposes a stringent limit on the pump and signal frequency stability when FBAs are used in some phase-sensitive detection schemes. Figure 12.10 shows the phase change ϕ against the detuning $\Delta\omega_S/2\pi$ for several values of the pump power, $P_S(L) = 100\,\text{nW}$, and $L = 5\,\text{km}$. If the Brillouin amplifier is used to perform a self-homodyne coherent receiver [53] for ASK signals, a phase stability of about 0.1 rad for the amplified signal carrier may be required. From Fig. 12.10, it can be seen that this is satisfied for a detuning $<200\,\text{kHz}$.

The nonlinear phase shift can be used with advantage in some self-homodyne schemes that require some specific adjustment of the carrier phase. In the case of a self-homodyne coherent receiver for PSK signals, for example, a quadrature phase correction is required [55]. Figure 12.10 shows that it is achieved with a detuning $\Delta\omega_S/2\pi \approx 4\,\text{MHz}$. The phase stability concerning variations of the signal or pump

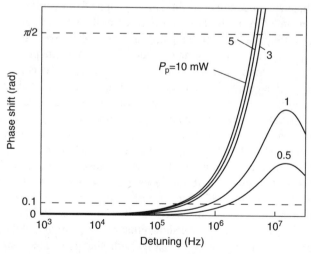

Figure 12.10   Nonlinear phase shift ϕ against the signal detuning $\Delta\omega_S/2\pi$ for several values of the pump power, $P_S(L) = 100\,\text{nW}$, and $L = 5\,\text{km}$.

powers is clearly improved by operation of the Brillouin amplifier in the saturated regime, as can be observed from the curves corresponding to higher pump powers in Fig. 12.10.

The narrow linewidth associated with the Brillouin gain can be used to advantage to realize a tunable narrowband filter for channel selection in a densely packed multichannel communication system [56]. A channel can be selectively amplified through Brillouin amplification by launching a pump beam at the receiver end so that it propagates inside the fiber in a direction opposite to that of the multichannel signal. Different channels can be selectively amplified by tuning the pump laser. This technique was demonstrated in a 1986 experiment [56], where 45 Mb/s channels as closely spaced as 140 MHz were successfully demultiplexed. Dispersion-shifted fibers with triangular core dopant profiles and a pump power of 14 mW were used to achieve a Brillouin gain bandwidth of 100 MHz and 25 dB of gain at 1550 nm. Modulating the pump wave can increase the bandwidth of the SBS gain in order to accommodate high bit rate WDM channels [47,57]. The adjustable shape and bandwidth, as well as the high out-of-band rejection, are interesting features that can be used to advantage in this approach [9].

Another important application of SBS occurs in the area of optical fiber sensors. Since the Brillouin frequency shift given by Eq. (12.5) depends on the fiber refractive index, it changes with every alteration of the refractive index. This alteration can be caused by local environmental influences such as temperature or a mechanical distortion of the fiber. Therefore, by monitoring changes of the Brillouin shift it is possible to get information on the temperature or mechanical stress along the fiber. This provides the possibility of making distributed fiber sensors capable of sensing temperature and strain changes over relatively long distances [11,12,58–69]. Such sensors can be used, for example, to provide information about the temperature distribution along streets or to monitor the mechanical stress of bridges or pipelines.

In a distributed temperature or strain sensor, the light from a pulsed pump laser and that from a tunable CW probe laser are injected at the opposite ends of a fiber. If the frequency difference between the pump and the probe waves corresponds to the Brillouin frequency shift (BFS) at any position in the fiber, the probe wave will be amplified while the pump wave is weakened. The time delay between the transmitted pulse and the received amplification of the probe signal indicates the position where Brillouin amplification occurs. The distribution of temperature or strain over the entire fiber length can be obtained by changing the frequency difference between the pump and probe waves. In a 1993 experiment, two diode-pumped 1319 nm Nd:YAG lasers were used as the pump and probe sources [58]. A spatial resolution of 10 m and a temperature resolution of 1°C were achieved with a 22 km long fiber.

The Brillouin frequency shift increases linearly with both temperature and strain. In one experiment, a BFS increase of 1.36 MHz per °C and of 594.1 MHz per percent elongation were measured at 1320 nm in a standard single-mode fiber [61]. Experimental results have shown that both the strain and the temperature coefficients of the BFS decrease linearly with the germania concentration in the fiber core while in slightly different slopes: -1.48% versus -1.61% per unit mole percentage, respectively [17]. The sensitivity to temperature or strain variations can be quantitatively enhanced by using fluorine-doped instead of germania-doped fibers [17,62].

Several different schemes based on correlation-domain [63], time-domain [64], or frequency-domain [65] techniques have been developed to realize SBS-based distributed sensors. However, all these techniques suffer a common physical difficulty in discriminating the response to strain from that to temperature, since the BFS is sensitive to both. One technique to distinguish strain from temperature consists in measuring the Brillouin gain coefficient or the Brillouin linewidth together with BFS [66]. The linewidth decreases and the gain peak increases with increasing temperature, whereas they remain approximately unchanged with strain [61]. Another method to discriminate the strain and temperature effects uses the resonance frequencies from two or more acoustic modes in a special fiber [67,68]. This technique seems particularly promising, since the measurement accuracy of BFS can be significantly enhanced [67].

12.8 SBS SLOW LIGHT

The velocity at which a pulse of light propagates through a medium (group velocity) is given by

$$v_g = \frac{c}{n + \omega(\mathrm{d}n/\mathrm{d}\omega)} \tag{12.50}$$

Equation (12.50) shows that the group velocity depends not only on the refractive index but also on the dispersion (i.e., $\mathrm{d}n/\mathrm{d}\omega$). The group velocity can exceed c if $\mathrm{d}n/\mathrm{d}\omega < 0$, which corresponds to anomalous dispersion. This case occurs when the frequency ω is close to the absorption frequency of the medium. On the contrary, the group velocity can be made much smaller than c if $\mathrm{d}n/\mathrm{d}\omega$ is large and positive, corresponding to a normal dispersion regime. This situation is achieved when the frequency ω is close to the amplifying resonance frequency of the medium. Such slow light has attracted much attention in the scientific communities since the first experimental demonstration of the phenomenon in the early 1980s [69]. As the pulse is slowed in the medium, nonlinear interactions are greatly enhanced due to compression of the local energy density, which provides the occurrence of nonlinear processes at much lower operating powers than in more common conditions. Some important applications for slow light have been proposed, including optical buffering, data synchronization, optical memories, and signal processing [70–74]. Among the various methods that have been used to realize the slow light are the electromagnetically induced transparency (EIT) [75], the coherent population oscillation (CPO) [76], and, more recently, the optical resonances associated with SBS [13,14,77], SRS [78], and parametric amplification [79] in optical fibers.

In an SBS slow-light experiment, a continuous-wave pump beam, with angular frequency ω_p, propagates through an optical fiber giving rise to amplifying and absorbing resonances due to the process of electrostriction. A counterpropagating probe pulse experiences slow (fast) light propagation when its carrier frequency is set to the amplifying (absorbing) resonance, which occurs in the vicinity of the Stokes (anti-Stokes) frequency. Fiber-based SBS slow light offers several advantages compared to other approaches: the use of an optical fiber provides long interaction lengths, the involved resonance can be created at any wavelength by changing the pump wavelength, and the process can be achieved at room temperature [73,74,77]. The SBS method should be, in principle, limited to data rates of a few tens of Mb/s due to the narrow Brillouin resonance width, which is about 30 MHz in standard single-mode fibers. However, recent experiments have successfully demonstrated the possibility of increasing the SBS slow-light bandwidth by broadening the spectrum of the pump field [14,80].

The complex SBS gain function can be written from Eq. (12.15) as

$$g = g_0 \frac{\Gamma_B}{\Gamma_B - i2(\omega - \omega_p + \omega_a)} \tag{12.51}$$

where $g_0 = g_{B0}/2$ is the peak value. Considering the propagation of a pulse along a fiber of length L, the difference between the transit times of the pulse with and without SBS is given by [77]

$$\Delta t_d = \frac{G_0}{\Gamma_B} \left\{ 1 - 3 \left[\frac{2(\omega - \omega_{S0})}{\Gamma_B} \right]^2 \right\} \tag{12.52}$$

where $G_0 = g_0 I_{p0} L$ is the gain parameter and I_{p0} is the pump intensity. The maximum delay occurs at the peak of the Brillouin gain and is equal to G_0/Γ_B. On the other hand, the signal pulse broadening factor b is given by [77]

$$b = \frac{t_{out}}{t_{in}} = \left[1 + \frac{16 \ln 2 G_0}{t_{in}^2 \Gamma_B^2} \right]^{1/2} \tag{12.53}$$

where t_{in} (t_{out}) is the duration (full width at half maximum) of the Gaussian-shaped Stokes pulses at the input (output). Equations (12.52) and (12.53) show that the SBS-induced delay can be controlled through the intensity of the pump field and that such a delay is always accompanied by some pulse distortion. The relative time delay for a fixed value of b is given by

$$\frac{\Delta t_d}{t_{in}} = \left[\frac{b^2 - 1}{16 \ln 2} \right]^{1/2} \sqrt{G_0} \tag{12.54}$$

In practice, amplified spontaneous scattering from thermal phonons causes the saturation of the pump field when $G_0 > 25$ [77]. Considering $G_0 = 25$ and $b = 2$, we find from Eq. (12.54) that $\Delta t_d/t_{in} = 2.6$.

As mentioned before, the width of the SBS gain curve can be increased by using a broadband pump. In such a case, the complex SBS gain function is given by the convolution of the intrinsic SBS gain spectrum and the power spectrum of the pump field. Considering a pump source with a Gaussian power spectrum of width $\Delta\omega_p$ smaller than Brillouin frequency shift ω_a, the complex gain function is approximately given by [14]

$$g(\omega) = g_0 I_{p0} \sqrt{\pi}\eta \exp(-\xi^2)\mathrm{erfc}(-i\xi) \qquad (12.55)$$

where erfc is the complementary error function, $\xi = (\omega + \omega_a - \omega_{p0})/\Delta\omega_p$, and $\eta = \Gamma_B/(2\Delta\omega_p)$. In obtaining Eq. (12.55), it was assumed that $\eta \ll 1$. The width of the gain profile is now given by $\Gamma = 2\sqrt{\ln 2}\Delta\omega_p$, instead of Γ_B, whereas the gain parameter is given by $G = \sqrt{\pi}\eta G_0$. The SBS slow-light delay at the gain peak is now given by [14]

$$\Delta t_d = \frac{2\sqrt{\ln 2}}{\sqrt{\pi}}\frac{G}{\Gamma} \approx 0.94\frac{G}{\Gamma} \qquad (12.56)$$

When the pump spectral width $\Delta\omega_p$ is larger than the Brillouin shift ω_a, the Stokes gain is cancelled by the anti-Stokes absorption, which shifts the effective SBS gain peak to lower frequencies and, hence, reduces the slow-light delay. However, it was found that a partial overlap of the Stokes and anti-Stokes resonances can actually lead to an enhancement of the slow-light delay–bandwidth product when $\Delta\omega_p \sim 1.3\omega_a$ [14]. Using this approach, a Brillouin slow-light bandwidth of 12.6 GHz was achieved using a 2 km long highly nonlinear fiber (HNLF), thus demonstrating that it can be integrated into existing data systems operating over 10 Gb/s [14]. Figure 12.11a shows the measured SBS gain spectrum, which is well fitted by the prediction of Eq. (12.55). Figure 12.11b–d shows the experimental results for the pulse delay, pulse width, and pulse waveforms in the case of an input pulse width of 75 ps. A 47 ps SBS slow-light delay is achieved at a pump power of ~580 mW that is coupled into the HNLF, which gives a gain of about 14 dB. The pulse delay scales linearly with the gain, demonstrating the ability to control all-optically the slow-light delay.

As discussed above, the SBS slow-light pulse delay Δt_d is proportional to G/Γ. The decrease in G that accompanies the increase in the SBS slow-light bandwidth needs to be compensated by increasing the fiber length or pump power, and/or using highly nonlinear optical fibers. In these circumstances, the use of fibers made of highly nonlinear glasses is a good option to enhance the slow-light generation. The use of a 2 m bismuth oxide HNLF was reported to generate 29 dB Brillouin gain and the resultant optical delay of 46 ns with 410 mW pump power [81]. In another experiment, a Brillouin gain of 43 dB was achieved with only 60 mW pump power in a 5 m long As_2O_3 chalcogenide fiber, which leads to an optical time delay of 37 ns [82].

Besides generally offering a higher value of the Brillouin gain coefficient, highly nonlinear glasses also present higher background losses than silica. When considering the background loss, the fiber length L must be replaced by the effective length

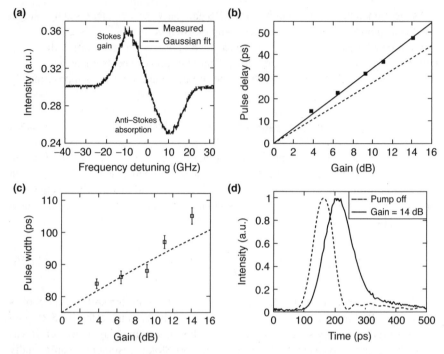

Figure 12.11 (a) Measured SBS gain with a dual Gaussian fit, (b) pulse delay, and (c) pulse width as a function of SBS gain. In (b), the solid line is the linear fit of the measured data (solid squares), and the dashed line is obtained with Eq. (12.56). (d) Pulse waveforms at 0 and 14 dB SBS gain. (After Ref. [14]; © 2007 IEEE.)

$L_{\text{eff}} = (1-\exp(-\alpha L))/\alpha$ when writing the gain parameter G_0. In this case, the maximum SBS-induced time delay per unit power can be written as

$$\frac{\Delta t_d}{P_{p0}} = \frac{g_0 K}{A_{\text{eff}} \Gamma_B}(1-\exp(-\alpha L))/\alpha \qquad (12.57)$$

The value of $(g_0 K)/(A_{\text{eff}} \Gamma_B)$ for the cases of silica, bismuth oxide, tellurite, and chalcogenide fibers is 0.000336, 0.067, 0.09257, and 0.2 ns/(mW m), respectively, whereas the background loss for the same fibers is 0.0005, 0.8, 0.02, and 1 dB/m, respectively [26]. The dependence of the time delay per unit power on the real fiber length is illustrated in Fig. 12.12. It can be seen that, among the four types of fibers, the largest time delay per unit power ($\sim$19.9 ns/mW) can be obtained from tellurite fiber due to its low background loss and relatively higher Brillouin gain coefficient.

12.9 FIBER BRILLOUIN LASERS

A Brillouin fiber laser (BFL) can be built by placing the fiber into the cavity of a resonator. A BFL with very narrow linewidth can be designed for virtually any

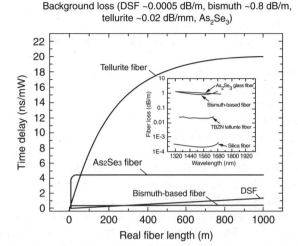

Background loss (DSF ~0.0005 dB/m, bismuth ~0.8 dB/m, tellurite ~0.02 dB/mm, As$_2$Se$_3$)

Figure 12.12 Dependence of SBS-induced time delay per unit power on the real fiber length for four types of fibers. *Inset*: The background losses of the four types of fibers. (After Ref. [26]; © 2008 IEEE.)

wavelength, provided that the required pump laser is available, thus showing a high potential for applications in metrology and spectroscopy [83,84]. Typically, most BFLs operate in a ring configuration, but there have also been some reports on Fabry–Perot cavity BFLs. In the Fabry–Perot geometry, both forward and backward propagating components are present, associated with the pump and Stokes waves. With sufficient pump intensities, higher order Stokes wave generation is possible through a cascaded SBS process, whereas four-wave mixing between copropagating pump and Stokes waves provides the generation of anti-Stokes waves [85]. Moreover, the performance of the output (supress) pump power is generally higher than the Brillouin laser peak power [86]. Some of these problems were solved in a recent experiment, in which a BFL with a Fabry–Perot cavity with improved characteristics was demonstrated [87]. Such a laser provided the possibility of achieving a peak power higher than the transmitted Brillouin pump power, besides avoiding the generation of higher order anti-Stokes and Stokes waves.

BFLs using a ring cavity are often preferred in order to avoid the generation of higher order Stokes waves through cascaded SBS. Using such a configuration, submilliwatt thresholds are easily achievable if the pump signal coincides with one of the resonator modes [88]. For optimum operation of the laser, a servo loop is generally used for frequency locking of the pump wave to a resonant frequency of the fiber ring resonator [89]. Moreover, due to its narrow bandwidth and small Brillouin gain coefficient, BFLs usually require the use of an active stabilization mechanism to achieve a continuous and stable operation [90].

Ring cavity BFLs can be used as high-performance laser gyroscopes, the so-called Brillouin fiber-optic gyroscopes (FOGs) [91–93]. The advantage of the Brillouin FOG over the interferometric and resonant FOGs is its direct readout of rotation rate, measured as a beat frequency between counterpropagating lasing waves inside the fiber resonator. However, as happens with other types of fiber-optic gyroscopes, the performance of Brillouin FOGs is affected by the Kerr effect, which causes a nonreciprocity between the two counterpropagating light waves in the fiber coil, resulting in a spurious rotation signal [93].

Hybrid Brillouin erbium fiber lasers (BEFLs) have been the object of much attention since the 1990s, considering their operation either in a single wavelength or at several wavelengths simultaneously, which can be tuned over a wide range [8,94–98]. The BEFL presents two gain media, which contribute to different aspects of the overall operation. The linear gain of the erbium-doped fiber (EDF) allows for large output power generation and contributes the majority of the output power. The SBS process provides additional gain at the characteristic Stokes-shifted frequency from the injected Brillouin pump, to determine the wavelength of operation.

Figure 12.13 shows the schematic diagram of a ring-type BEFL [8]. A Brillouin pump, P, is used to generate Brillouin gain in the single-mode optical fiber (SMOF) in the clockwise direction. The additional gain in the clockwise direction over that produced by the EDF occurs in a small bandwidth at the Stokes-shifted frequency from the Brillouin pump. The gain produced by the EDF is just less than the threshold gain, and the addition of the Brillouin gain allows laser action at the Stokes-shifted frequency. Figure 12.14 shows the evolution of the Brillouin signal as a function of the 980 nm pumping level. Also shown is the Brillouin pump that was kept at a constant level. The figure clearly indicates that the BEFL output signal can become larger than the Brillouin pump at certain levels of the 980 nm pump.

With sufficient pump intensities, multiwavelength operation of a BEFL is possible through a cascaded SBS process. Multiwavelength fiber laser sources working at the room temperature can find applications in many fields, such as precise spectroscopy, photonics component characterization, optical sensing, and dense wavelength division multiplexing (DWDM) optical communication [99]. In the case of a ring-type BEFL, its output contains only redshifted Stokes lines, each propagating in a single direction only, and hence even- and odd-order Stokes lines are observed in the clockwise and counterclockwise directions, respectively. This problem can be

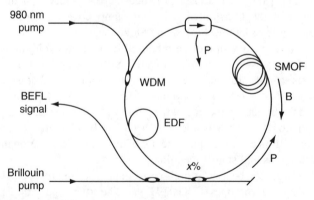

Figure 12.13 Schematic diagram of the Brillouin erbium fiber laser. (After Ref. [8]; © 1997 IEEE.)

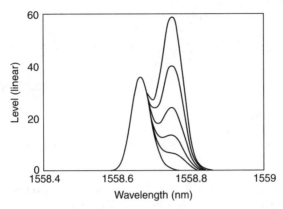

Figure 12.14   Optical spectrum of the BEFL output mixed with the Brillouin pump for six 980 nm pump powers (0, 8, 14, 23, 35, and 50 mW). (After Ref. [8]; © 1997 IEEE.)

avoided by using a Fabry–Perot cavity, since in this case all waves propagate in both directions. A multiwavelength lasing scheme was demonstrated in which a fiber Sagnac loop was used as one of the mirrors of a Fabry–Perot-type BEFL, whereas the other mirror was constituted by a 100% reflecting metal coating [95]. The Sagnac loop permitted the simultaneous presence of SBS pump and Stokes lines within the loop and thus generated higher order Stokes and anti-Stokes waves through four-wave mixing (FWM) process. A total of 34 spectral lines were generated through the SBS and FWM processes with 1.5 mW SBS pump power at 1561 nm and 80 mW EDF pump power.

Tuning of the frequency comb generated inside a BEFL is important in some applications. A tuning range of up to 60 nm was achieved with a Fabry–Perot-type BEFL, in which two fiber loops containing optical circulators acted as mirrors [98]. This laser included a 10 m long EDFA, while the Brillouin gain was achieved in a 8.8 km long SMF pumped with an external cavity laser tunable from 1520 to 1620 nm.

The main inconvenience for practical application of the BEFL reported in Ref. 98 is the need of an external Brillouin pump and the necessity to adjust the wavelength of such Brillouin pump accordingly in the process of wavelength tuning. An interesting configuration that avoids this problem is a self-seeded BEFL using an internally self-excited Brillouin pump [100,101]. Figure 12.15 shows a schematic of the linear cavity self-seeded BEFL. It includes a 16 m long EDFA, which is pumped by a 980 nm laser diode, a 12.5 km long SMF, a high-birefringence fiber Sagnac loop mirror, and an optical circulator (OC). The two-port connected OC and the Sagnac loop mirror serve as two reflectors. The Sagnac loop mirror also serves as a tunable wideband reflection filter.

More compact BEFLs can be realized by using fibers made of highly nonlinear glasses instead of several km long conventional fibers. In a 2008 experiment, a BELF was constructed using a tellurite fiber with a length of only 200 m in a ring cavity [26]. The laser setup included an 18 m long EDFA, which was pumped by a power of 330 mW emitted by a 974 nm laser diode. The maximum unsaturated power of

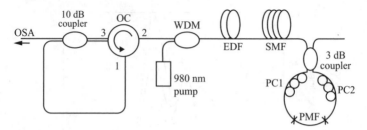

Figure 12.15 Schematic diagram of the self-seeded linear cavity Brillouin fiber laser. (After Ref. [101]; © 2006 OSA.)

54.6 mW at 1550 nm was achieved. The emission spectrum of the Brillouin comb laser consisted of 26 wavelengths spaced by 7.97 GHz, which is the Brillouin shift of tellurite fiber.

PROBLEMS

12.1 Describe the similarities and the main differences between SBS and SRS. Explain the origin of these differences.

12.2 Obtain an expression for the Brillouin frequency shift starting from the phase matching condition. Explain why SBS occurs only in the backward direction.

12.3 Consider an optical fiber whose core is constituted by silica glass doped with germania (GeO_2), with a refractive index $n = 1.47$ and an acoustic velocity $v_a = 5.6$km/s. Calculate the Brillouin frequency shift of the Stokes wave in such a fiber, assuming a pump wavelength $\lambda_p = 1.55$µm.

12.4 Starting from Eq. (12.11) and assuming that the rate of change of the two electric fields is much lower than the acoustic wave damping rate, obtain the result given by Eq. (12.12) for the amplitude of the displacement wave.

12.5 Estimate the SBS threshold at a pump wavelength $\lambda_p = 1.55$µm of a 50 km long standard single-mode fiber ($K_B = 1.5$) with an effective core area $A_{eff} = 50\,\mu m^2$, an attenuation constant of $\alpha = 0.2$ dB/km, and a gain coefficient $g_{B0} = 5 \times 10^{-11}$ m/W. Obtain the SBS threshold of the same fiber at $\lambda_p = 1.3$µm, assuming an attenuation constant $\alpha = 0.6$ dB/km.

12.6 Consider a 60 km long polarization maintaining fiber ($K_B = 1$) with an effective core area $A_{eff} = 50\,\mu m^2$, an attenuation constant of $\alpha = 0.2$ dB/km, and a gain coefficient $g_{B0} = 5 \times 10^{-11}$ m/W. Obtain the power of the Stokes wave at the fiber input assuming a pump power of 3 mW and an initiating Stokes power of 0.5 nW.

12.7 Prove that Eqs. (12.39)–(12.42) constitute indeed an approximate solution of Eqs. (12.20) and (12.21). Verify the validity of the results given by Eqs. (12.48) and (12.49).

REFERENCES

1. E. P. Ippen and R. H. Stolen, *Appl. Phys. Lett.* **21**, 539 (1972).

2. R. H. Stolen, *Proc. IEEE* **68**, 1232 (1980).

3. D. Cotter, *J. Opt. Commun.* **4**, 10 (1983).

4. Y. Aoki, *J. Lightwave Technol.* **6**, 1225 (1988).

5. F. Forghieri, R. W. Tkach, and A. R. Chraplyvy, in I. P. Kaminow and T. L. Koch (Eds.), *Optical Fiber Telecommunications*, Vol. IIIA, Academic Press, New York, 1997, p. 196.

6. R. W. Tkach and A. R. Chraplyvy, *Opt. Quantum Electron.* **21**, S105 (1989).

7. M. F. Ferreira, J. F. Rocha, and J. L. Pinto, *Opt. Quantum Electron.* **26**, 35 (1994).

8. D. Y. Stepanov and G. J. Cowle, *IEEE J. Sel. Top. Quantum Electron.* **3**, 1049 (1997).

9. A. Zadok, A. Eyal, and M. Tur, *J. Ligtwave Technol.* **25**, 2168 (2007).

10. S. M. Massey and T. H. Russel, *IEEE J. Sel. Top. Quantum Electron.* **15**, 399 (2009).

11. M. Nikles, L. Thevenaz, and P. A. Robert, *Opt. Lett.* **21**, 738 (1996).

12. K. Hotate and M. Tanaka, *IEEE Photon. Technol. Lett.* **14**, 179 (2002).

13. K. Y. Song, M. G. Herraez, and L. Thevenaz, *Opt. Express* **13**, 82 (2005).

14. Z. Zhu, A. M. Dawes, D. J. Gauthier, L. Zhang, A. E. Willner, *J. Lightwave Technol.* **25**, 201 (2007).

15. Z. Zhu, D. J. Gauthier, and R. W. Boyd, *Science* **318**, 1748 (2007).

16. R. W. Tkach, A. R. Chraplyvy, and R. M. Derosier, *Electron. Lett.* **22**, 1011 (1986).

17. W. Zou, Z. He, and K. Hotate, *J. Lightwave Technol.* **26**, 1854 (2008).

18. K. Shiraki, M. Ohashi, and M. Tateda, *J. Lightwave Technol.* **14**, 50 (1996).

19. A. Yeniay, J.-M. Delavaux, and J. Toulouse, *J. Lightwave Technol.* **20**, 1425 (2002).

20. J. Stone and A. R. Chraplyvy, *Electron. Lett.* **19**, 275 (1983).

21. N. Shibata, R. G. Waarts, and R. P. Braun, *Opt. Lett.* **12**, 269 (1987).

22. Y. Koyamada, S. Sato. S. Nakamura, H. Sotobayashi, and W. Chujo, *J. Lightwave Technol.* **22**, 631 (2004).

23. J. H. Lee, T. Tanemura, K. Kikuchi, T. Nagashima, T. Hasegawa, S. Ohara, and N. Sugimoto, *Opt. Lett.* **30**, 1698 (2005).

24. C. K. Jen, A. Saffai-Jazi, and G. W. Farnell, *IEEE Trans. Ultrason. Ferroelectr. Freq. Control* **33**, 634 (1986).

25. A. Yariv, *Quantum Electronics*, 3rd ed., Wiley, New York, 1989.

26. G. Qin, H. Sotobayashi, M. Tsuchiya, A. Mori, T. Suzuki, and Y. Ohishi, *J. Lightwave Technol.* **26**, 492 (2008).

27. K. S. Abedin, *Opt. Express* **13**, 10266 (2005).

28. E. Lichtman, A. A. Friesem, R. G. Waarts, and H. H. Yaffe, *J. Opt. Soc. Am. B* **4**, 1397 (1987).

29. R. G. Smith, *Appl. Opt.* **11**, 2489 (1972).

30. M. O. Van Deventer and A. J. Boot, *J. Lightwave Technol.* **12**, 585 (1994).

31. P. Bayvel and P. M. Radmore, *Electron. Lett.* **26**, 434 (1990).

32. R. D. Esman and K. J. Williams, *Proceedings of OFC'96* 1996, Paper ThF5, p. 227.

33. X. P. Mao, R. W. Tkach, A. R. Chraplyvy, R. M. Jopson, and R. M. Derosier, *IEEE Photon. Technol. Lett.* **4**, 66 (1992).

34. N. Yoshizawa, *J. Lightwave Technol.* **11**, 1518 (1993).

35. S. L. Zhang and J. J. O'Reilly, *IEEE Photon. Technol. Lett.* **5**, 537 (1993).

36. C. N. Pannel, P. St. J. Russel, and T. P. Newson, *J. Opt. Soc. Am. B* **10**, 684 (1993).

37. M. F. Ferreira, *Electron. Lett.* **30**, 40 (1994).

38. M. F. Ferreira, *J. Lightwave Technol.* **13**, 1692 (1995).

39. A. Kobyakov, M. Mehendale, M. Vasilyev, S. Tsuda, and A. F. Evans, *J. Lightwave Technol.* **20**, 1635 (2002).

40. M. F. Ferreira, *Electron. Lett.* **31**, 1182 (1995).

41. M. O. van Deventer, J. J. van der Tol, and A. J. Boot, *IEEE Photon. Technol. Lett.* **6**, 291 (1994).

42. Y. Aoki, K. Tajima, I. Mito, *J. Lightwave Technol.* **6**, 710 (1988).

43. A. R. Chraplyvy, *J. Lightwave Technol.* **8**, 1548 (1990).

44. E. Lichtman, R. G. Waarts, A. A. Friesem, *J. Lightwave Technol.* **7**, 171 (1989).

45. N. A. Olsson and J. P. van der Ziel, *Appl. Phys. Lett.* **48**, 1329 (1986).

46. N. A. Olsson and J. P. van der Ziel, *Electron. Lett.* **22**, 488 (1986).

47. N. A. Olsson, and J. P. van der Ziel, *J. Lightave Technol.* **5**, 147 (1987).

48. J. H. Lee, T. Tanemura, K. Kikuchi, T. Nagashima, T. Hasegawa, S. Ohara, and N. Sugimoto, *Opt. Lett.* **30**, 1698 (2005).

49. K. S. Abedin, *Opt. Lett.* **31**, 1615 (2006).

50. K. S. Abedin, *Opt. Express* **14**, 11766 (2006).

51. A. Mori, H. Masuda, K. Shikano, and M. Shimizu, *J. Lightwave Technol.* **21**, 1300 (2003).

52. H. Masuda, A. Mori, K. Shikano, and M. Shimizu, *J. Lightwave Technol.* **24**, 504 (2006).

53. C. G. Atkins, D. Cotter, D. W. Smith, and R. Wyatt, *Electron. Lett.* **22**, 556 (1986).

54. R. G. Waarts, A. A. Friesem, and Y. Hefetz, *Opt. Lett.* **13**, 152 (1988).

55. D. Cotter, D. W. Smith, C. G. Atkins, and R. Wyatt, *Electron. Lett.* **22**, 671 (1986).

56. A. R. Chraplyvy and R. W. Tkach, *Electron. Lett.* **22**, 1084 (1986).

57. T. Tanemura, Y. Takushima, and K. Kikuchi, *Opt. Lett.* **27**, 1552 (2002).

58. X. Bao, D. J. Webb, and D. A. Jackson, *Opt. Lett.* **18**, 1561 (1993).

59. H. H. Kee, G. P. Lees, and T. P. Newson, *Opt. Lett.* **25**, 695 (2000).

60. Y. T. Cho, M. N. Alahbabi, G. Brambilla, and T. P. Newson, *IEEE Photon. Technol. Lett.* **17**, 1256 (2005).

61. M. Niklès, L. Thévenaz, and Ph. Robert, *J. Lightwave Technol.* **15**, 1842 (1997).

62. W. Zou, Z. He, and K. Hotate, *Opt. Express* **16**, 18804 (2008).

63. K. Hotate and M. Tanaka, *IEEE Photon. Technol. Lett.* **14**, 197 (2002).

64. M. N. Alahbabi, Y. T. Cho, and T. P. Newson, *J. Opt. Soc. Am. B* **22**, 1321 (2005).

65. D. Garus, T. Gogolla, K. Krebber, and F. Schliep, *J. Lightwave Technol.* **15**, 654 (1997).

66. M. N. Alahbabi, Y. T. Cho, and T. P. Newson, *Opt. Lett.* **29**, 26 (2004).

67. W. Zou, Z. He, M. Kishi, and K. Hotate, *Opt. Lett.* **32**, 600 (2007).

68. L. Zou, X. Bao, S. Afshar, and L. Chen, *Opt. Lett.* **29**, 1485 (2004).

69. S. Chu and S. Wong, *Phys. Rev. Lett.* **48**, 738 (1982).

70. D. J. Gauthier, A. L. Gaeta, and R. W. Boyd, *Photon. Spectra* **40**, 44 (2006).

71. R. W. Boyd, D. J. Gauthier, and A. L. Gaeta, *Opt. Photon. News* **17**, 19 (2006).

72. F. Xia, L. Sekaric, and Y. Vlasov, *Nat. Photon.* **1**, 65 (2007).

73. E. Mateo, F. Yaman, and G. Li, *Opt. Lett.* **33**, 488 (2008).

74. J. Liu, T. H. Cheng, Y. K. Yeo, Y. Wang, L. Xue, W. Rong, L. Zhou, G. Xiao, D. Wang, and X. Yu, *J. Lightwave Technol.* **27**, 1279 (2009).

75. J. Zhang, G. Hernandez, and Y. Zhu, *Opt. Lett.* **33**, 46 (2008).

76. P. Palinginis, F. Sedgwick, S. Crankshaw, M. Moewe, and C. Chang-Hasnain, *Opt. Express* **13**, 9909 (2005).

77. Y. Okawach, M. Bigelow, J. Sharping, Z. Zhu, A. Schweinsberg, D. J. Gauthier, R. W. Boyd, and A. L. Gaeta, *Phys. Rev. Lett.* **94**, 153902 (2005).

78. J. E. Sharping, Y. Okawachi, and A. L. Gaeta, *Opt. Express* **13**, 6096 (2005).

79. D. Dahan and G. Eisenstein, *Opt. Express* **13**, 6234 (2005).

80. M. G. Herráez, K. Y. Song, and L. Thévenaz, *Opt. Express* **14**, 1395 (2006).

81. C. Jauregui, H. Ono, P. Petropoulos, and D. J. Richardson, *Optical Fiber Communication Conference and Exposition (OFC 2006)* 2006, Paper PDP2.

82. K. Y. Song, K. S. Abedin, K. Hotate, M. G. Herráez, and L. Thévenaz, *Opt. Express* **14**, 5860 (2006).

83. J. Boschung, L. Thevenas, and P. A. Robert, *Electron. Lett.* **30**, 1488 (1994).

84. M. R. Shirazi, S. W. Harun, M. Biglary, K. Thambiratnam, and H. Ahmad, *ISATS Trans. Electron. Signal Proc.* **1**, 30 (2007).

85. K. O. Hill, B. S. Kawasaki, and D. C. Johnson, *Appl. Phys. Lett.* **28**, 608 (1976).

86. H. Ahmad, M. R. Shirazi, M. Biglary, and S. W. Harun, *Microwave Opt. Technol. Lett.* **50**, 265 (2008).

87. M. R. Shirazi, S. W. Harun, M. Biglary, and H. Ahmad, *Opt. Lett.* **33**, 770 (2008).

88. L. F. Stokes, M. Chodorov, and H. J. Shaw, *Opt. Lett.* **7**, 509 (1976).

89. R. Carrol, C. D. Coccoli, D. Cardarelli, and G. T. Coate, *Proc. SPIE* **719**, 169 (1986).

90. D. R. Ponikvar and S. Ezekiel, *Opt. Lett.* **6**, 398 (1981).

91. R. K. Kadjwar and I. P. Giles, *Electron. Lett.* **25**, 1729 (1989).

92. F. Zarinatchi, S. P. Smith, and S. Ezekiel, *Opt. Lett.* **16**, 229 (1991).

93. S. Huang, K. Thévenaz, K. Toyama, B. Y. Kim, and H. J. Shaw, *IEEE Photon. Technol. Lett.* **5**, 365 (1993).

94. S. Yamashita and G. J. Cowle, *IEEE Photon. Technol. Lett.* **10**, 796 (1998).

95. M. K. Abd-Rahman, M. K. Abdullah, and H. Ahmad, *Opt. Commun.* **181**, 135 (2000).

96. J. C. Yong, L. Thévenaz, and B. Y. Kim, *J. Lightwave Technol.* **21**, 546 (2003).

97. M. H. Al-Mansoori, M. K. Abd-Rahman, F. R. M. Adikan, and M. A. Mahdi, *Opt. Express* **13**, 3471 (2005).

98. D. S. Lim, H. K. Lee, K. H. Kim, S. B. Kang, J. T. Ahn, and M. Y. Jeon, *Opt. Lett.* **23**, 1671 (1998).

99. Y. S. Hurh, G. S. Hwang, J. Y. Jeon, K. G. Lee, K. W. Shin, S. S. Lee, K. Y. Yi, and J. S. Lee, *IEEE Photon. Technol. Lett.* **17**, 696 (2005).

100. Y. J. Song, I. Zhan, J. H. Ji, Y. Su, Q. H. Ye, and Y. X. Xia, *Opt. Lett.* **30**, 486 (2005).

101. L. Zhan, J. H. Li, J. Xia, S. Y. Luo, and Y. X. Xia, *Opt. Express* **14**, 10233 (2006).

13

HIGHLY NONLINEAR AND MICROSTRUCTURED FIBERS

In standard single-mode fibers, the nonlinear parameter γ, defined in Eq. (4.42), has a typical value of only $\gamma \approx 1.3 \ \text{W}^{-1} \ \text{km}^{-1}$. Such value is too small for some applications requiring highly efficient nonlinear processes. As a consequence, a fiber length of several kilometers is generally necessary in such applications. This long fiber length poses serious limitations in its use for processing optical signals owing to the practical issues of the size and stability of the setup, as well as low stimulated Brillouin scattering (SBS) threshold. To shorten the length of interaction, highly nonlinear (HNL) silica fibers with a smaller effective mode area, and hence a larger nonlinear coefficient ($\gamma \approx 11 \ \text{W}^{-1} \ \text{km}^{-1}$), have been widely used [1]. About 1 km of this type of fiber is usually used for the demonstration of different nonlinear effects. The fiber nonlinearity can be further enhanced by tailoring either the fiber structure or the material from which the fiber is made, or a combination of both. Different types of silica-based photonic crystal fibers have been designed to address that purpose [2]. On the other hand, new types of HNL fibers using materials with higher nonlinear characteristics have been developed during recent years. Using such fibers, the required fiber length for several nonlinear processing applications was impressively reduced to the order of centimeters.

This chapter consists of six sections and can be divided into two main parts. The first part, consisting of Sections 13.1–13.3, provides a review of the more representative types of HNL fibers, made from silica or from other highly nonlinear materials. The second part, consisting of Sections 13.4–13.6, is dedicated to the description of novel nonlinear phenomena exhibited by such HNL fibers.

Nonlinear Effects in Optical Fibers. By Mário F. S. Ferreira.
Copyright © 2011 John Wiley & Sons, Inc. Published 2011 by John Wiley & Sons, Inc.

13.1 THE NONLINEAR PARAMETER IN SILICA FIBERS

As defined in Chapter 4, the nonlinear parameter γ is given by

$$\gamma = \frac{2\pi}{\lambda} \frac{n_2}{A_{\text{eff}}} \qquad (13.1)$$

where λ is the light wavelength, n_2 is the nonlinear index coefficient, and A_{eff} is the effective mode area, given by Eq. (4.43). In practice, the value of γ can be determined using any of the three major nonlinear effects occurring inside optical fibers: self-phase modulation (SPM), cross-phase modulation (XPM), and four-wave mixing (FWM). The SPM technique, for example, makes use of the pulse spectral broadening and measures the maximum value of the nonlinear phase shift, ϕ_{NL}, which is related to γ as given by Eq. (5.27) [3]. The XPM technique uses the pump-probe configuration. When the pump light is modulated, the probe spectrum develops frequency-modulated (FM) sidebands because of the XPM-induced phase shift, from which the nonlinear parameter γ can be determined [4]. Concerning the FWM approach, it uses the fact that both the amplitudes and frequencies of the generated sidebands also depend on the nonlinear parameter γ, as seen in Chapter 6 [5].

Equation (13.1) shows that, for a fixed wavelength, since n_2 is determined by the material from which the fiber is made, the more practical way of increasing the nonlinear parameter γ is to reduce the effective mode area A_{eff}. However, the nonlinear parameter γ can also be enhanced using the dopant dependence of the nonlinear refractive index n_2. This possibility is illustrated in Fig. 13.1, which shows the relationship between relative index difference Δ_d and the nonlinear refractive index n_2 in bulk glass [6]. The relative index difference is defined as $\Delta_d = (n_d^2 - n_0^2)/2n_d^2$,

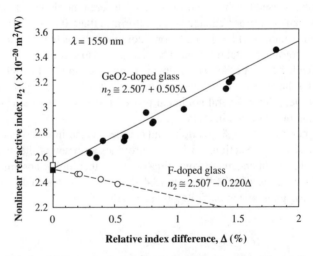

Figure 13.1 Relationship between the relative index difference Δ and the nonlinear refractive index n_2 of GeO$_2$- (closed circles) and F- (open circles) doped bulk glass. (After Ref. [6]; © 2002 IEEE.)

where n_0 and n_d denote the refractive index of pure silica and doped glass, respectively. The closed and open circles correspond to GeO_2- and F-doped bulk glasses, respectively, whereas the squares show the n_2 of pure silica glasses. The dopant dependence can be approximated using a linear relation, as shown in Fig. 13.1. It can be observed that the nonlinear refractive index n_2 increases with the relative index difference for the GeO_2-doped bulk glass. On the other hand, the dopant dependence of n_2 for the F-doped bulk glass is the inverse and about 2.3 times smaller than that for the GeO_2-doped case.

In order to enhance the nonlinear parameter γ, the best option will be provided by a fiber heavily doped with GeO_2 in the core and having a large refractive index difference between the core and the cladding, because such fibers exhibit a large nonlinear refractive index n_2 as well as a small effective area A_{eff} [7]. Figure 13.2 shows the schematic refractive index profiles of a HNLF and a standard single-mode fiber (SSMF) [8]. In typical HNLFs, the relative refractive index difference of the core to the outer cladding (Δn^+) is around 3%, while the core diameter is around 4 μm. The W-cladding profile with a fluorine-doped depressed cladding having a refractive index difference of Δn^- allows the single-mode operation in the communication bands as well as flexibility in designing of chromatic dispersion [9]. Using such an approach, new kinds of HNLFs with specific dispersive properties, namely, dispersion-shifted fibers, dispersion-compensating fibers, dispersion-decreasing fibers, and dispersion-flattened fibers, have been developed [7,10,11]. The values of the nonlinear parameter γ for most of these fibers are in the range $10–20 \, W^{-1} \, km^{-1}$ [11]. Increasing γ much above these values is not possible in this type of fibers, since the optical mode confinement is lost when the core diameter is further reduced.

The normalized frequency V, defined in Eq. (3.43), is given by

$$V = k_0 a (n_1^2 - n_c^2)^{1/2} \quad (13.2)$$

where k_0 is the vacuum wave number, a is the core radius, and n_1 (n_c) is the core (cladding) refractive index. Equation (13.2) can be approximated as

$$V = k_0 a n_1 \sqrt{2\Delta} \quad (13.3)$$

where $\Delta = (n_1 - n_c)/n_1$ is the relative index difference between the core and the cladding of the fiber. A single-mode fiber is achieved when $V < 2.405$. If Δ is kept

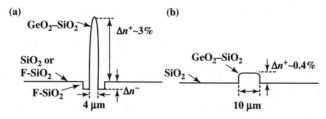

Figure 13.2 Schematic refractive index profiles of (a) an HNLF and (b) an SSMF. (After Ref. [8]; © 2009 IEEE.)

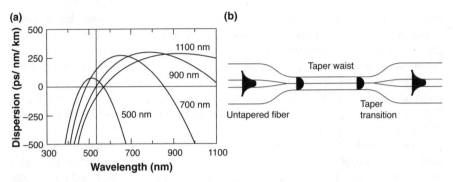

Figure 13.3 (a) Calculated dispersion spectra of taper waists for different values of the diameter. (b) Schematic of a tapered fiber. (After Ref. [19]; © 2004 OSA.)

constant and the core radius a is reduced, the normalized frequency V decreases and the mode confinement is lost. Equation (13.3) shows that, in order to maintain a given value of V when decreasing a, Δ must increase such that $a^2\Delta$ remains constant.

The maximum value of Δ is achieved when the cladding material is replaced by air ($n_c \approx 1$). In such a case, the mode remains confined to the core even if the core diameter is close to 1 µm. Narrow-core fibers with air cladding have been produced by tapering standard optical fibers, with an original cladding diameter of 125 µm [12–14]. Tapering a fiber involves heating and stretching it to form a narrow waist connected to untapered fiber by taper transitions (see Fig. 13.3b). If the transitions are gradual, light propagating along the fiber suffers very little loss.

The nonlinear parameter γ is significantly enhanced when the core diameter is reduced to values of the order of 2 µm. Considering the Gaussian approximation for the mode profile, and using $\lambda = 1$ µm for the wavelength of light transmitted through the fiber, $n_1 = 1.45$ for the silica core and $n_c = 1$ for the air cladding, it was found that a maximum value $\gamma \approx 370\ W^{-1}\ km^{-1}$ occurs for $V \approx 1.85$ [15].

In recent years, there has been a great interest in tapered fibers with a core diameter below 1 µm [16–18]. Numerical results, obtained without using the Gaussian approximation, have shown that the mean field diameter of the optical mode of such a nanoscale fiber with silica core and air cladding achieves a minimum value when the core diameter is 0.74λ [17]. The nonlinear parameter γ then attains a maximum value, which scales with λ^{-3}. Using $n_2 = 2.6 \times 10^{-20}\ m^2/W$, a value $\gamma = 662\ W^{-1}\ km^{-1}$ is obtained at $\lambda = 0.8$ µm. The dispersive properties of a tapered fiber are very sensitive to the core size and can be adjusted by changing the core diameter [17,19]. The zero dispersion wavelength (ZDW) can be shifted to the visible range and, in some cases, a second ZDW appears at longer wavelengths, defining a wavelength window between the two ZDWs, which shows anomalous group velocity dispersion (GVD). This is illustrated in Fig. 13.3a, which shows the calculated dispersion spectra of taper waists for several values of the waist diameter [19]. We observe that, as the diameter of a taper waist decreases below 1 µm, the peak in dispersion shifts to shorter wavelengths, giving a flattened-at-zero response for diameters of ~500 nm.

13.2 MICROSTRUCTURED FIBERS

Microstructured fibers (MFs) represents a new class of optical fibers that are characterized by the fact that the silica cladding presents an array of embedded air holes. They are also referred to as photonic crystal fibers (PCFs). This designation is justified since they were first realized in 1996 in the form of a photonic crystal cladding with a periodic array of air holes [20]. Photonic crystals were first reported in 1987 [21,22] and consist of transparent material periodically structured. This structure can exhibit photonic band gaps, such that light with a wavelength within those bands cannot propagate in the crystal. Actually, there are many kinds of photonic crystal fibers, even if some of them do not use the photonic band gap effect. Therefore, the designation microstructured fibers appears as more general.

A common technique for fabricating microstructured fibers is stacking of circular capillaries. Typically, the preform is a meter long and 20 mm in diameter and contains several hundred capillaries [23]. The capillaries are stacked horizontally in an appropriately shaped jig to form the planned microstructure. After being inserted into a jacketing tube, the completed preform will then be heated and drawn down on a fiber drawing tower. Control of pressure and vacuum during the draw provides to some extent the possibility to adjust the structural parameters, namely, the d/Λ value, where d is the air hole diameter and Λ is the hole-to-hole spacing.

An alternative way to fabricate microstructured fibers is known as *extrusion technique* [24], which permits the formation of structures that are not readily made by stacking. In this case, the preform with the required pattern of holes is produced by extruding material selectively from a solid glass rod with a diameter of about 2 cm. The structured preform is then drawn down on a fiber drawing tower to the dimensions of a fiber. While not suitable for silica, this technique is useful for making MFs from compound silica glasses, tellurites, chalcogenides, and polymers.

Microstructured fibers can be divided into two main types, which are represented in Fig. 13.4. One class of fibers, first proposed in 1999 [25], have a central region containing air. Such fibers are usually called hollow-core MFs and the light propagating in them is confined to the core by the photonic band gap effect. Extremely low losses are expected from this type of fibers, since the light travels predominantly in empty space. Although the best attenuation reported to date for this type of MFs is 1.2 dB/km [26], values well below 0.2 dB/km seem to be possible with further technological advances. The nonlinear effects are also strongly reduced and the dispersion becomes negligible in air-core MFs. Therefore, they offer the possibility to solve in the future some limitations of actual communication systems over optical fibers [27].

The second type of microstructured fibers has a solid core, in which the light is guided due to total internal reflection, since the core has a higher refractive index than the cladding. In such fibers, the periodic nature of air holes in the cladding is not important as long as they provide an effective reduction of its refractive index below that of the silica core [27,28]. This helps to concentrate the mode field in a very small area. Due to the resulting high intensity, the fiber can show a high nonlinear behavior. Moreover, when an asymmetric structure is used, such fiber also

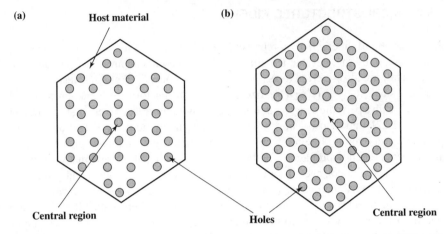

Figure 13.4   Two main microstructured fiber configurations with (a) hollow core and (b) solid core.

exhibits a high birefringence, which provides for the polarization maintaining of the propagated field.

Solid-core MFs exhibit generally high losses, which can exceed 1000 dB/km [29]. This occurs particularly when the core diameter is reduced in order to increase the nonlinearity parameter γ. Such losses are essentially due to the nature of the mode confinement, which is provided by the air holes embedded in the silica cladding. The large index contrast greatly enhances scattering at the air/silica interfaces, causing additional scattering losses that are typically negligible in conventional fibers. Increasing the core diameter reduces the fiber losses, but generally only at the expense of a simultaneous reduction of the nonlinearity parameter [30]. Recently, a 100 km long solid-core MF was fabricated with a low loss of 0.3 dB/km at 1550 nm [31]. The mode-field diameter (MFD) and the effective core area at this wavelength were 7.8 μm and 50 μm^2, respectively. This fiber was used in the first MF-based penalty-free dispersion-managed soliton transmission at 10 Gb/s [31,32]. The slightly higher attenuation of this MF compared to conventional SMF is attributed to roughness at the glass–air interfaces [33].

The number of guided modes in a conventional fiber depends on the normalized frequency V given by Eq. (13.2), which depends on the wavelength. In conventional fibers, the single-mode condition $V < 2.405$ is satisfied only if the wavelength is in the infrared region. For lower values of the wavelength, the fiber becomes multimode. However, in an MF the normalized frequency is given by a slightly different expression from Eq. (13.2), with a replaced by Λ and n_c replaced by the effective cladding index of the microstructured cladding. Such effective refractive index depends significantly on the ratios d/Λ and λ/Λ [34] and asymptotically approaches the core index in the short-wavelength limit. As a consequence, both the effective index difference and the numerical aperture go to zero in this limit.

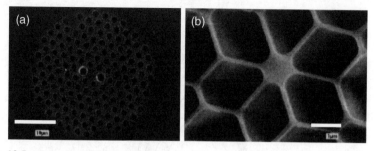

Figure 13.5 Scanning electron micrographs of two different MFs: (a) birefringent MF and (b) MF with very small core (diameter 800 nm) and zero dispersion wavelength 560 nm. (After Ref. [2]; © 2006 IEEE.)

Surprisingly, in an MF it is verified that the behavior of the cladding effective index cancels out the wavelength dependence of the normalized frequency V, such that it approaches a constant value as $\Lambda/\lambda \to \infty$. As a consequence, it was observed that an MF shows a single-mode behavior at all wavelengths if $d/\Lambda < 0.43$; such a fiber is known as the *endlessly single-mode* (ESM) fiber [35,36]. Furthermore, because of the scale invariance of MFs, the ESM behavior can be achieved for all core diameters.

If the core of an MF is deliberately distorted so as to become twofold symmetric, extremely high values of birefringence can be achieved. For example, by introducing capillaries with different wall thickness above and below a solid glass core (Fig. 13.5a), values of birefringence some 10 times larger than those in conventional fibers can be obtained [37]. Moreover, experiments show that such birefringence is some 100 times less sensitive to temperature variations than in conventional fibers [38].

The microstructured cladding offers greatly enhanced design flexibility and can manipulate the dispersion characteristics by controlling structural parameters such as the hair hole diameter d and the hole-to-hole spacing Λ [34]. In fact, the dispersive properties of MFs are quite sensitive to these parameters and can be tailored by changing appropriately each of them. Figure 13.6 shows the chromatic dispersion characteristics of an MF with perfect hexagonal symmetry as a function of wavelength for d/Λ ranging from 0.2 to 0.9 in steps of 0.1. Dispersion engineering allows the MF to be designed with a ZDW in the visible or near-infrared regions. An MF with a large normal dispersion in the 1550 nm wavelength region can be obtained by using a large hair hole diameter d and a very small hole-to-hole spacing Λ. Moreover, Fig. 13.6c shows that a nearly zero dispersion-flattened behavior can be achieved with $\Lambda \approx 2.5\,\mu m$ and $d/\Lambda \approx 0.25$.

In solid-core MF, as the holes get larger, the core becomes more and more isolated, until it resembles an isolated strand of silica glass. The MFs with larger cores exhibit semi-infinite anomalous dispersion above the ZDW. By decreasing the core size, the ZDW tends to be shifted to a shorter wavelength, leading to the anomalous dispersion at near-infrared and visible wavelengths. For example, the MF in Fig. 13.5b has zero dispersion at 560 nm [2]. When the core size is decreased further, a second ZDW

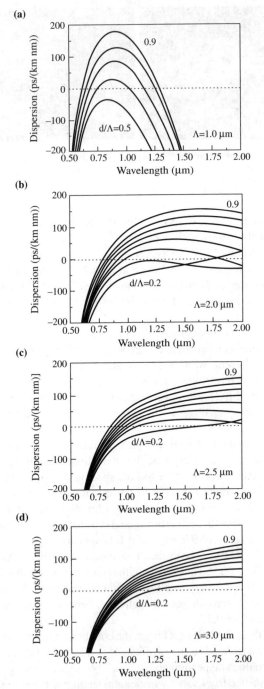

Figure 13.6 Chromatic dispersion characteristics of MFs for (a) $\Lambda = 1.0\,\mu m$, (b) $\Lambda = 2.0\,\mu m$, (c) $\Lambda = 2.5\,\mu m$, and (d) $\Lambda = 3.0\,\mu m$. (After Ref. [34]; © 2005 IEEE.)

Figure 13.7 MF cores of diameter 3.1 (untapered), 0.7, and 0.5 μm (left to right, different scales). (After Ref. [19]; © 2004 OSA.)

arises in the longer wavelength side, such that the GVD is anomalous in the spectral window between the two ZDWs and normal outside it, as illustrated in Fig. 13.3a. Submicron-diameter MF cores with low dispersion at 532 nm have been fabricated using a conventional tapering process [19]. SEM images of core diameters of 700 and 500 nm obtained by tapering from a 3.1 μm MF are shown in Fig. 13.7. This technique allows to preserve the nanostructure in MFs and circumvents the apparent inability in making such submicron core MFs by standard fiber drawing.

In the case of microstructures fibers and to take into account the different proportions of light in glass and air, the nonlinear coefficient γ must be redefined as follows [39]:

$$\gamma = k_0 \sum_i \frac{n_2^i}{A_{eff}^i} = k_0 \frac{n_2^{eff}}{A_{core}} \qquad (13.4)$$

In Eq. (13.4) n_2^i is the nonlinear refractive index of material i ($2.9 \times 10^{-23} \mathrm{m^2/W}$ for air and $2.6 \times 10^{-20} \mathrm{m^2/W}$ for silica), A_{eff}^i is the effective area for the light in material i, and n_2^{eff} is the effective nonlinear index for fiber, with a core area A_{core}.

A solid-core MF similar to the one in Fig. 13.5b but with a core diameter of 1 μm has a nonlinear coefficient $\gamma = 240 \, \mathrm{W^{-1} \, km^{-1}}$ at 850 nm, which must be compared with the highest values $\sim 20 \, \mathrm{W^{-1} \, km^{-1}}$ obtainable in conventional single-mode fibers. Considering a peak power of $P_0 = 5 \, \mathrm{kW}$, the nonlinear length for the same solid-core MF will be $L_{NL} < 1 \, \mathrm{mm}$. For dispersion values in the range $-400 < \beta_2 < 400 \, \mathrm{ps^2/km}$ and pulse durations $t_0 = 300 \, \mathrm{fs}$, we have a dispersion length $L_D = t_0^2/|\beta_2| > 0.2 \, \mathrm{m}$. Considering also typical values of loss (1–100 dB/km), we conclude that both the effective fiber length L_{eff} (given by Eq. (4.29)) and the dispersion length L_D are much longer than the nonlinear length L_{NL}, which means that strong nonlinear effects will be readily observable in solid-core MFs.

Concerning the air-core MFs, their nonlinear properties are very weak due to the small overlap between the light and glass. An effective nonlinear refractive index $n_e^{eff} = 8.6 \times 10^{-23} \, \mathrm{m^2/W}$ (about 300 times smaller than that in silica glass) and a nonlinear coefficient $\gamma = 0.023 \, \mathrm{W^{-1} \, km^{-1}}$ were reported recently for a fiber of this type [40]. However, the nonlinear effects in this type of MFs can be greatly enhanced if air is replaced with a suitable gas or liquid [41]. For example, in a 2002 experiment stimulated Raman scattering was observed to occur in a hydrogen-filled hollow-core MF at very low threshold pulse energies [42].

13.3 NON-SILICA FIBERS

The value of the nonlinear parameter γ can be enhanced if the optical fiber is made using glasses with higher nonlinearities than silica, such as lead silicate, tellurite, bismuth oxide, and chalcogenide glasses. Such possibility is illustrated in Fig. 13.8a, which shows the nonlinear index coefficient n_2 against the linear refractive index, n, for several optical glasses [43,44]. As expected, n_2 increases with n, since both depend

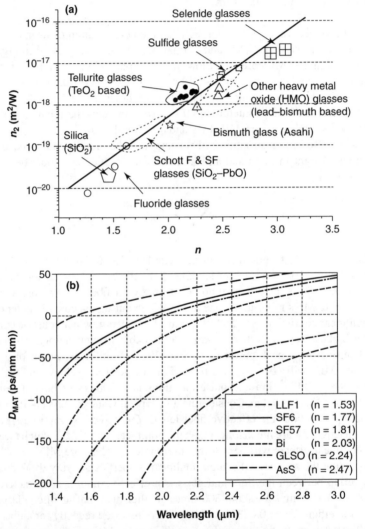

Figure 13.8 (a) Nonlinear index coefficient n_2 against the linear refractive index for several optical glasses. (b) Material dispersive curves of Schott glasses (LLF1, SF6, SF57), bismuth oxide glass (Bi), and chalcogenide glass (GLSO). The number in the legend indicates the linear index at $1.06\,\mu m$. (After Ref. [44]; © 2007 IEEE.)

on the density of the material. Introducing heavy atoms or ions with a large ionic radius (i.e., using chalcogen elements S, Se, and Te to replace oxygen) act to increase the polarizability of the components in the glass matrix and also increase the nonlinear index n_2. Figure 13.8a shows that the nonlinear index can be increased by a factor of almost 1000 compared to its value for silica fibers using some types of highly nonlinear glasses. The ZDW of a glass shifts to longer wavelengths with increasing linear refractive index, as shown in Fig. 13.8b. Heavy metal oxide glasses (lead silicate, bismuth oxide, tellurite) have linear indices in the range 1.8–2.0, nonlinear indices ~10 times higher than that of silica, and material ZDWs of 2–3 μm. Chalcogenide glasses (GLS, As_2S_3) have linear indices of 2.2–2.4, nonlinear indices significantly higher than those of the oxide glasses, and material ZDWs larger than 4 μm.

Lead silicate and tellurite glasses can offer values of n_2 that are about 10–20 times larger than that of silica. For example, a value $n_2 \approx 2.5 \times 10^{-19} \mathrm{m^2/W}$ has been measured for tellurite glasses [45]. More recently, a value $n_2 = 4.1 \times 10^{-19} \mathrm{m^2/W}$ at 1.06 μm was measured for Schott SF57 glass [46]. Both types of materials have been used for making microstructured fibers. The extrusion technique for fiber preform manufacture is particularly suitable when using compound glasses, since they exhibit low soft temperatures of ~500 °C, as opposed to ~2000 °C for silica. Besides being simpler than the more common stacking technique developed for silica MFs, extrusion allows the realization of cladding structures consisting of mostly air, as required to increase the fiber nonlinearity.

Figure 13.9 shows the measured nonlinear phase shift induced through self-phase modulation of a continuous-wave dual-frequency optical beat signal propagated through a 144 cm long MF made of lead silicate glass (SF57) and presenting an effective mode area $A_{\mathrm{eff}} = 1.1$ μm^2 [46]. From the slope of the linear fit, a nonlinear

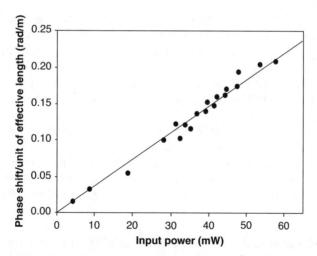

Figure 13.9 Measurement of the nonlinear phase shift induced by self-phase modulation as a function of power launched in lead silicate fiber of length 144 cm. (After Ref. [46]; © 2006 IEEE.)

parameter $\gamma = 1860\ W^{-1}\ km^{-1}$ at 1.55 µm was obtained. Higher values of the nonlinear parameter are expected at shorter wavelengths due to a combination of the $1/\lambda$ dependence of γ and the fact that shorter wavelengths can be confined to the core down to smaller core dimensions.

Much attention has been paid in recent years to fibers based on bismuth oxide (Bi_2O_3). Among the different types of nonlinear fibers, Bi-NLFs have a relatively mature technology and the fibers can be spliced to standard single-mode fibers. They also possess good chemical, mechanical, and thermal stability and have a low photosensitivity compared to chalcogenide fibers. Depending on the concentration of bismuth in the fiber, the nonlinear refractive index n_2 can be varied from 30×10^{-20} to $1.1 \times 10^{-18}\ m^2/W$, representing a 10–40 enhancement as compared to that of silica [47–49]. By 2008, a bismuth oxide fiber with a mode field diameter of 1.97 µm exhibited a nonlinear parameter $\gamma \approx 1100\ W^{-1}\ km^{-1}$ at 1550 nm. Only 35 cm length of such a highly nonlinear fiber was needed to demonstrate several applications on nonlinear processing of optical signals [49]. The large Brillouin gain offered by the Bi-NLF has also found important applications in slow light. A 2 m long Bi-NLF was successfully used to produce a fourfold reduction in the group velocity of 180 ns pulses using only 400 mW input power to provide the Brillouin gain [50].

A nonlinear parameter $\gamma = 2450\ W^{-1}\ km^{-1}$ was measured in a 2004 experiment for an 85 cm long chalcogenide fiber with a core diameter of 7 µm [51]. This corresponds to a nonlinear refractive index $n_2 = 2.4 \times 10^{-17}\ m^2/W$, which is larger by a factor of about 1000 compared to the silica case. The same fiber also exhibited a Raman gain parameter that was nearly 800 times larger than that of silica fibers [51]. This allows the use of either a smaller fiber length or a lower pump power to attain amplifier gain characteristics similar to those of silica-based devices. Due to their high nonlinearity, chalcogenide fibers have been considered as a good option to enhance several nonlinear effects.

In recent years, PCF structures have been fabricated in various types of chalcogenide glasses for several applications. Employment of photonic crystal geometry in the cladding of As_2Se_3 fibers can give rise to a strong overlap between pump and signal, providing higher Raman gain efficiency than conventional fibers, as demonstrated for silica-based PCF structures [52,53]. It was shown that the Raman gain efficiency in As_2Se_3 PCFs can be improved by a factor of more than 4 compared to conventional As_2Se_3 fibers [54].

Compared to the bismuth fiber, a chalcogenide fiber has an intrinsically higher material nonlinearity (up to 10 times that of bismuth oxide). However, because the core area of available chalcogenide fibers is higher than that of bismuth fiber, the resulting nonlinear parameter γ is similar in both cases. It is expected that further progress in fabrication methods will lead to chalcogenide fibers with core areas comparable to the current bismuth fiber, which will provide nonlinear parameters $\gamma > 10,000\ W^{-1}\ km^{-1}$ [55].

Both the bismuth oxide fiber and the chalcogenide fiber present a relatively high normal dispersion at 1550 nm communication window: typically −300 ps/(nm km) in the first case and −500 ps/(nm km) in the second one. Such large values can be useful in some applications but can cause limitations in other applications. Recently, air cladding type dispersion-shifted bismuth oxide PCFs have been successfully fabricated [56], which allows the control of dispersion characteristics. It is expected that the progress in fabrication technology will offer in the near future new opportunities for nonlinear applications of this type of fiber, namely, the realization of efficient wideband FWM-based wavelength converters and compact fiber-optic parametric amplifiers [49].

13.4 SOLITON SELF-FREQUENCY SHIFT

It was shown in Section 8.3.2 that intrapulse Raman scattering is responsible for the soliton shift toward lower frequencies. Such phenomenon was reported for the first time in 1986 [57] and it is referred to as *soliton self-frequency shift* (SSFS). Due to Raman gain, the blue portion of the soliton spectrum pumps the red portion of the spectrum, causing a continuous redshift in the soliton spectrum. This effect is responsible for the fission of higher order solitons and the generation of the so-called Raman solitons.

Since its discovery in conventional single-mode fibers, the SSFS effect has also been observed in other types of fibers, including tapered microstructured air-silica fiber [58], solid-core MFs [59,60], and air-core MFs [61]. The SSFS effect can be significantly enhanced in some highly nonlinear fibers, where it has been used during the recent years for producing femtosecond pulses [62–65], as well as to realize several signal processing functions [66–68].

The Raman-induced SSFS given in Eq. (8.31) can be written in real world units as

$$\Delta f = -\frac{4t_R |\beta_2| z}{15\pi t_0^4} \tag{13.5}$$

where $t_R \approx 5$ fs is the Raman parameter. From Eq. (13.5) it can be seen that the SSFS varies inversely with the fourth power of the soliton width t_0. Moreover, using the condition $N = \gamma P_0 t_0^2 / |\beta_2| = 1$, which is valid for a fundamental soliton, Eq. (13.5) can be rewritten in the form

$$\Delta f = -\frac{4t_R (\gamma P_0)^2 z}{15\pi |\beta_2|} \tag{13.6}$$

Since the SSFS effect is proportional to (γP_0),^{2} it will be enhanced if short pulses with high peak pulses are propagated in highly nonlinear fibers.

Figure 13.10 shows the optical spectra of the output pulses from a 60 cm long photonic crystal fiber, obtained in a 2002 experiment [62]. In this experiment, 70 fs pulses with a 48 MHz repetition rate at a wavelength of 782 nm were launched at the fiber input. As the fiber input power is increased above a given value, the center wavelength of the generated Raman soliton is shifted toward the longer wavelength side continuously due to the SSFS effect. The total shift at the fiber output achieves a maximum value of 900 nm when the fiber input power is 3.6 mW. The spectral shape

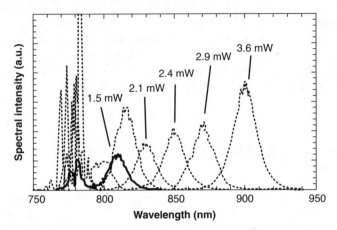

Figure 13.10   Optical spectra of generated wavelength-tunable femtosecond soliton pulse as the fiber input power was changed. (After Ref. [62]; © 2002 IEEE.)

of the Raman pulses is almost *sech*^{2} and their spectral width has an almost constant value of 18 nm.

It can be observed from Fig. 13.10 that a second Raman soliton is generated when the input power is 3.6 mW. In fact, for this pump pulse, the corresponding soliton order exceeds 2. Two fundamental solitons are then created through the fission process experienced by the higher order soliton initially propagating inside the fiber.

The Raman solitons created through the fission process are generally perturbed by the third- and higher order dispersion of the fiber, emitting dispersive waves, which constitute the so-called *nonsoliton radiation* (NSR) [69] or *Cherenkov radiation* [70]. Such radiation was observed in a 2001 experiment using 110 fs pulses from a laser operating at 1556 nm, which were launched into a polarization-maintaining dispersion-shifted fiber with a GVD $\beta_2 = -0.1$ ps^2/km at the laser wavelength [71]. In this experiment, the nonsoliton radiation appeared in the normal GVD regime of the fiber. Moreover, it was observed that the NSR overlaps temporally with the Raman soliton, which is due to the phenomenon of soliton trapping.

The NSR is emitted at a frequency such that its phase velocity matches that of the soliton. Neglecting the fourth- and higher order dispersions, the frequency shift $\Omega_d = \omega_d - \omega_s$ between the frequency of dispersive waves (ω_d) and that of the soliton (ω_s) is given approximately by [70]

$$\Omega_d \approx -\frac{3\beta_2}{\beta_3} + \frac{\gamma P_0 \beta_3}{3\beta_2^2} \qquad (13.7)$$

where γ is the nonlinear parameter, P_0 is the peak power of the Raman soliton, and β_j ($j = 2, 3$) are the dispersion parameters, calculated at the soliton central frequency ω_s. For solitons propagating in the anomalous GVD regime, we have $\beta_2 < 0$. In these circumstances, Eq. (13.7) shows that $\Omega_d > 0$ when $\beta_3 > 0$, in which case the NSR is emitted at wavelengths shorter than that of the soliton. On the contrary, if $\beta_3 < 0$ the NSR is emitted at a longer wavelength than that of the soliton.

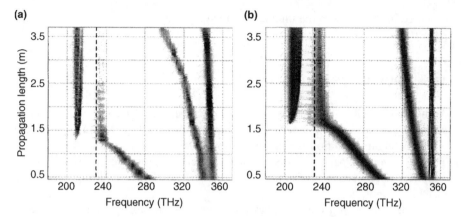

Figure 13.11 (a) Experimental and (b) numerical spectra as a function of propagation distance when 200 fs pulses with a peak power of 230 W are launched into the MF with two ZDWs. The dashed vertical line corresponds to the second frequency with zero dispersion. (After Ref. [73]; 2003 AAAS.)

As seen in previous sections, some HNLFs exhibit two ZDWs, one falling in the visible region and the other in the infrared region, such that GVD is anomalous in the spectral window between them. The third-order dispersion parameter β_3 is positive near the first ZDW and negative near the second one. Near the first ZDW the NSR is always blueshifted, so that the recoil and SSFS act in the same spectral direction. However, the opposite behavior occurs near the second ZDW. As the soliton loses energy to the NSR, it becomes slightly wider, as a result of energy conservation. On the other hand, momentum conservation determines that, as NSR is emitted in the normal GVD regime, the soliton should recoil further into the anomalous GVD regime, providing the suppression of SSFS [72].

Figure 13.11a shows the cancellation of the SSFS observed in a 2003 experiment [73]. These results were obtained using a silica-core MF with ZDWs at 600 and 1300 nm, in which 200 fs pulses at a wavelength of 860 nm and with a peak power of 230 W were launched. These experimental conditions provided the initial formation of a pulse corresponding to the fourth-order soliton solution of the NLSE. In the first stage, the pulse splits into two Raman shifting solitons, together with some residual radiation, which retains the pump frequency. The more intense soliton acquires a stronger Raman shift. It passes the minimum of GVD and enters the spectral region with negative dispersion slope, where redshifted NSR quickly approaches the central part of the soliton spectrum. This leads to the amplification of the NSR, which stabilizes the soliton frequency at ≈ 1270 nm, through the recoil mechanism. Figure 13.11b shows the corresponding modeling results, which are in excellent agreement with experiment.

The effect of SSFS cancellation described above cannot be observed in conventional optical fibers, since for the commonly used frequencies the radiation is always blueshifted, so that the recoil and Raman effects act in the same spectral direction. Such cancellation is an example of the unexpected effects that can be observed in PCFs due to the unique combination of their dispersive and nonlinear properties.

13.5 FOUR-WAVE MIXING

As discussed in Chapter 6, new frequencies can be generated in conventional optical fibers by parametric FWM when they are pumped close to the ZDW. In this case, the phase matching condition is sensitive to the exact shape of the dispersion curve. Thus, we must expect that the peculiar dispersive properties of some highly nonlinear fibers have a profound impact on the FWM process through the phase matching condition.

As seen in Section 6.3, the phase matching condition for the FWM process requires that dispersive effects compensate nonlinear ones through the following equation:

$$\beta_2 \Omega_s^2 + \frac{\beta_4}{12} \Omega_s^4 + 2\gamma P_p = 0 \qquad (13.8)$$

where β_2 and β_4 are, respectively, the second- and fourth-order dispersion terms at the pump frequency, Ω_s is the frequency shift between the signal or idler frequencies and the pump frequency, γ is the nonlinear parameter, and P_p is the pump peak power. The third and all-odd dispersion orders play no role in Eq. (13.8) since they cancel out from the degenerate phase matching condition $2\beta(\omega_p) - \beta(+\Omega_s) - \beta(-\Omega_s) = 0$, where $\beta(\omega)$ is the exact wave vector.

The impact of the fourth-order dispersion term (β_4) on parametric processes was experimentally studied a few years ago in standard telecommunications fibers [74] and in MFs [75]. In the anomalous dispersion regime ($\beta_2 < 0$), the β_2 term dominates in Eq. (13.8) and the frequency shift is given by Eq. (6.26). However, modulation instability can also occur in the normal GVD regime ($\beta_2 > 0$). In this case, the positive nonlinear phase mismatch $2\gamma P$ is compensated by the negative value of the linear phase mismatch due to β_4.

In order to study the dependence of signal and idler frequencies with the pump wavelength, it is necessary to use the actual dispersion curve to find the involved dispersion parameters. Figure 13.12a shows the variation of GVD with wavelength of an MF used in a 2008 experiment [76]. The zero dispersion wavelength of such a fiber is located at 1092 nm. By solving Eq. (13.8), the phase matching diagram in Fig. 13.12b for the signal and idler wavelengths was obtained using a pump power $P_p = 2\,\text{kW}$, corresponding to the peak power experimentally launched into the MF. Crosses depict phase-matched FWM wavelengths experimentally observed.

As can be seen from Fig. 13.12b, the frequency shift between the pump and the FWM sidebands is much larger than the Raman gain band (of 13.2 THz) when the pump wavelength is below the ZDW. Therefore, no significant spectral overlap is possible between Raman and FWM gain bands in this situation. Such feature is particularly important to reduce the Raman-related noise and allows the use of the MF

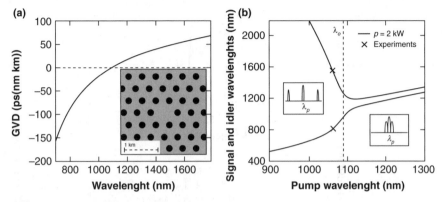

Figure 13.12 (a) GVD curve of the MF in the inset and (b) phase matching diagram calculated with Eq. (13.8). Crosses depict phase-matched FWM wavelengths experimentally observed. Insets in (b) represent typical gain spectra corresponding to both dispersion regimes. (After Ref. [76]; © 2008 OSA.)

as a compact bright tunable single-mode source of entangled photon pairs, with wide applications in quantum communications [77–79] (see Section 6.8).

FWM in highly nonlinear and microstructured fibers can be used to realize both fiber-optic parametric amplifiers (FOPAs) and optical parametric oscillators (OPOs) [80–84]. The physics of parametric amplification in HNLFs is similar to that of standard optical fibers. Differences arise, first, from the enhanced nonlinear parameter γ. Moreover, as seen in Section 13.2, some MFs can exhibit single transverse mode propagation over a wide range of wavelengths, which leads to excellent spatial overlap between propagating modes at widely different wavelengths. Finally, the peculiar dispersive characteristics of HNLFs can enhance the phase matching condition of the FWM process.

As seen in Chapter 6, the small-signal gain of the signal field in a single-pump fiber parametric amplifier is given by

$$G_s = \frac{P_s(L)}{P_s(0)} = 1 + \left[\frac{\gamma P_p}{g} \sinh(gL)\right]^2 \qquad (13.9)$$

where P_s is the signal power, P_p is the pump power, and the parametric gain coefficient is $g = [-\Delta k((\Delta k/4) + \gamma P_p)]^{1/2}$.

In order to achieve an efficient gain, the phase matching condition $\Delta k + 2\gamma P_p = 0$ must be satisfied. As shown in Fig. 13.12b, if the pump is located in the normal dispersion region, the phase-matched wavelengths are relatively far apart. However, they become closer when the pump wavelength is tuned into the region of anomalous dispersion. On the other hand, the width of the phase matching bands for pumping in the anomalous region is relatively large, compared to that observed for pumping in the normal dispersion regime. The region around the ZDW is of particular interest, because it provides both broad phase-matched bands and widely spaced matching wavelengths [80].

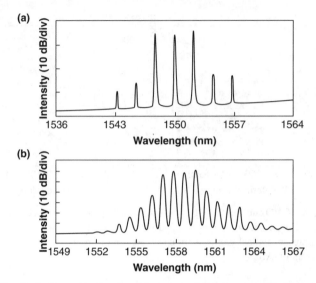

Figure 13.13 Optical spectra showing (a) three and (b) four wavelengths output from an erbium-doped fiber laser stabilized by FWM in a 35 cm Bi-NLF. (After Ref. [49]; © 2008 IEEE.)

In a 2001 experiment, a 6.1 m long MF was used to realize a single-pump parametric amplifier, providing a gain of 13 dB over 30 nm bandwidth with a pump peak power of only 6 W [85]. Higher gains can be achieved using materials with higher nonlinearity. For example, a peak gain as high as 58 dB was recently obtained with a single-pump FOPA using a 2 m long bismuth oxide-based fiber [86]. Due to the large values of β_2 and β_4, this FOPA was very narrowband and its gain peak wavelength was tunable in proportion to the pump wavelength.

FWM in HNLFs can be used to stabilize the output of multiwavelength erbium-doped fiber lasers, through the continuous annihilation and creation of photons at the wavelengths of interest [87]. A stable tunable dual-wavelength output over 21.4 nm was demonstrated using a 35 cm long HNL bismuth oxide fiber [88]. The number of oscillating wavelengths can be increased simply by applying a stronger FWM in the fiber. In such a case, more wavelength components are involved in the energy exchange process to stabilize the laser output. Using the same 35 cm long HNL bismuth oxide fiber and a 30 dBm EDFA output power, up to four different wavelengths were produced recently [49]. Figure 13.13 shows the spectra of the FWM-stabilized lasing output for three and four wavelengths.

In addition to its use in generating continuous-wave (CW) multiwavelength laser output, FWM in HNL bismuth oxide fiber has also been used in several other applications, namely, to produce wavelength and width-tunable optical pulses [89,90], frequency multiplication of a microwave photonic carrier [91], and wavelength conversion of 40 Gb/s polarization-multiplexed ASK-DPSK data signals [49].

13.6 SUPERCONTINUUM GENERATION

The supercontinuum (SC) generation is essentially a nonlinear phenomenon and corresponds to an extremely wide spectrum achieved by an optical pulse while propagating in a nonlinear medium. It generally results from the synergy between several fundamental nonlinear processes, such as SPM, XPM, SRS, and FWM. The spectral locations and powers of the pumps, as well as the nonlinear and dispersive characteristics of the medium, determine the relative importance and the interaction between these nonlinear processes. The first observation of supercontinuum was realized in 1970 by Alfano and Shapiro in bulk borosilicate glass [92]. SC in fibers occurred for the first time in a 1976 experiment, when 10 ns pulses with more than 1 kW peak power were launched in a 20 m long fiber, producing a 180 nm wide spectrum [93].

A supercontinuum source can find applications in the area of biomedical optics, where it allows the improvement of longitudinal resolution in optical coherence tomography by more than an order of magnitude [94,95]; in optical frequency metrology [96,97]; in all kinds of spectroscopy; and as multiwavelength source in the telecommunications area. The last case will be discussed in Chapter 14.

13.6.1 Basic Physics of Supercontinuum Generation

SPM is one of the nonlinear effects that are responsible for the spectral broadening of picosecond pulses. As seen in Chapter 5, SPM-induced spectral broadening depends on the propagation distance, power density, and pulse shape. Its role is not significant at relatively low input powers and for relatively long pulses (within picosecond and nanosecond range), so that the rate of power variation with time, dP/dt, is small. In fact, early experiments aiming the generation of supercontinuum in fibers revealed that it was mainly due to SRS and FWM [93,98]. Multimode propagation through the fiber provided the necessary phase matching conditions for the FWM process. Concerning the SRS process, its contribution for the supercontinuum generation is effective only on the Stokes (long-wavelength) side of the spectrum. Hence, such contribution tends to make the overall spectrum asymmetric.

The dispersive properties of the fiber become especially important in the case of the FWM process. If the required phase matching conditions are satisfied, FWM generates sidebands on the short- and long-wavelength sides of the pulse spectrum. This process can produce a wideband supercontinuum even in the normal GVD region ($\beta_2 > 0$) of an optical fiber, where the positive nonlinear phase mismatch can be compensated by the negative linear phase mismatch due to β_4, as seen in Section 13.5. In a 1998 experiment, an SC spectrum over 140 nm was generated when 0.9 ps pulses were launched on a 1.7 km long dispersion-flattened fiber having a normal GVD $\beta_2 = 0.1 \text{ ps}^2/\text{km}$ at 1569 nm [99].

In conventional dispersion-shifted fibers, the pulses are usually launched in the 1550 nm window and the resulting supercontinuum spectrum extends at most from 1300 to 1700 nm. However, as seen in Section 13.2, microstructures fibers can be designed with a ZDW of anywhere from 1550 down to 565 nm [2,100]. Using such

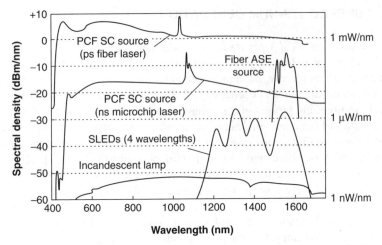

Figure 13.14   Comparison of brightness of various broadband light sources (SLED: super-luminescent light-emitting diode; ASE: amplified spontaneous emission; SC: supercontinuum). The microchip laser SC spectrum was obtained by pumping at 1064 nm with 600 ps pulses. (After Ref. [2]; © 2006 IEEE.)

fibers, it becomes possible to generate a supercontinuum extending from the visible to the near-infrared region. Supercontinuum generation has been achieved in different MFs at 532 [101], 647 [102], 1064 [103], and 1550 nm [104]. The advantages of using MF-based SC sources can be seen from Fig. 13.14, which offers a comparison of the bandwidth and spectrum available from different broadband light sources [2]. Compact MF-based SC sources pumped by inexpensive microchip lasers at 1064 or 532 nm are useful in several areas of science.

The remarkable features of microstructured fibers provide a plethora of nonlinear mechanisms of spectral broadening that may differ depending on waveguide dispersion profile, power, wavelength, and duration of pumping pulses. In particular, the ZDW of an MF can set near the pump wavelength that one wants to use, permitting intrinsic phase matching condition for parametric processes. Using pump pulses of subkilowatt peak power and a duration of some tens of ps, a spatially single-mode SC with more than 600 nm wide can be generated by the interplay of SRS and FWM [105,106]. In a 2008 experiment, a flat octave-spanning supercontinuum was generated from the visible to the near infrared, by pumping an MF at 1064.5 nm, corresponding to the normal dispersion regime, with 0.6 ns pulses with a 7 kHz repetition rate and ~60 mW average power [76]. Figure 13.15 shows the output spectra for L increasing from 1 to 100 m, for a peak power of 2 kW launched into the fiber. For $L = 1$ m (bottom of Fig. 13.15), the signal and idler waves generated by degenerate FWM are clearly visible. They are located, respectively, at 810 and 1548 nm, which is in excellent agreement with Eq. (13.8). By increasing the fiber length, the spectral width of these two waves and that of the pump increase. The broadening of the pump spectrum is expected since the injected peak power is much higher than the Raman threshold. As a consequence, the phase matching relation of

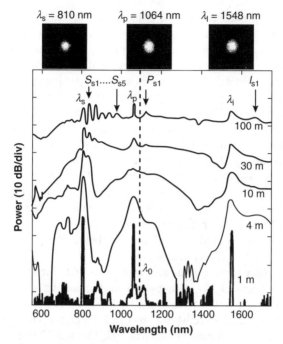

Figure 13.15 Evolution of the output spectrum as a function of the fiber length for a pump peak power of 2 kW launched into the MF. The vertical line represents the ZDW. (After Ref. [76]; © 2008 OSA.)

Eq. (13.8) can be satisfied between additional wavelengths located close to the pump and other wavelengths located around the signal and idler waves. For a fiber length of 30 m, the signal and idler waves start generating their own Raman cascade. For a fiber length of 100 m, the power of the signal wave becomes strong enough to generate up to five Raman Stokes orders located in the normal GVD region, very well defined and labeled S_{s1} to S_{s5} in Fig. 13.15. In contrast, the Raman orders generated by the pump and idler waves fall in the anomalous regime and broaden rapidly, evolving into a continuum. This behavior can be attributed to the modulation instability combined with solitonic effects.

Supercontinuum can also be generated using sufficiently high power from CW lasers. The CW-pumped SCs are generally initiated by the modulation instability process, which can lead to the formation of a train of solitons if one ensures that the pump light is launched in the anomalous dispersion region. These pulses are not stable through the action of higher order dispersion and nonlinear effects. For example, in fibers with positive dispersion slopes ($\beta_3 > 0$), these solitonic pulses shed energy to phase-matched NSR located on the short-wavelength side of the pump [107], as discussed in Section 13.4. Simultaneously, they are redshifted through the combined effects of intrapulse Raman scattering and spectral recoil. Experimentally, this phenomenon has been proved to be a primary importance to generate short wavelengths in the pulse pumping regime. It has also been combined with dispersion-engineered MFs to further extend SCs to the UV in nanosecond and picosecond pumping schemes [108]. This idea consists of modifying the dispersion curve of the fiber so that group velocity matching conditions for trapped dispersive waves continuously evolve along propagation. This leads to the generation of new wavelengths as the ZDW decreases along propagation.

In a 2008 experiment, an SC ranging from 670 to 1350 nm with 9.55 W output power was achieved in the CW pumping regime, using a 200 m long ZDW decreasing MF pumped by a 20 W ytterbium fiber laser at 1.06 μm [109]. The fiber was composed of a 100 m long section with a constant dispersion followed by a 100 m long section with linearly decreasing ZDW. The computed group index and dispersion curves at the fiber input and output are represented in Fig. 13.16a by the solid and dashed curves, respectively. To decrease the ZDW along propagation, the outer diameter of the second section fiber was linearly reduced from an initial value of 125 μm to a final value of 80 μm. The longitudinal evolution of the ZDW is represented in Fig. 13.16b. The initial ZDW was located at 1053 nm, just below the pump wavelength and decreased to 950 nm in a quasilinear way.

Figure 13.17 shows the output spectrum measured for three different input powers of 8.2, 11.3, and 13.5 W. As expected, the broadening of the output spectrum increases on both sides with increasing launch power. The spectrum is limited at long wavelengths by solitons redshifted by the SSFS effect. The trapped dispersive wave appears as a power increase located at 763, 751, and 720 nm, indicated by the arrows in Fig. 13.17. The obtained results confirm that the extension of the SC toward short wavelengths is mainly due to the trapping of dispersive waves by redshifted solitons [110,111].

Pumping an MF by femtosecond pulses with a normal dispersion wavelength below ZDW, where soliton formation is not allowed, leads only to spectral broadening through SPM [112,113]. As the intensity of the input pulse is increased, the spectrum is increasingly broadened. As referred before, one of the most prominent features of MFs is the presence of anomalous dispersion region in visible wavelengths, which has

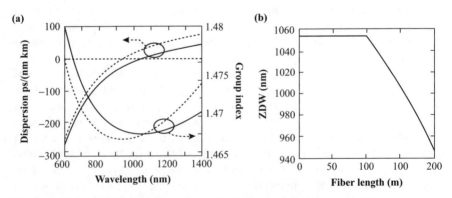

Figure 13.16 (a) Calculated dispersion curves at the input (solid curve) and output (dashed curves) of the MF. (b) Evolution of the ZDW along the fiber length. (After Ref. [109]; © 2008 OSA.)

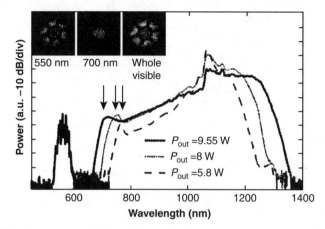

Figure 13.17 Output spectra for output powers of 5.8 (dashed curve), 8 (dotted curve), and 9.55 W (bold curve). The inset shows the far-field output spot at 550 nm, 700 nm, and in the whole visible range. (After Ref. [109]; © 2008 OSA.)

a profound effect on the nonlinear optical behavior within the fiber, leading to the formation of solitons. An interpretation of SC arising from pumping fs pulses inside the anomalous dispersion region is attributed to the fission of higher order solitons into redshifted fundamental solitons associated with phase-matched blueshifted NSR [114,115].

Figure 13.18 shows the experimental results for the initial SC spectrum provided by a 1 m length of fiber as a function of input power, for different degrees of anomalous dispersion [116]. The ZDW of the used fiber was 780 nm and the pumping wavelength was tuned at (a) 790, (b) 805, and (c) 850 nm. A pump power much higher that that needed to form a fundamental soliton excites higher order solitons. For example, a bounded soliton around 860 nm pumped at 850 nm with 0.4 kW peak power corresponds to $N = 4.87$. A pronounced pulse splitting and the SSFS by intrapulse Raman scattering are observed for higher order solitons. After the fission of higher order solitons, the blueshifted peak is generated as indicated by arrows in Fig. 13.18a–c.

Figure 13.19 shows the input power dependence of SC spectra at higher excitation power observed for the same cases considered in Fig. 13.18. SC obtained in the vicinity of the ZDW shows smooth SC with a relatively low input power of 5 kW (Fig. 13.19b). Peaks in the proximity of the ZDW serve as parametric pumps, which generate further peaks between blueshifted and SSFS peaks. At higher pump energies, all those peaks merge to give a continuous spectrum, whose bandwidth increases by increasing the input power. XPM followed by all the generated frequencies also made the resultant spectrum smooth. The generation of SC at deep anomalous dispersion (Fig. 13.19c) exhibits substantial broad SC at relatively low powers, but the generated SC profile is less smooth.

The use of multicomponent glasses allows the balance of nonlinearity and dispersion to be adjusted as well as offering extended transparency into the IR [117]. SC generation over 1200–1800 nm has been demonstrated in a 2 cm long highly

(a)

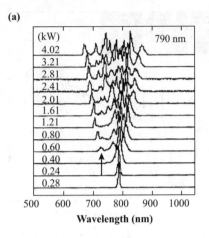

(b)

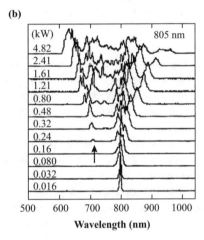

(c)

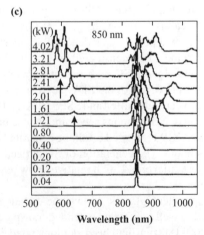

Figure 13.18   Initial SC spectra obtained by tuning the input wavelength at (a) 790, (b) 805, and (c) 850 nm as a function of input power for a 1 m long MF fiber. (After Ref. [116]; © 2004 IEEE.)

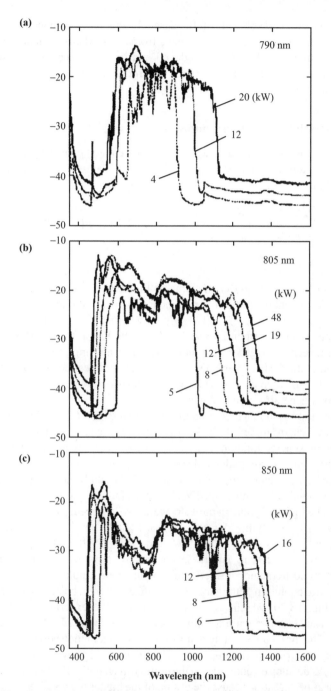

Figure 13.19 Growing processes of SC by tuning the input wavelength at (a) 790, (b) 805, and (c) 850 nm as a function of input power for a 1 m long MF fiber. (After Ref. [116]; © 2004 IEEE.)

nonlinear bismuth oxide fiber [118]. Because this fiber had a normal dispersion at the communication window, such SC originated from the SPM effect. Hence, the spectral variation and noise in the SC was reduced and the chirp accumulated during the propagation was readily compensated to produce a femtosecond compressed pulse [118].

13.6.2 Modeling the Supercontinuum

Modeling the supercontinuum generation can be realized considering a generalized NLSE that includes higher order dispersion effects, as well as intrapulse Raman scattering [44,119–122]. Such equation can be written as

$$\frac{\partial U}{\partial z} - i \sum_{k \geq 2} \frac{i^k \beta_k}{k!} \frac{\partial^k U}{\partial \tau^k} + \frac{\alpha(\omega)}{2} U = i\gamma \left(1 + \frac{i}{\omega_0} \frac{\partial}{\partial t}\right) \left(U(z,t) \int_{-\infty}^{+t} R(t') |U(z,t-t')|^2 dt'\right)$$

$$(13.10)$$

where β_k are the dispersion coefficients at the center frequency and $\alpha(\omega)$ is the frequency-dependent fiber loss. The nonlinear response to the applied field $U(z,t)$ has been written as $R(t) = (1-f_R)\delta(t) + f_R h(t)$, where the δ-function represents the instantaneous electron response (responsible for the Kerr effect), $h(t)$ represents the delayed ionic response (responsible for the Raman scattering), and f_R is the fractional contribution of the delayed Raman response to the nonlinear polarization, in which a value $f_R = 0.18$ is often assumed [123]. The split-step Fourier method discussed in Section 4.6 is generally used to solve Eq. (13.10).

The simulation of the supercontinuum was realized in Ref. 122 using fiber and laser parameters taken from Ref. 119. Such parameters represent typical experimental conditions for supercontinuum generation in MFs employing femtosecond lasers and pumping in the anomalous dispersion regime. An input pulse $U(0,\tau) = \sqrt{P_0} \operatorname{sech}(\tau/t_0)$, is assumed, where $P_0 = 10\,\text{kW}$ and the width $t_0 = 28.4\,\text{fs}$, which corresponds to an intensity FWHM of 50 fs. The pulse is initially centered at 850 nm. The following fiber parameters were also assumed: $\beta_2 = -12.76\,\text{ps}^2/\text{km}$, $\beta_3 = 8.119 \times 10^{-2}\,\text{ps}^3/\text{km}$, $\beta_4 = -1.321 \times 10^{-4}\,\text{ps}^4/\text{km}$, $\beta_5 = 3.032 \times 10^{-7}\,\text{ps}^5/\text{km}$, $\beta_6 = -4.196 \times 10^{-10}\,\text{ps}^6/\text{km}$, $\beta_7 = 2.570 \times 10^{-13}\,\text{ps}^7/\text{km}$, $\gamma = 45$ W^{-1}/km, and $L = 10\,\text{cm}$. For these conditions, the soliton order is around 5. $N = 2^{13}$ time and frequency discretization points were employed.

The spectral evolution of the supercontinuum at different positions along the MF is shown in Fig. 13.20, which reproduces very well the experimental observations reported in Ref. 119. Dramatic spectral broadening is observed in the first centimeter of the fiber. The initial broadening, mainly due to the interaction between the SPM and GVD, is symmetric. Afterward, the spectrum becomes asymmetric and distinct spectral peaks develop on both sides of the pump wavelength as higher order dispersive and nonlinear perturbations cause the fission of the higher order soliton. The resulting fundamental solitons then undergo a continuous self-frequency shifting to longer wavelengths because of intrapulse Raman scattering. The emergence and self-shifting

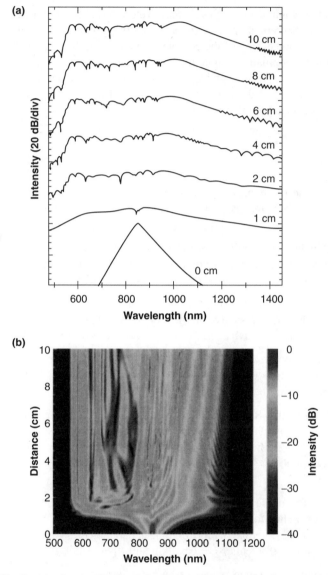

Figure 13.20 Results from a numerical simulation of supercontinuum generation in an MF. (a) Spectra at discrete locations along the fiber and (b) density plot as a function of the propagation distance. (After Ref. [122]; © 2007 IEEE.)

of one of these fundamental solitons on the long wavelength side is clearly seen in Fig. 13.20b. In this process, each Raman soliton sheds some of its energy in the form of nonsolitonic radiation on the short-wavelength side of the pump, which also leads to the appearance of discrete spectral components in this region of the spectrum.

In spite of the significant progress in the development of SC sources and successful application of theoretical models to optimize the generation of broad radiation sources, the comprehension of the complex nonlinear mechanism of SC generation is not complete. Actually, a quantitative agreement between experimental and numerical simulation results of SC generation is rare. Among other aspects, the polarization effects and the coherence properties of SC are two issues, which are attracting a considerable attention at present.

PROBLEMS

13.1 Determine the nonlinear Kerr coefficient of a tapered fiber, which has an effective core area $A_{eff} = 3\,\mu m^2$ and a nonlinear parameter at 1550 nm $\gamma = 100\,W^{-1}\,km^{-1}$.

13.2 Explain how the nonlinear phenomenon of SPM can be used to determine the nonlinear parameter γ for an optical fiber. Why is it important to know the exact shape of the input pulses used in such measurement?

13.3 Describe the phenomenon of nonsolitonic radiation and explain the conditions under which it can occur. Comment on the new features of NSR that could be observed in fibers exhibiting two zero dispersion wavelengths.

13.4 Give a physical explanation for the self-frequency shift effect occurring during the propagation of an ultrashort soliton. Explain how this effect can be suppressed in fibers that exhibit two ZDWs.

13.5 Determine the Raman-induced frequency shift experienced by a 100 fs (FWHM) input pulse, which propagates as a fundamental soliton, at the output of a 20 m long highly nonlinear fiber with $\beta_2 = -30\,ps^2/km$. Assume the typical value $t_R = 3$ fs. Considering that such fiber has a nonlinear parameter $\gamma = 50\,W^{-1}\,km^{-1}$, determine the pulse peak power.

13.6 Derive the phase matching condition given by Eq. (13.8) and obtain the frequency shift for a pulse with a peak power $P_p = 100$ W propagating in a highly nonlinear fiber with a nonlinear parameter $\gamma = 200\,W^{-1}\,km^{-1}$ and dispersion coefficients $\beta_2 = 6\,ps^2/km$ and $\beta_4 = -2 \times 10^{-4}\,ps^4/km$.

13.7 Describe the physical processes involved in the supercontinuum generation when a CW beam is launched into a highly nonlinear fiber.

REFERENCES

1. G. Berrettini, G. Meloni, A. Bogoni, and L. Poti, *IEEE Photon. Technol. Lett.* **18**, 2439 (2006).
2. P. S. J. Russel, *J. Lightwave Technol.* **24**, 4729 (2006).

3. R. H. Stolen, W. A. Reed, K. S. Kim, and G. T. Harvey, *J. Lightwave Technol.* **16**, 1006 (1998).

4. T. Kato, Y. Suetsugu, M. Takagi, E. Sasaoka, and M. Nishimura, *Opt. Lett.* **29**, 988 (1995).

5. L. Prigent and J. P. Hamaide, *IEEE Photon. Lett.* **5**, 1092 (1993).

6. K. Nakajima and M. Ohashi, *IEEE Photon. Technol. Lett.* **14**, 492 (2002).

7. T. Okuno, M. Onishi, T. Kashiwada, S. Ishikawa, and M. Nishimura, *IEEE J. Sel. Top. Quantum Electron.* **5**, 1385 (1999).

8. M. Hirano, T. Nakanishi, T. Okuno, and M. Onishi, *J. Sel. Top. Quantum Electron.* **15**, 103 (2009).

9. M. Takahashi, R. Sugizaki, J. Hiroishi, M. Tadakuma, Y. Taniguchi, and T. Yagi, *J. Lightwave Technol.* **23**, 3615 (2005).

10. K. Mori, H. Takara, and S. Kawanishi, *J. Opt. Soc. Am. B* **18**, 1780 (2001).

11. T. Okuno, M. Hirano, T. Kato, M. Shigemats, and M. Onishi, *Electron. Lett.* **39**, 972 (2003).

12. J. M. Harbold, F. O. Ilday, F. W. Wise, T. A. Birks, W. J. Wadsworth, and Z. Chen, *Opt. Lett.* **27**, 1558 (2000).

13. F. Lu and W. H. Knox, *Opt. Express* **12**, 347 (2004).

14. M. A. Foster and A. L. Gaeta, *Opt. Express* **12**, 3137 (2004).

15. A. Zheltitikov, *J. Opt. Soc. Am. B* **22**, 1100 (2005).

16. L. M. Tong, J. Y. Lou, and E. Mazur, *Opt. Express* **12**, 1025 (2004).

17. M. A. Foster, K. D. Moll, and A. L. Gaeta, *Opt. Express* **12**, 2880 (2004).

18. C. M. Cordeiro, W. J. Wadsworth, T. A. Birks, and P. J. St. Russel, *Opt. Lett.* **30**, 1980 (2005).

19. S. G. Leon-Saval, T. A. Birks, W. J. Wadsworth, and P. J. St. Russel, *Opt. Express* **12**, 2864 (2004).

20. J. C. Knight, T. A. Birks, P. J. St. Russel, and D. M. Atkin, *Opt. Lett.* **21**, 1547 (1996).

21. E. Yablonovitch, *Phys. Rev. Lett.* **58**, 2059 (1987).

22. S. John, *Phys. Rev. Lett.* **58**, 2486 (1987).

23. J. C. Knight, *Nature* **424**, 847 (2003).

24. V. V. Kumar, A. K. George, W. H. Reeves, J. C. Knight, P. J. St. Russel, F. G. Omenetto, and A. J. Taylor, *Opt. Express* **10**, 1520 (2002).

25. R. F. Cregan, B. J. Mangan, and J. C. Knight, *Science* **285**, 1537 (1999).

26. P. J. Roberts, F. Couny, H. Sabert, et al., *Opt. Express* **13**, 236 (2005).

27. J. Broeng, D. Mogilevstev, S. E. Barkou, and A. Bjarklev, *Opt. Fiber Technol.* **5**, 305 (1999).

28. B. J. Eggleton, C. Kerbage, P. S. Westbrook, R. S. Windeler, and A. Hale, *Opt. Express* **9**, 698 (2001).

29. V. Finazzi, T. M. Monro, and D. J. Richardson, *J. Opt. Soc. Am. B* **20**, 1427 (2003).

30. K. Nakajima, J. Zhou, K. Tajima, K. Kurokawa, C. Fukai, and I. Sankawa, *J. Lightwave Technol.* **17**, 7 (2005).

31. K. Turokwa, K. Tajima, K. Tsujikawa, and K. Nakagawa, *J. Lightwave Technol.* **24**, 32 (2006).

32. K. Kurokawa, K. Tajima, and K. Kakajima, *J. Lightwave Technol.* **25**, 75 (2007).

33. P. J. Roberts, F. Couny, H. Sabert, et al., *Opt. Express* **13**, 7779 (2005).

34. K. Saito and M. Koshiba, *J. Lightwave Technol.* **23**, 3580 (2005).

35. T. A. Birks, J. C. Knight, and P. J. St. Russel, *Opt. Lett.* **22**, 961 (1997).

36. L. Dong, H. McKay, and L. Fu, *Opt. Lett.* **33**, 2440 (2008).

37. A. Ortigosa-Blanch, J. C. Knight, W. J. Wadsworth, J. Arriaga, B. J. Mangan, T. A. Birks, and P. J. St. Russel, *Opt. Lett.* **25**, 1325 (2000).

38. D. Kim and J. U. Kang, *Opt. Express* **12**, 4490 (2004).

39. J. Laegsgaard, N. A. Mortenson, J. Riishede, and A. Bjarklev, *J. Opt. Soc. Am. B* **20**, 2046 (2003).

40. F. Luan, J. C. Knight, P. J. St. Russel, S. Campbell, D. Xiao, D. T. Reid, B. J. Mangan, D. P. Wiliams, and P. J. Roberts, *Opt. Express* **12**, 835 (2004).

41. P. J. St. Russell, *Science* **299**, 358 (2003).

42. F. Benabid, G. Antonopoulos, J. C. Knight, and P. J. St. Russel, *Science* **298**, 5592 (2002).

43. X. Feng, A. K. Mairaj, D. W. Hewak, and T. M. Monro, *J. Lightwave Technol.* **23**, 2046 (2005).

44. J. H. V. Price, T. M. Monro, H. Ebendorff-Heidepriem, F. Poletti, P. Horak, V. Finazzi, J. Y. Y. Leong, P. Petropoulos, J. C. Flanagan, G. Brambilla, X. Feng, and D. J. Richardson, *IEEE J. Sel. Top. Quantum Electron.* **23**, 738 (2007).

45. J. S. Wang, E. M. Vogel, and E. Snitzer, *Opt. Mater.* **3**, 187 (1994).

46. J. Y. Y. Leong, P. Petropoulos, J. H. V. Price, H. Ebendorff-Heidepriem, S. Asimakis, R. C. Moore, K. E. Frampton, V. Finazzi, X. Feng, T. M. Monro, and D. J. Richardson, *J. Lightwave Technol.* **24**, 183 (2006).

47. J. H. Lee, K. Kikuchi, T. Nagashime, T. Hasegawa, S. Ohara, and N. Sugimoto, *Opt. Express* **13**, 3144 (2005).

48. J. H. Lee, T. Nagashima, T. Hasegawa, S. Ohara, N. Sugimoto, and K. Kikuchi, *J. Lightwave Technol.* **24**, 22 (2006).

49. P. Fok and C. Shu, *J. Sel. Top. Quantum Electron.* **14**, 587 (2008).

50. C. Jauregui, H. Ono, P. Petropoulos, and D. J. Richardson, *Proceedings of Optical Fiber Communications Conference* 2006, Paper PDP2.

51. R. E. Slusher, G. Lenz, J. Hodelin, J. Sanghera, L. B. Shaw, and I. D. Aggarwal, *J. Opt. Soc. Am. B* **21**, 1146 (2004).

52. M Fuochi, F. Poli, A. Cucinotta, and L. Vincetti, *J. Lightwave Technol.* **21**, 2247 (2003).

53. S. K. Varshney, T. Fujisawa, K. Saito, and M. Koshiba, *Opt. Express* **13**, 9516 (2005).

54. S. K. Varshney, K. Saito, K. Iizawa, Y. Tsuchida, M. Koshiba, and R. K. Sinha, *Opt. Lett.* **33**, 2431 (2008).

55. S. Radic, D. J. Moss, and B. J. Eggleton, in I. Kaminow, T. Li,and A. E. Willner (Eds.), *Optical Fiber Telecommunications. VA. Components and Subsystems.* Elsevier Inc., 2008.

56. H. Ebendorff-Heidepriem, P. Petropoulos, S. Asimadis, V. Finazzi, R. Moore, K. Frampton, F. Koizumi, D. Richardson, and T. Monro, *Opt. Express* **12**, 2082 (2004).

57. F. M. Mitschke and L. F. Mollenauer, *Opt. Lett.* **11**, 659 (1986).

58. X. Liu, C. Xu, W. H. Knox, J. K. Chandalia, B. J. Eggleton, S. G. Kosinski, and R. S. Windler, *Opt. Lett.* **26**, 358 (2001).

59. B. R. Washburn, S. E. Ralph, P. A. Lacourt, J. M. Dudley, W. T. Rhodes, R. S. Windeler, and S. Coen, *Electron. Lett.* **37**, 1510 (2001).

60. I. G. Cormack, D. T. Reid, W. J. Wadsworth, J. C. Knight, and P. J. St. Russel, *Electron. Lett.* **38**, 167 (2002).

61. D. G. Ouzounov, F. R. Ahmad, D. Muller, N. Venkataraman, M. T. Gallagher, M. G. Thomas, J. Silcox, K. W. Koch, and A. L. Gaeta, *Science* **301**, 1702 (2003).

62. N. Nishizawa, Y. Ito, and T. Goto, *IEEE Photon. Technol. Lett.* **14**, 986 (2002).

63. A. Efimov, A. J. Taylor, F. G. Omenetto, and E. Vanin, *Opt. Lett.* **29**, 271 (2004).

64. K. S. Abedin and F. Kubota, *IEEE J. Sel. Top. Quantum Electron.* **10**, 1203 (2004).

65. J. H. Lee, J. Howe, C. Xu, and X. Liu, *IEEE J. Sel. Top. Quantum Electron.* **14**, 713 (2008).

66. N. Nishizawa and T. Goto, *Opt. Express* **11**, 359 (2003).

67. M. Kato, K. Fujiura, and T. Kurihara, *Electron. Lett.* **40**, 381 (2004).

68. S. Oda and A. Maruta, *Opt. Express* **14**, 7895 (2006).

69. A. V. Husakou and J. Herrmann, *Phys. Rev. Lett.* **87**, 203901 (2001).

70. N. Akhmediev and M. Karlsson, *Phys. Rev. A* **51**, 2602 (1995).

71. N. Nishizawa and T. Goto, *Opt. Express* **8**, 328 (2001).

72. F. Biancalana, D. V. Skryabin, and A. V. Yulin, *Phys. Rev. E* **70**, 016615 (2004).

73. D. V. Skryabin, F. Luan, J. C. Knight, and P. J. St. Russel, *Science*, **301**, 1705 (2003).

74. S. Pitois and G. Millot, *Opt. Commun.* **226**, 415 (2003).

75. W. Wadsworth, N. Joly, Knight, T. Birks, F. Biancala, and P. Russel, *Opt. Express* **12**, 299 (2004).

76. A. Kudlinski, V. Pureur, G. Bouwrnans, and A. Mussot, *Opt. Lett.* **33**, 2488 (2008).

77. J. G. Rarity, J. Fulconis, J. Duligall, W. J. Wadsworth, and P. J. St. Russell, *Opt. Express* **13**, 534 (2005).

78. J. Fan, A. Migdall, and L. J. Wang, *Opt. Lett.* **30**, 3368 (2005).

79. J. Fulconis, O. Alibart, W. J. Wadsworth, P. J. St. Russell, and J. G. Rarity, *Opt. Express* **13**, 7572 (2005).

80. J. E. Sharping, *J. Lightwave Technol.* **26**, 2184 (2008).

81. J. E. Sharping, M. Fiorentino, P. Kumar, and R. S. Windeler, *Opt. Lett.* **27**, 1675 (2002).

82. Y. Deng, Q. Lin, F. Lu, G. Agrawal, and W. Knox, *Opt. Lett.* **30**, 1234 (2005).

83. J. E. Sharping, M. A. Foster, A. L. Gaeta, J. Lasri, O. Lyngnes, and K. Vogel, *Opt. Express* **15**, 1474 (2007).

84. J. E. Sharping, J. R. Sanborn, M. A. Foster, D. Broaddus, and A. L. Gaeta, *Opt. Express* **16**, 18050 (2008).

85. J. E. Sharping, M. Fiorentino, A. Coker, P. Kumar, and R. S. Windeler, *Opt. Lett.* **26**, 1948 (2001).

86. K. Seki and S. Yamashita, *Opt. Express* **16**, 13871 (2008).

87. X. M. Liu, X. F. Yang, F. Y. Lu, J. H. Ng, X. Q. Zhou, and C. Lu, *Opt. Express* **13**, 142 (2005).

88. M. P. Fok and C. Shu, *Opt. Express* **15**, 5925 (2007).

89. C. Yu, L. S. Yan, T. Luo, Y. Wang, Z. Pan, and A. E. Wilner, *IEEE Photon. Technol. Lett.* **17**, 636 (2005).

90. M. P. Fok and C. Shu, *Proceedings of Fiber Communication Conference*, 2006, Paper OWI 34

91. A. J. Seeds and K. J. Williams, *J. Lightwave Technol.* **24**, 4628 (2006).

92. R. R. Alfano and S. L. Shapiro, *Phys. Rev. Lett.* **24**, 584 (1970).

93. C. Lin and R. H. Stolen, *Appl. Phys. Lett.* **28**, 216 (1976).

94. A. F. Fercher, W. Drexler, C. K. Hitzenberger, and T. Lasser, *Rep. Prog. Phys.* **66**, 239 (2003).

95. G. Humbert, W. J. Wadsworth, S. G. Leon-Saval, J. C. Knight, T. A. Birks, P. S. J. Russell, M. J. Lederer, D. Kopf, K. Wiesauer, E. I. Breuer, and D. Stifter, *Opt. Express* **14**, 1596 (2006).

96. M. Bellini and T. W. Hänsch, *Opt. Lett.* **25**, 1049 (2000).

97. H. Hundertmark, D. Wandt, C. Fallnich, N. Haverkamp, and H. Telle, *Opt. Express* **12**, 770 (2004).

98. P. L. Baldeck and R. R. Alfano, *J. Lightwave Technol.* **5**, 1712 (1987).

99. Y. Takushima, F. Futami, and K. Kikuchi, *IEEE Photon. Technol. Lett.* **10**, 1560 (1998).

100. W. H. Reeves, D. V. Skryabin, F. Biancalana, J. C. Knight, P. J. St. Russel, F. G. Omenetto, A. Efomov, and A. J. Taylor, *Nature*, **424**, 511 (2003).

101. S. G. Leon-Saval, T. A. Birks, W. J. Wadsworth, P. S. J. Russel, and M. W. Mason, *Opt. Express* **12**, 2864 (2004).

102. S. Coen, A. H. L. Chau, R. Leonhardt, J. D. Harvey, J. C. Knight, W. J. Wadsworth, and P. S. J. Russell, *J. Opt. Soc. Am. B* **19**, 753 (2002).

103. W. J. Wadsworth, N. Joly, J. C. Knight, T. A. Birks, F. Biancalana, and P. J. L. Russell, *Opt. Express* **12**, 299 (2004).

104. V. V. Ravi, K. Kumar, A. K. George, W. H. Reeves, J. C. Knight, P. S. J. Russell, F. G. Omenetto, and A. J. Taylor, *Opt. Express* **10**, 1520 (2002).

105. S. Coen, A. H. Chau, R. Leonhardt, J. D. Harvey, J. C. Knight, W. J. Wadsworth, and P. J. St. Russel, *J. Opt. Soc. Am. B* **19**, 753 (2002).

106. J. M. Dudley, L. Provino, N. Grossard, H. Maillotte, R. S. Windeler, B. J. Eggleton, and S. Coen, *J. Opt. Soc. Am. B* **19**, 765 (2002).

107. N. Akhmediev and M. Karlsson, *Phys. Rev. A* **51**, 2602 (1995).

108. A. Kudlinski, A. K. George, J. C. Knight, J. C. Travers, A. B. Rulkov, S. V. Popov, and J. R. Taylor, *Opt. Express* **14**, 5715 (2006).

109. A. Kudlinski and A. Mussot, *Opt. Lett.* **33**, 2407 (2008).

110. A. V. Gorbach, D. V. Skryabin, J. M. Stone, and J. C. Knight, *Opt. Express* **14**, 9854 (2006).

111. A. V. Gorbach and D. V. Skryabin, *Nat. Photon.* **1**, 653 (2007).

112. A. Ortigosa-Blanch, J. C. Knight, and P. J. St. Russel, *J. Opt. Soc. Am. B* **19**, 2567 (2002).

113. G. Geny, M. Lehtonen, H. Ludvigsen, J. Broeng, and M. Kaivola, *Opt. Express* **10**, 1083 (2002).

114. A. V. Husakou and J. Herrmann, *Appl. Phys. B* **77**, 227 (2003).

115. T. Schreiber, J. Limpert, H. Zellmer, A. Tunnermann, and K. P. Hansen, *Opt. Commun.* **228**, 71 (2003).

116. S. Sakamaki, M. Nakao, M. Naganuma, and M. Izutsu, *IEEE J. Sel. Top. Quantum Electron.* **10**, 876 (2004).

117. F. G. Omenetto, N. A. Wolchover, M. R. Wehner, M. Ross, A. Efimov, A. J. Taylor, V. V. Ravi, K. Kumar, A. K. George, J. C. Knight, N. Y. Joly, and P. J. St. Russell, *Opt. Express* **14**, 4928 (2006).

118. J. T. Gopinath, H. M. Shen, H. Sotobayashi, E. P. Ippen, T. Hasegawa, T. Nagashima, and N. Sugimoto, *J. Lightwave Technol.* **23**, 3591 (2005).

119. J. M. Dudley and S. Coen, *J. Sel. Top. Quantum Electron.* **8**, 651 (2002).

120. A. L. Gaeta, *Opt. Lett.* **27**, 924 (2002).

121. A. V. Husakou and J. Herrmann, *J. Opt. Soc. Am. B* **19**, 2171 (2002).

122. J. Hult, *J. Lightwave Technol.* **25**, 3770 (2007).

123. R. H. Stolen, J. P. Gordon, W. J. Tomlinson, and H. A. Haus, *J. Opt. Soc. Am. B* **6**, 1159 (1989).

14

OPTICAL SIGNAL PROCESSING

Extensive research has recently been done aiming at ultrafast optical networks at the bit rate over 100 Gb/s, where a single optical carrier is employed [1]. For such increasing bit rates, electronic devices are not suitable, due to their practical speed limits. All-optical signal processing appears, therefore, as a key and promising technology for improving the flexibility and increasing the capacity of future photonic networks [2].

The third-order $\chi^{(3)}$ optical nonlinearity in silica-based single-mode fibers is one of the most important effects that can be used for all-optical signal processing [3]. This happens not only because the third-order nonlinearity has a response time typically <10 fs, but also because it is responsible for a wide range of phenomena, as discussed in previous chapters of this book.

Using conventional optical fibers for nonlinear optical processing, a length of several kilometers is usually required due to its relatively small nonlinear parameter [4]. Such long fibers pose some practical limitations, concerned namely with the size and stability of the system. In order to minimize these problems, highly nonlinear silica fibers with a smaller effective mode area, and hence a larger nonlinear parameter ($\gamma \sim 11$ W^{-1} km^{-1}), were developed and used [5]. However, in spite of the fact that the required fiber length for some applications was thus reduced to about 1 km, the issues related to the long fiber lengths were not completely resolved. As seen in Chapter 13, microstructured fibers (MFs) with an extremely small effective core area and exhibiting significantly enhanced nonlinear characteristics have been fabricated in the recent years [6,7]. Wavelength conversion via four-wave mixing (FWM) was achieved using a 64 m long silica MF [8], which means a reduction in fiber length by one order of magnitude.

Another main step to improve the efficiency of the nonlinear effects is to realize highly nonlinear fibers (HNLFs) using materials with a nonlinear refractive index higher than that of the silica glass, namely, lead silicate, tellurite, bismuth glasses, and chalcogenide glasses [9–12]. Using such HNLFs, the required fiber length for nonlinear processing can be dramatically reduced to the order of centimeters, representing a further improvement by several orders of magnitude.

A wide range of fiber-based all-optical signal processing devices have been investigated during the recent years by using self-phase modulation (SPM) [13,14], cross-phase modulation (XPM) [15–18], and four-wave mixing [19–22] in some of the above-mentioned highly nonlinear fibers. In this chapter, we provide a brief account of this vast and fast developing area.

14.1 NONLINEAR SOURCES FOR WDM SYSTEMS

Actual wavelength division multiplexing (WDM) transmission systems use a large number of DFB lasers. Each DFB laser corresponds to a wavelength channel and is generally used together with a modulator and a wavelength-stabilizing system. This approach is costly and becomes impractical when the number of channels becomes large.

One approach to realize a practical WDM source uses the technique of spectrum slicing. In this approach, the light of a broadband source is sliced into several narrow bands centered at the wavelengths of the WDM channels. A multipeak optical filter can be used to realize the spectrum slicing. Since some of the components are shared among all the wavelengths, the number of active elements in such multiwavelength transmitters is greatly reduced. Spectrum slicing for WDM applications has been reported employing amplified spontaneous emission (ASE) from light-emitting diodes [23,24], superluminescent diodes [25,26], or erbium-doped fiber amplifiers [27]. However, the transmission capacity of such incoherent spectrum-sliced systems is generally limited by the spontaneous emission beat noise. This limitation can be alleviated using coherent light sources, such as mode-locked lasers [28,29] and supercontinuum (SC) generators [30–33]. The coherent nature of the broadband radiation originating from these sources is crucial to achieve high transmission capacities in both multiwavelength optical time division multiplexing (OTDM) and dense wavelength division multiplexing (DWDM) systems.

The spectral width of femtosecond pulses emitted by a mode-locked laser is already sufficiently large to begin with and can be used for frequency slicing. In fact, according to the Fourier theorem for bandwidth-limited pulses, the temporal duration and the bandwidth are in inverse proportion to each other. For example, a laser producing a train of 50 fs transform-limited Gaussian pulses has an optical bandwidth of 70 nm at 1550 nm. The bandwidth of such pulses can be further broadened by chirping them using 10–15 km of a standard telecommunication fiber. A major advantage of this spectrum slicing technique is its limited hardware requirement for generating and encoding a large number of WDM channels.

Supercontinuum generation and spectrum slicing is another practical possibility of reducing the cost and complexity of multiwavelength transmitters for both OTDM and WDM systems. Figure 14.1 illustrates a typical implementation of such a

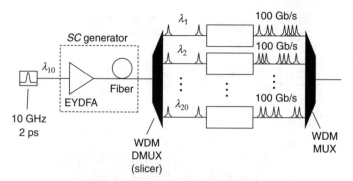

Figure 14.1 Schematic representation of a multiwavelength OTDM system using a spectrum-sliced supercontinuum transmitter. EYDFA denotes an erbium–ytterbium-doped fiber amplifier. (After Ref. [34]; © 2001 Elsevier.)

transmitter for a multiwavelength OTDM system [34]. Picosecond pulses are obtained at repetition rates around 10 GHz from an actively mode-locked laser [32,33], a gain-switched laser [35], or an electroabsorption-modulated DFB laser [36]. The pulses are amplified and propagated in a fiber, to generate a supercontinuum. A passive multipeak optical filter, such as a WDM demultiplexer, slices the supercontinuum and creates multiple 10 GHz pulse trains having different carrier wavelengths. These pulses are then sent to OTDM stages where data encoding and optical interleaving is realized.

Three characteristics of the supercontinuum are particularly important in the context of its application as a WDM system transmitter. First, its spectrum should ideally coincide with the transmission window. Second, it should be smooth enough, in order to provide similar powers to each channel, resulting in uniform impact of the nonlinear effects. Third, the amplitude noise of the filtered SC should be sufficiently reduced. This last characteristic becomes especially important for high-capacity WDM systems, since the intensity variations among pulses at a given wavelength have a negative impact on the performance of that channel.

The properties of the supercontinuum depend significantly on the type of fiber used for the nonlinear expansion as well as on the seed pulses. As discussed in Section 13.6, the SC phenomenon is the result of a rather complicated combination of nonlinear physical processes, such as SPM, XPM, and FWM. Furthermore, if the generated bandwidth exceeds the Raman gain spectrum, stimulated Raman scattering must also be considered. On the other hand, if the fiber is operated in the anomalous regime, higher order soliton generation and fission, together with the emission of nonsolitonic radiation, represent important mechanisms in the spectral broadening process.

Early in 1995, 3.5 ps pulses from a mode-locked fiber laser were broadened spectrally up to 200 nm through supercontinuum generation by exploiting the nonlinear effects in a 3 km long optical fiber, resulting in a 200-channel WDM source [37]. Two years later, a supercontinuum source was used to demonstrate data transmission at a bit rate of 1.4 Tb/s using seven WDM channels, each operating at 200 Gb/s, with 600 GHz spacing [38]. By 2002, a 200 nm wide supercontinuum was

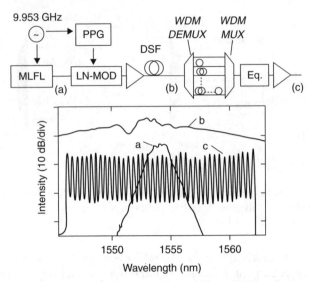

Figure 14.2 Long-distance 400 Gb/s SC-DWDM transmitter layout. MLFL denotes the mode-locked fiber laser, PPG the pulse pattern generator, LN-MOD the lithium niobate modulator, DSF the dispersion-shifted fiber, and Eq. a dynamic power equalizer. The inset shows the optical spectrum at three points within the transmitter. (After Ref. [34]; © 2001 Elsevier.)

used to create a WDM source with 4200 channels with only 5 GHz spacing [39]. Such a WDM source was later used to create 50 GHz spaced channels in a spectral range between 1425 and 1675 nm [40].

Figure 14.2 shows a schematic of a spectrum-sliced SC generator used in a long-distance WDM transmission experiment [33]. A 10 GHz harmonically mode-locked fiber laser producing a train of 2 ps was used to generate SC light through SPM in the normal dispersion regime. A WDM demux/mux combination was used to realize the spectrum slicing, producing 40 channels between 1546.2 and 1562 nm, with a channel spacing of 50 GHz. A power equalizer was also included to equalize the power levels of all channels, which were afterward propagated over 544 km of SMF for a total capacity of 400 Gb/s.

The structure of the cavity modes present in the original laser output is preserved when a normal dispersion fiber is used to generate the supercontinuum [41]. This enables the generation of an ultrabroadband frequency comb, in which the separation between peaks corresponds to the microwave mode-locking frequency of the source laser. Each peak can then be considered as a potential transmission channel. Using this property, more than 1000 optical frequency channels with a channel spacing of 12.5 GHz were generated in 2000 from a single SC source [42]. Following the same principle, 124 nm seamless transmission of 3.13 Tb/s (10-channel DWDM × 313 Gb/s) over 160 km, with a channel spacing of 50 GHz, was reported [43]. Raman amplification in hybrid tellurite/silica fiber was used in this experiment to improve gain flatness. More recently, a field demonstration of 1046-channel ultra-DWDM transmission over 126 km was realized using an SC multicarrier source spanning

1.54–1.6 µm, which was mainly generated through SPM-induced spectral broadening. The channel spacing was 6.25 GHz and the signal data rate per channel was 2.67 Gb/s [44].

14.2 OPTICAL REGENERATION

In high-speed optical networks, optical signals are affected by several effects, namely, ASE from optical amplifiers, chromatic dispersion, polarization mode dispersion, and nonlinear phenomena in transmission fibers. Therefore, all-optical regenerators including reamplification and reshaping (2R) or reamplification, reshaping, and retiming (3R) functions are key devices for such high-speed optical networks. Compared to conventional techniques, all-optical signal regeneration offers much higher response speeds and significantly lower costs, since they avoid optoelectronic/electrooptic (OE/EO) conversion and demultiplexing.

2R optical regeneration based on SPM with subsequent spectral filtering was first proposed by Mamyshev [13]. These regenerators have been the object of a lot of attention, since they present a number of key features. Among them are the simplicity, a bandwidth that is limited only by the intrinsic material nonlinear response, and the possibility to achieve a direct bit error rate (BER) improvement [45,46]. Up to 1 million kilometers of unrepeatered transmission was achieved using such regenerators [47].

The principle of operation of a Mamyshev regenerator is as follows. A noisy input RZ (return to zero) signal is phase modulated by its own waveform (SPM), while it propagates through a nonlinear fiber. The temporal phase variation induces the instantaneous frequency shift and the signal spectrum is broadened. The intensity slope point of the original signal is spectrally shifted and can be selected using a bandpass filter (BPF) offset from the input center wavelength. Furthermore, above a certain level, any variation in the signal intensity will translate into a change in the width of the SPM-induced broadened spectrum. Since the bandwidth of the optical filter is fixed, the intensity of the filtered output is insensitive to those variations of the signal power, which results in an effective suppression of the amplitude noise at bit "one" level. On the other hand, optical noise at bit "zero" is low and does not induce any spectral shift, being filtered out by the BPF. This results in an improvement of the optical signal-to-noise ratio (SNR).

The signal processing based on the above principle of operation is illustrated in Fig. 14.3. Such a process can be characterized by three parameters: the instantaneous frequency shift induced by phase modulation, f_{PM}, the BPF offset frequency relative to the original signal center frequency, f_{BPF}, and the BPF bandwidth, Δf_{BPF}. The requirement for f_{PM} in order to guarantee a proper operation is given by [48]

$$f_{PM} > 2\Delta f_{in} \tag{14.1}$$

where Δf_{in} is the bandwidth of the incoming optical signal. The maximum value of SPM-induced frequency shift depends on the pulse shape and can be obtained from

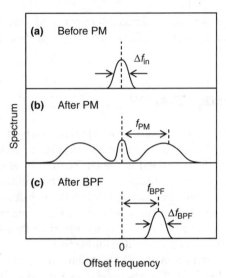

Figure 14.3 Optical spectra of the data signal before (a) and after (b) phase modulation, and after bandpass filtering (c). (After Ref. [48]; © 2008 IEEE.)

the results presented in Section 5.1. Assuming that Gaussian pulses with a peak power P_0 are launched into an optical fiber of length L, Eq. (14.1) provides the following condition [48]:

$$\gamma P_0 L > \frac{4\pi}{3.45} \tag{14.2}$$

where γ is the nonlinear parameter of the fiber. Considering a 500 m long silica-based HNLF, such that $\gamma = 20 \, W^{-1} \, km^{-1}$, the required peak power is $P_0 \approx 0.4 \, W$.

2R regenerators based on SPM with subsequent spectral filtering have been demonstrated using different types of highly nonlinear fibers [45–56]. Since logical "ones" (e.g., pulses) and logical "zeros" (e.g., ASE noise) are processed with distinct transfer functions, such regenerators are able to directly improve the BER of a noisy signal. A BER improvement from 5×10^{-9} to 4×10^{-12} for an optical signal-to-noise ratio of 23 dB, or a sensitivity improvement of 6 dB at a BER of 10^{-10}, was achieved using a 851 m long highly nonlinear silica fiber, together with a BPF offset 1.2 nm from the signal center wavelength and with a bandwidth of 0.56 nm [46]. A more compact version of such a 2R regenerator was realized in the same experiment using a 2.8 m long highly nonlinear As_2Se_3 chalcogenide fiber presenting an MFD of ~6.4 µm at 1550 nm, an average core/cladding refractive index of 2.7, and a numerical aperture of 0.18. The Kerr coefficient of the fiber is about 400 times greater than that of the silica. Figure 14.4a–d shows the experimental and simulation results for the output pulse spectra obtained with no filter present considering different coupled peak powers. As the peak power increases, the SPM broadens the signal spectrum. Figure 14.4e presents the bandpass filter transmission spectrum, offset by

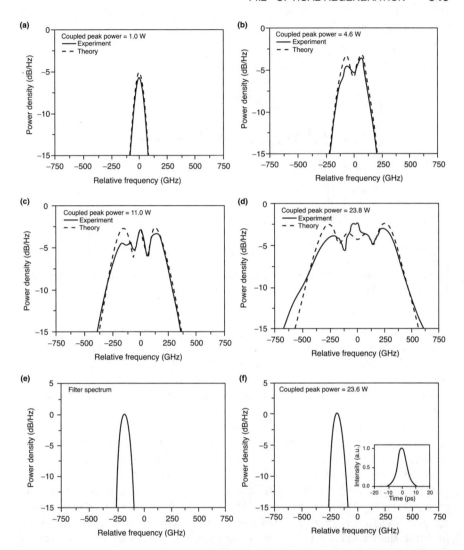

Figure 14.4 (a–d) Measured (solid curves) and theoretical (dashed curves) pulse spectra at the output of the chalcogenide fiber for different input peak powers. (e) The bandpass filter transmission spectrum, with a 3 dB bandwidth of 0.56 nm (70 GHz), offset by 1.3 nm from the input center frequency. (f) Output pulse spectrum with 23.8 W of peak power. Inset in (f) shows pulse autocorrelation. (After Ref. [46]; © 2006 IEEE.)

1.3 nm from the input center frequency, and with a 3 dB bandwidth of 70 GHz (0.56 nm), whereas Fig. 14.4f shows the output pulse spectrum after filtering for a peak pulse power of 23.8 W.

Figure 14.5 illustrates the experimental and theoretical power transfer functions of the same regenerator. Taking into account the two-photon absorption (TPA)

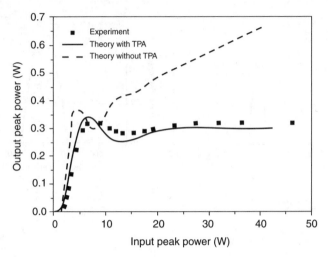

Figure 14.5 Experimental and theoretical regenerator transfer functions for a filter offset of 1.3 nm. (After Ref. [46]; © 2006 IEEE.)

effect seems to be essential to properly describe the experimental results. The TPA affects the shape and contributes to the flattening of the upper level of the step-like transfer function. The curves exhibit a threshold at 1 W and an output power limiting function at ∼8 W peak input power. The first feature is important to suppress the noise in the "zeros", whereas the second feature helps in suppressing the noise in the "ones."

The As_2Se_3 fiber used in the regenerator of Fig. 14.5 has a linear normal dispersion of 504 ps/(nm km) at 1550 nm. This large value of the dispersion proves to be important in smoothing out the nonlinear transfer curves at high peak power levels [46]. The total dispersion of 1–2 ps/nm present in the device is similar to that of typical silica HNLF with one to several kilometers used in some regeneration experiments [51]. However, in the case of the As_2Se_3 fiber, such a level of dispersion corresponds to a very short length of a few meters. Besides the advantages resulting from the short length, a reduction of the operating powers by over an order of magnitude is expected in the future by reducing the mode field diameter as well as possibly increasing the intrinsic nonlinearity of the chalcogenide glass.

2R regeneration of an amplitude shifted keying (ASK) data signal was also achieved recently using a 35 cm long highly nonlinear bismuth oxide fiber (Bi-NLF) [55]. In this experiment, a fiber laser was harmonically mode locked at 10 GHz at a center wavelength of 1550.66 nm. The output was externally modulated using an electrooptic modulator, which was biased in order to intentionally degrade the signal extinction ratio. The degraded signal was then amplified with an EDFA and afterward launched to the Bi-NLF. Figure 14.6a shows the measured eye diagram of the distorted 10 Gb/s RZ input signal, which presents an extinction ratio of only 7.4 dB. After undergoing SPM in the Bi-NLF, the signal degradation still remains, as observed from Fig. 14.6b. Using a BPF placed 2 nm away from the spectral peak

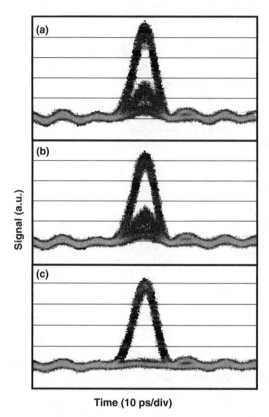

Time (10 ps/div)

Figure 14.6   Measured eye diagrams in an SPM-based optical regeneration experiment using a Bi-NLF: (a) distorted RZ input signal; (b) RZ signal after the Bi-NLF; (c) spectrally filtered RZ output signal. (After Ref. [55]; 2008 IEEE.)

of the input signal, a regenerated signal is achieved, as shown by the eye diagram in Fig. 14.6c. The extinction ratio of the regenerated signal was improved to 15.5 dB. The BER measurement results of the RZ data signal are depicted in Fig. 14.7 and show a sensitivity improvement of 5 dB at a BER of 10^{-9}.

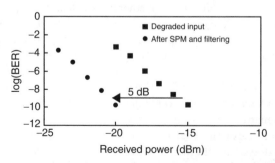

Figure 14.7   BER measurement results of the 10 Gb/s RZ data signals without (squares) and with (circles) optical regeneration. (After Ref. [55]; © 2008 IEEE.)

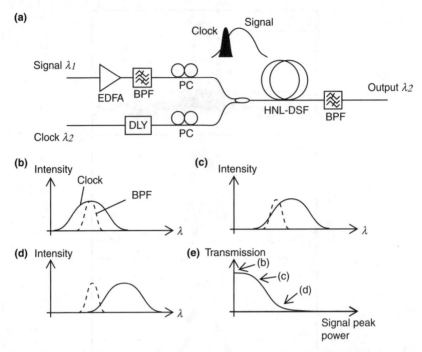

Figure 14.8 Schematic and principle of the 3R regenerator based on XPM-induced spectral shift followed by filtering. The peak power of the signal pulse is increased from (b) to (d). (After Ref. [58]; © 2005 IEEE.)

In spite of all the advantages offered by 2R regeneration, it is not able to retime the signal and therefore cannot correct for timing jitter. Full regeneration (3R) can be achieved using XPM [57–59] and offers many of the advantages of the 2R scheme, while including the retiming function in an all-optical manner.

The principle of operation of a 3R regenerator, first proposed by Suzuki et al. [58], is illustrated in Fig. 14.8. An impaired signal with a center wavelength λ_1 is strongly pumped by an EDFA. Such a signal is launched into a highly nonlinear dispersion-shifted fiber (HNL-DSF) together with a clean clock pulse train at λ_2. A BPF with a center wavelength λ_2 is placed behind the HNL-DSF. As the signal bits and clock pulses temporally overlap in the HNLF, the center wavelength of the clock pulse shifts from λ_2 toward the longer side through XPM. The amount of the wavelength shift is proportional to the slope of the signal pulse, which depends on the peak power, as represented in Fig. 14.8b–d. The output power of the clock pulse from the BPF is then a nonlinear function of the signal peak power, as shown in Fig. 14.8e. This figure shows that two levels appear in the output: fluctuated "zeros" in the original signal are converted to "one" bits, while the fluctuated original "one" bits are absorbed. As a result, the noise imposed on the signal is redistributed and regeneration is realized. Although this regeneration architecture inverts the bits, a noninverting mode of operation can be realized by offsetting the center wavelength

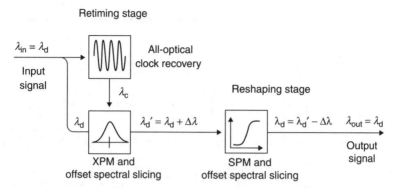

Figure 14.9 Block diagram of polarization-independent all-optical 3R regenerator. (After Ref. [65]; © 2008 IEEE.)

of the BPF behind the HNLF away from λ_2 to transmit only XPM-generated frequencies.

A common challenge of nonlinear effects such as four-wave mixing and cross-phase modulation is their inherent sensitivity to the state of polarization (SOP) of the input signal. Since the SOP of a data signal varies randomly with time in practical systems, it is important that the all-optical retiming function be polarization independent. Several techniques have been proposed to address this problem, namely, depolarization of the clock signal [60], polarization diversity [61], fiber twisting for circularly polarizing fiber [62], and the use of fiber birefringence [63]. However, these techniques often require additional complexity and/or careful control and stability of the clock signal SOP.

Polarization-independent all-optical retiming without additional complexity has been demonstrated recently using XPM in an HNLF and offset spectral filtering [64,65]. This technique is based on the fact that there exist wavelength regions for which the XPM-broadened spectrum is polarization independent. A block diagram of an all-optical 3R regenerator using such an approach and demonstrated in a 2008 experiment [65] is shown in Fig. 14.9. In the retiming stage, an optical clock signal is recovered and used to induce XPM on the input data signal in an HNLF, followed by offset spectral filtering. A tunable optical delay controls the temporal alignment of the clock and data signals at the input of the HNLF. The retimed signal is obtained by filtering the broadened spectrum of the data signal with a BPF offset from the carrier wavelength of the data signal by $\Delta\lambda$. The subsequent reshaping stage uses SPM in an HNLF and spectral filtering with an offset of $-\Delta\lambda$, thus preserving the signal wavelength. With proper design, high-performance polarization-independent all-optical 3R regeneration can be achieved with such setup [65].

14.3 OPTICAL PULSE TRAIN GENERATION

Besides the regeneration function described in the previous section, the SPM-based pulse reshaping in HNLFs can also be used with advantage in the generation of optical

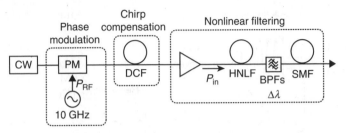

Figure 14.10   Configuration of an optical pulse train generator. (After Ref. [48]; © 2008 IEEE.)

pulse trains. This application is particularly important for 160 Gb/s OTDM trans-mitters, where the generation of 10 GHz picosecond optical pulse trains is one main issue. High-quality pulse trains are required for such OTDM transmitters, namely, a reduced pulse width, a low timing jitter, a long-term stability, and wavelength and repetition rate tunability.

The configuration of a pulse train generator using SPM-based pulse reshaping in an HNLF is shown in Fig. 14.10 [48]. The repetition rate f_{rep} of the pulse train is 10 GHz. First, a CW light is sinusoidally phase modulated by an LiNbO$_3$ (LN) phase modulator driven by a 10 GHz clock with a power of P_{RF}. The chirp induced by the phase modulation is compensated by a dispersion-compensating fiber (DCF) as a dispersive medium. The phase-modulated wave is then converted to a picosecond pulse train at the repetition rate of f_{rep}, which is generally accompanied with a large pedestal [48]. Such pedestal is eliminated by the pulse reshaping based on SPM in an optical fiber [13]. Since the spectrum of the pedestal remains in the vicinity of the center wavelength of the input spectrum, it can be suppressed using a BPF offset from this frequency. Finally, a standard single-mode fiber (SMF) is used to compensate for the residual chirp of the reshaped pulse. Highly stable generation of a 10 GHz 2 ps optical pulse train tunable over the entire C band was achieved using a pulse generator with the configuration of Fig. 14.10, incorporating a 500 m long highly nonlinear silica fiber [48]. The applicability of such a pulse generator to the 160 Gb/s transmitter was also demonstrated in the same experiment.

14.4   WAVELENGTH CONVERSION

In future optical communication networks, each local network would have an appropriate carrier wavelength depending on the demand for the network. Therefore, wavelength conversion over a wide range of wavelengths will be one of the key functions necessary to maximize information transfer efficiency between the different networks.

All-optical wavelength conversion can be realized using mainly two kinds of devices: semiconductor optical amplifier (SOA)-based converters [66,67] and fiber-based converters. Wavelength converters based on SOAs exhibit many attractive features but are generally limited to signal rates of $\sim$40 Gb/s owing to carrier recombination lifetimes or intraband dynamics [68]. By contrast, converters based on nonlinear fiber effects can support almost unlimited bit rates and are transparent to signal modulation format due to the femtosecond response time of the fiber nonlinearity [69]. Previous ultrahigh-speed wavelength conversion of return-to-zero data has been demonstrated using SPM in fiber [13], XPM in fiber [16], FWM in fiber [70], and XPM in nonlinear optical loop mirror (NOLM) [71]. In this section, we will discuss the use of XPM and FWM effects in fibers to realize the wavelength conversion.

14.4.1 Wavelength Conversion with FWM

Considering the case of degenerate FWM and assuming that the intensity of the pump wave is much higher than that of the signal and idler waves, Eqs. (6.8)–(6.11) with $U_1 = U_2 = U_p$, $U_3 = U_s$, and $U_4 = U_i$, give

$$\frac{\partial U_p}{\partial z} = i\gamma |U_p|^2 U_p \tag{14.3}$$

$$\frac{\partial U_s}{\partial z} = i2\gamma \left[|U_p|^2 U_s + U_p^2 U_i^* \, e^{-i\Delta kz} \right] \tag{14.4}$$

$$\frac{\partial U_i}{\partial z} = i2\gamma \left[|U_p|^2 U_i + U_p^2 U_s^* \, e^{-i\Delta kz} \right] \tag{14.5}$$

where γ is the nonlinear parameter and $\Delta k = \beta_s + \beta_i - 2\beta_p$ is the phase mismatch of the wave vectors. As seen in Chapter 6, the effective phase mismatch with induced nonlinearity is given by

$$\kappa = \Delta k + 2\gamma P_p \tag{14.6}$$

where P_p is the pump power. Using the results obtained in Section 6.5, the efficiency of the FWM converter can be written as

$$G_i = \frac{P_i(L)}{P_s(0)} = \left(\frac{\gamma P_p}{g} \right)^2 \sin \mathrm{h}^2(gL) \tag{14.7}$$

where L is the fiber interaction length and the parametric gain coefficient g is given by

$$g^2 = \left[(\gamma P_p)^2 - (\kappa/2)^2 \right] = -\Delta k \left(\frac{\Delta k}{4} + \gamma P_p \right) \tag{14.8}$$

The parameter g represents real gain over a conversion bandwidth corresponding to $-4\gamma P_p < \Delta k < 0$.

The derivation of Eq. (14.7) ignores the pump depletion, fiber loss, competing nonlinear processes, and walk-off between the pump and the signal pulses. Since these effects would lower the parametric gain by lowering the pump power or reduce the interaction length, the conversion efficiency given by Eq. (14.7) corresponds effectively to an optimum value.

From Eq. (14.8) it can be seen that the maximum gain for the parametric process is $g_{max} = \gamma P_p$ and occurs if $\kappa = 0$. Hence, fibers with a high nonlinear parameter γ will be advantageous for efficient wavelength conversion. On the other hand, in order to fulfill the phase matching condition, fibers must be operated in their dispersion minimum. Fibers with a high value of γ and exhibiting the ZDW around 1.5 μm are called HNL-DSFs. In a 1998 experiment, a peak conversion efficiency of 28 dB was achieved over a bandwidth of 40 nm, using a 720 m long DSF and a pulsed pump with a peak power of 600 mW [70]. Wavelength band conversion was also demonstrated in a 2001 experiment [72], through which the existing WDM sources in C-band were wavelength converted to the S-band. More than 30 nm conversion bandwidth with greater than 4.7 dB conversion efficiency was measured in a 315 m long HNL-DSF by using 860 mW pump power at ~1532 nm [72].

When the GVD $|\beta_2|$ is small, we must take into account the fourth-order term in the expansion of Δk, which becomes

$$\Delta k = \beta_2 \Omega_s^2 + \frac{\beta_4}{12} \Omega_s^4 \qquad (14.9)$$

where $\Omega_s = \omega_p - \omega_s = \omega_i - \omega_p$. Equation (14.9) can be modified and written in terms of the involved wavelengths as

$$\Delta k = \left[2\pi c \left(\frac{1}{\lambda_p} - \frac{1}{\lambda_s} \right) \right]^2 \left[\beta_2 + \frac{1}{3} \beta_4 \pi^2 c^2 \left(\frac{1}{\lambda_p} - \frac{1}{\lambda_s} \right)^2 \right] \qquad (14.10)$$

For low pump powers, we can neglect the nonlinear term in Eq. (14.6) and the phase matching condition is given simply by $\Delta k = 0$. In order to realize this condition, Eq. (14.10) shows that the relation

$$\frac{1}{\lambda_p} - \frac{1}{\lambda_s} = \pm \frac{1}{\pi c} \sqrt{\frac{-3\beta_2}{\beta_4}} \qquad (14.11)$$

should be satisfied. As seen from Eq. (14.10), when $\beta_2 = 0$ the conversion bandwidth is proportional to $|\beta_4|^{-4}$. Hence, depending on the value of β_4, FWM-based wavelength conversion can be classified into three types [73]. The first one corresponds to conventional conversion and uses an HNLF with a moderate value $|\beta_4| \approx 1 \times 10^{-55}$ s^4/m. The second one corresponds to large values of β_4 and provides narrowband conversion. The third one is realized using HNLFs with a small absolute value of β_4, which provides a broadband conversion. An HNLF with a nonlinear parameter $\gamma = 25$ W^{-1} km^{-1} and a reduced value $\beta_4 = 0.2 \times 10^{-55}$ s^4/m has been fabricated recently [73]. Figure 14.11 shows the measured and calculated results for the FWM wavelength conversion efficiencies obtained using such an HNLF with a length of 100 m. Also shown are the results for a conventional HNLF

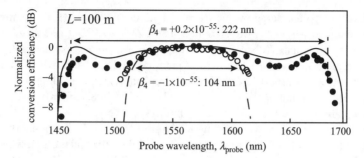

Figure 14.11 Normalized conversion efficiency against the signal wavelength. Solid circles and solid curve: measured and calculated results for an HNLF with $\beta_4 = +0.2 \times 10^{-55}$ s^4/m, respectively. Open circles and dotted curve: measured and calculated results for an HNLF with $\beta_4 = -1 \times 10^{-55}$ s^4/m, respectively. (After Ref. [73]; © 2009 IEEE.)

with $\beta_4 = -1 \times 10^{-55}$ s^4/m. The pump wavelength was set to 1562.5 nm, slightly longer than the ZDW of 1562.1 nm, in order to have a small negative value of -1×10^{-29} s^2/m for β_2. The wavelength conversion in the reduced-β_4 HNLF can be realized within an extremely broad 3 dB bandwidth of 222 nm from 1460 to 1682 nm. This result together with other measurements obtained with different fiber lengths has confirmed that the bandwidths obtained for the reduced-β_4 HNLF are almost a factor of 2 larger compared to those reported for conventional HNLFs [74–76].

In addition to the broadband wavelength conversion, there is also a growing interest in developing wavelength selective devices from multiple signals, such as reconfigurable optical add–drop multiplexers. In order to realize a selective wavelength conversion, Δk should be small in a narrow bandwidth around λ_s, as determined by Eq. (14.11). Therefore, besides β_2/β_4 being negative, we find from Eq. (14.10) that both $|\beta_2|$ and $|\beta_4|$ should be large. The desired value of β_2/β_4 and a tunable wavelength conversion can be realized by tuning the pump wavelength near the ZDW. Figure 14.12 shows the measured conversion efficiencies against λ_s in a 100 m

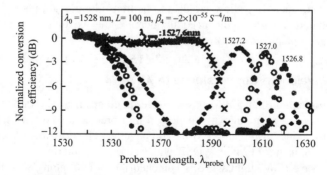

Figure 14.12 Measured normalized conversion efficiency against the signal wavelength for different pump wavelengths, using a 100 m long HNLF with $\beta_4 = -2 \times 10^{-55}$ s^4/m. (After Ref. [73]; © 2009 IEEE.)

long HNLF with an enlarged value $\beta_4 = -2 \times 10^{-55}$ s^4/m [73]. The pump wavelength is detuned at a shorter wavelength from the ZDW of 1528.0 nm. As λ_p is shifted to a shorter wavelength, β_2 becomes larger and a selective conversion is realized. The results depicted in Fig. 14.12 show that pumping at 1527.6 nm, which is near the ZDW, provides a broad wavelength conversion bandwidth, extending from 1530 to 1580 nm. However, tuning λ_p to 1527.2, 1527.0, and 1526.8, narrowband conversion is realized around 1600, 1610, and 1620 nm, respectively.

The performance of FWM wavelength converters can be affected if the spectrum of the converted signal (idler) is broadened by the dithering of the pump, which is required to avoid stimulated Brillouin scattering (SBS). In a one-pump FOPA, this broadening can reach several GHz, which is too large for many practical applications. The use of two pumps phase modulated 180° out of phase can cancel out the broadening of the converted signal [77]. A similar effect, together with polarization-independent wavelength conversion, was achieved using binary phase shift keying modulation of the two orthogonally polarized pump waves [78]. It must be noticed, however, that the use of pump phase modulation to suppress SBS can introduce signal degradation and optical SNR penalties in the case of phase-modulated signals [79].

An effective method to increase the SBS threshold consists in using short lengths of HNLFs to realize wavelength conversion. In a 2004 experiment, a 64 m long silica photonic crystal fiber has been employed for wavelength conversion of communication signals by FWM [61], offering a reduction in fiber length by one order of magnitude. By 2005, only 1 m length of a bismuth oxide optical fiber (Bi-NLF) with an ultrahigh nonlinearity of ~ 1100 W^{-1} km^{-1} was used to make an FWM-based wavelength converter capable of operating at 80 Gb/s [80]. One main advantage of the Bi-NLF is the possibility of fusion splicing to conventional silica fibers. More recently, a 35 cm long Bi-NLF was also used to demonstrate FWM-based wavelength conversion of 40 Gb/s polarization-multiplexed ASK-DPSK signals [55]. Using a dual-pump configuration, a conversion range greater than 30 nm was achieved. This experiment demonstrated the true transparency of FWM wavelength conversion to modulation format. Signals in multilevel modulation format are particularly attractive in this context, since they allow the data to be transmitted at a higher bit rate than binary modulation without the need to increase the existing bandwidth of the electronic and optoelectronic components.

14.4.2 Wavelength Conversion with XPM

One main limitation of an FWM-based converter comes from the fact that it is not possible to convert an unknown wavelength to a predetermined wavelength. This limitation is avoided using an XPM-based converter. The principle of operation of such converter using XPM in a DSF with subsequent filtering is as follows. A continuous-wave (CW) light wave is launched into a DSF along with the control signal. Here, the XPM will act to broaden the spectrum of the CW light wave, where a mark is copropagated with it. In this way, red- and blueshifted sidebands are generated on the CW light wave. By filtering out one of these sidebands, the wavelength-

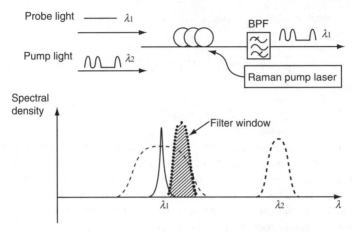

Figure 14.13 Schematic of an XPM-based Raman gain-enhanced wavelength converter. (After Ref. [83]; © 2005 IEEE.)

converted signals can be obtained [16,48,81]. The pulse width of the converted pulse based on XPM in the DSF is determined by the power of the control pulse, dispersion effect, and walk-off time between the control pulse and the CW light wave [82]. Broadband pulse width maintained wavelength conversion could be realized by use of a short HNL-DSF because dispersion and walk-off effects can be reduced [81].

The use of Raman gain can significantly enhance the performance of an XPM-based fiber wavelength converter [83,84]. Figure 14.13 shows a schematic of such Raman gain-enhanced wavelength converter (RE-WC). It is basically constituted by a nonlinear fiber, a Raman pump laser, and a bandpass filter. The data pump light at wavelength λ_2 is combined with the CW probe light at wavelength λ_1 and injected into the fiber. The Raman pump amplifies simultaneously the pump light and the probe light and significantly enhances the amount of XPM. At the output of the fiber, a BPF is used to extract one of the sidebands on the CW probe, generating an amplitude-modulated signal from the phase modulation of the CW probe. This amplitude-modulated signal will thus form a wavelength-converted replica of the original data signal.

In the absence of Raman gain and considering the pulsed pump light and the CW probe light propagating in the fiber with the same polarization, the nonlinear equations that govern their behaviors can be written from Eqs. (5.41) and (5.42), after neglecting the GVD and FWM terms, as follows:

$$\frac{\partial U_s}{\partial z} + \frac{1}{v_{gs}}\frac{\partial U_s}{\partial t} = -\frac{\alpha}{2}U_s + i\gamma(|U_s|^2 + 2|U_p|^2)U_s \qquad (14.12)$$

$$\frac{\partial U_p}{\partial z} + \frac{1}{v_{gp}}\frac{\partial U_p}{\partial t} = -\frac{\alpha}{2}U_p + i\gamma(|U_p|^2 + 2|U_s|^2)U_p \qquad (14.13)$$

where $U_j, j = s, p$, is the slowly varying complex amplitude, v_{gj} is the group velocity, α is the attenuation coefficient, and γ is the fiber nonlinear parameter. If the pulse dispersion is neglected, the solution of Eq. (14.12) at the output of the fiber with length L is given by

$$U_s(L,t) = U_s(0, t - L/v_{gs}) \exp(-\alpha L/2) \exp\{i\Delta\phi(L,t)\} \qquad (14.14)$$

where

$$\Delta\phi = \gamma L_{eff}(P_s + 2P_p) \qquad (14.15)$$

is the nonlinear phase shift, L_{eff} is the effective length of the interaction in the fiber, and $P_j = |U_j|^2$ is the power of the signal ($j = s$) or pump ($j = p$) waves. The first term in Eq. (14.15) is due to SPM, whereas the second term is due to the XPM effect. This second term determines the conversion efficiency of the XPM-based wavelength conversion. When Raman gain is present, the accumulated phase shift for a given signal power level is increased. This effect is due to the increase of effective interaction length, L_{eff}, which becomes

$$L_{eff} = \frac{1 - \exp\{-(\alpha - g)L\}}{\alpha - g} \qquad (14.16)$$

where g is the gain provided by the Raman amplification.

The Raman gain enhancement of a regenerative ultrafast all-optical XPM wavelength converter was experimentally demonstrated in Ref. 83 at 40 and 80 Gb/s. The XPM conversion of a 40 Gb/s OTDM signal was performed with $\sim$1 dB penalty in the receiver sensitivity, while the conversion of a strongly degraded 80 Gb/s signal produced $\sim$2 dB sensitivity improvement, demonstrating signal regeneration. The Raman gain greatly enhanced the wavelength conversion efficiency at 80 Gb/s by 21 dB at a Raman pump power of 600 mW using 1 km of highly nonlinear fiber. Figure 14.14 shows the simulated and experimental results for the XPM enhancement factor, E_{XPM}, defined as the ratio of $\Delta\phi_{XPM} = 2\gamma L_{eff}P_p$ with and without Raman gain, both for a 5 km long standard DSF and for a 1 km long HNLF.

From Eqs. (14.15) and (14.16) it can be seen that, in order to achieve a given phase shift, the signal launch power can be reduced in the presence of Raman gain. On the other hand, for a given CW launch power, more power will be available in the wavelength-converted output signal if it is amplified while passing through the fiber. As a consequence, less amplified spontaneous emission noise will be generated when the converted signal is amplified at the output of the converter. This is seen in Fig. 14.15, where the noise level after filtering and amplification is reduced by 3.5 dB while the signal power is increased by $\sim$2 dB [84]. In this 2007 experiment, Raman-assisted wavelength conversion by XPM was demonstrated using a 500 m long HNLF at the record high bit rate of 640 Gb/s.

Wavelength conversion using XPM in chalcogenide fiber [85] has been recently realized. Figure 14.16 shows the experimental setup used to demonstrate λ-conversion in a 1 m long As$_2$Se$_3$ fiber, which had a nonlinear parameter $\gamma \sim 1200$ W^{-1} km^{-1} and a dispersion $D = -560$ ps/(nm km) at 1550 nm. The optical data was amplified and a 1.3 nm tunable bandpass filter (TBF) was used to remove out-of-band ASE. The

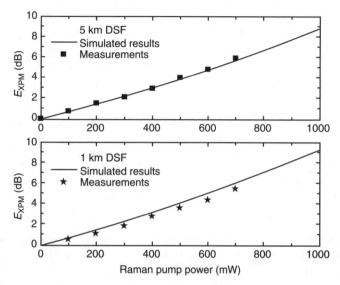

Figure 14.14 Calculated and measured results for XPM enhancement for both DSF and HNLF. (After Ref. [83]; © 2005 IEEE.)

pulses were combined with a CW probe from a wavelength tunable amplified laser diode and coupled into the As_2Se_3 fiber. In-line polarization controllers ensured that the polarization states of the pump and probe were aligned. The output of the As_2Se_3 fiber was then sent through a sharp 0.56 nm tunable grating filter offset to longer wavelengths by 0.55–0.70 nm to remove the pump and select a single XPM sideband. A second 1.3 nm TBF was used to remove out-of-band ASE of the amplified signal. An in-line, 200 pm wide, fiber Bragg grating (FBG) notch filter was used to further

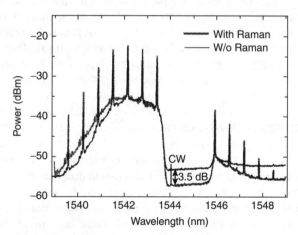

Figure 14.15 Optical spectra for the wavelength-converted signal showing the effect of applying Raman gain to the conversion process. (After Ref. [84]; © IEEE.)

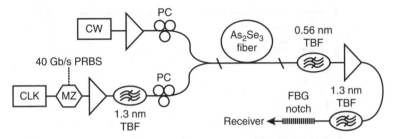

Figure 14.16 System setup demonstrating XPM-based wavelength conversion. CLK: 10 GHz actively mode-locked fiber laser; FBG notch: fiber Bragg grating notch filter; MZ: Mach–Zehnder modulator; PC: polarization controller; PRBS: pseudorandom bit sequence; TBF: tunable bandpass filter. (After Ref. [85]; © 2006 OSA.)

suppress the residual CW. Using such setup, error-free conversion was demonstrated at 10 Gb/s RZ near 1550 nm over 10 nm wavelength range with 7.1 ps pulses at 2.1 W peak power, achieving only 1.4 dB excess penalty.

XPM-based wavelength conversion was also demonstrated in a 1 m long bismuth oxide fiber over 15 nm wavelength range at 160 Gb/s [86]. In fact, the As_2Se_3 fiber presents the highest Kerr nonlinearity to date: $n_2 \sim 1.1 \times 10^{-17} m^2/W$, which is $\sim$400 times that of silica and an order of magnitude greater than Bi_2O_3. However, the much larger effective core area of the As_2Se_3 fiber used in Ref. 85 compared to that of the Bi_2O_3 fiber in Ref. 86 resulted in a similar nonlinear parameter γ in both cases. An effective core area of the As_2Se_3 fiber similar to that of the Bi_2O_3 fiber would result in a nonlinear parameter $\gamma = 11,100 \, W^{-1} \, km^{-1}$. Such reduction of the As_2Se_3 fiber core area will significantly reduce the device length, fiber losses, and dispersion-related impairments.

Pump-probe walk-off is an issue in XPM-based wavelength converters, since it limits the conversion bandwidth. A typical conversion bandwidth of 10 nm can be achieved for actual 1 m long Bi_2O_3 devices. In spite of the fact that the dispersion parameter of As_2Se_3 is higher than that of Bi_2O_3, it is expected that using shorter As_2Se_3 fiber lengths it will be possible to achieve a conversion bandwidth >40 nm. In contrast with silica DSFs, both As_2Se_3 and Bi_2O_3 fibers present a relatively constant dispersion over the whole communications band [87].

14.5 ALL-OPTICAL SWITCHING

An optical switch with fully transparent features in both time and wavelength domains is a key device for providing several functions required in optical signal processing. A practical optical switch should have a broad bandwidth over the entire transmission band and should be capable of ultrahigh-speed operation at a data rate of 100 Gb/s or higher. The third-order nonlinearity in optical fibers can provide such features, especially if highly nonlinear fibers are considered. In this section, both XPM- and FWM-based optical fiber switches are discussed.

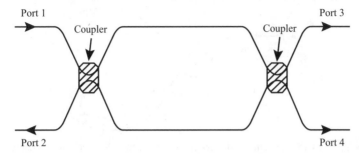

Figure 14.17 Schematic illustration of a Mach–Zehnder interferometer.

14.5.1 XPM-Induced Optical Switching

Interferometers offer the possibility to convert the XPM-induced phase shift into an amplitude change. Considering the particular case of the Mach–Zehnder interferometer (MZI), shown schematically in Fig. 14.17, the initial signal pulse is divided into two parts and both propagate in different branches of the interferometer. If both branches have the same properties and in the absence of any control pulse, the two partial pulses experience the same phase shift and interfere constructively at the end of the interferometer. However, if a control pulse at a different wavelength is injected into one branch of the interferometer, it changes the signal phase through XPM in that arm.

The output intensity of the interferometer is related to the input intensity as [88]

$$I_{\text{out}} = I_{\text{in}} \cos\left(\frac{\Delta\phi_{\text{L}} + \Delta\phi_{\text{NL}}}{2}\right) \qquad (14.17)$$

where $\Delta\phi_{\text{L}}$ is the initial phase difference between the two partial pulses and $\Delta\phi_{\text{NL}}$ is the nonlinear phase difference induced by the control pulse via XPM. Assuming that $\Delta\phi_{\text{L}} = 0$, a destructive interference occurs and the signal pulse will not be transmitted if the XPM-induced phase shift is π. From Eq. (14.15), this condition is given by

$$\Delta\phi_{\text{NL}} = 2\gamma P L_{\text{eff}} = \pi \qquad (14.18)$$

Considering a typical single-mode fiber with a nonlinear parameter $\gamma = 1.3\,\text{W}^{-1}$ km^{-1} and an effective length $L_{\text{eff}} = 1$ km, Eq. (14.18) gives an input power of 1.2 W for the control pulse.

Reducing the effective core area A_{eff} increases the nonlinear parameter γ and, consequently, reduces the power required for XPM-induced switching. For example, using fibers with $A_{\text{eff}} = 2\,\mu\text{m}^2$ in each arm of an MZI, an XPM-induced phase shift of $10°$ was achieved with a pump power of only 15 mW [89].

A Sagnac interferometer can also be used as an optical switch. Figure 14.18 shows a schematic representation of such an interferometer. The input beam is divided in the fiber coupler into two beams counterpropagating in the fiber loop. After one

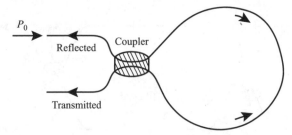

Figure 14.18   Schematic illustration of an all-fiber Sagnac interferometer.

round-trip, both beams arrive at the same time at the coupler, where they interfere coherently, according to their relative phase difference. If a 3 dB fiber coupler is used, the input beam is totally reflected and the Sagnac interferometer serves as an ideal mirror. For this reason, it is also known as a nonlinear optical loop mirror. This device was already discussed in Section 9.1.

Let us consider a control pulse that is injected into the Sagnac loop such that it propagates in only one direction. Due to the XPM effect in the fiber, the signal pulse that propagates in that direction experiences a phase shift while the counterpropagating pulse remains unaffected. The XPM-induced phase shift is again given by $\Delta\phi_{NL} = 2\gamma PL_{eff}$. As a result, the signal pulse is reflected in the absence of the control pulse but is transmitted when a control pulse is injected such that it produces a phase shift of π. The potential of XPM-induced switching in all-fiber Sagnac interferometers was demonstrated in several experiments [90–93].

Sagnac interferometers can be used for demultiplexing OTDM signals. The control signal consists of a train of optical pulses that is injected into the loop such that it propagates in a given direction. If the control (clock) signal is timed such that it overlaps with pulses belonging to a specific OTDM channel, the XPM-induced phase shift allows the transmission of this channel, whereas all the remaining channels are reflected. Different channels can be demultiplexed simultaneously by using several Sagnac loops [94]. An 11 km long Sagnac loop was used in 1993 to demultiplex a 40 Gb/s signal to individual 10 Gb/s channels [95].

The main limitation of a Sagnac interferometer used to provide XPM-induced switching and demultiplexing stems from the weak fiber nonlinearity. According to Eq. (14.18), fibers of several kilometers are necessary for a control pulse power in the mW range. In these circumstances, if the signal and control pulses have different wavelengths, the walk-off effects due to the group velocity mismatch must be taken into consideration. The control pulse width and the walk-off between the signal and control pulses determine the switching speed and the maximum bit rate for demultiplexing with a NOLM [96].

A walk-off-free NOLM was proposed that consisted of nine sections of short dispersion-flattened fibers with different dispersion values [97]. However, such device has a very complex configuration. The suppression of the walk-off effect can be achieved by using a fiber whose zero dispersion wavelength lies between the signal and the control wavelengths, such that both experience the same group velocity. A

Sagnac loop interferometer operating under these conditions was realized in 1990 using a 200 m long polarization maintaining fiber [98]. In a 2002 experiment, a 100 m long HNL-DSF with a nonlinear parameter $\gamma = 15\,W^{-1}\,km^{-1}$ and a ZDW at 1553.5 nm was used for error-free demultiplexing of 320 Gb/s TDM signals down to 10 Gb/s [99].

Another possibility to suppress the pulse walk-off is to use an orthogonally polarized pump at the same wavelength as that of the signal. In this case, the group velocity mismatch due to the polarization mode dispersion is relatively low. Moreover, it can be used to advantage by constructing a Sagnac loop consisting of multiple sections of polarization maintaining fibers that are spliced together in such a way that the slow and fast axes are interchanged periodically [100]. As a result, the pump and signal pulses are forced to collide multiple times inside the loop, and the XPM-induced phase shift is enhanced considerably.

Both the interaction length and the control power can be decreased using microstructured fibers. In a 2002 experiment, XPM-induced switching of 2.6 ps (FWHM) signal pulses in a Sagnac interferometer built with a 5.8 m long microstructured fiber was achieved [101]. Optical time division demultiplexing of 160 Gb/s OTDM signal based on XPM and subsequent optical filtering has also been demonstrated more recently using a 2 m long Bi_2O_3-based highly nonlinear microstructured fiber with $\gamma > 600\,W^{-1}\,km^{-1}$ and GVD $\sim50\,ps^2/km$ [48]. Using such a fiber, the long-term stability was drastically improved compared to silica fiber devices because the very short fiber length reduces the phase drift between the signal and control pulses.

14.5.2 Optical Switching Using FWM

Four-wave mixing can also be used to switch a given sequence of signal pulses propagating along an optical fiber. To realize such an operation, a control signal consisting of a train of optical pulses with a frequency (f_c) different from that of the signal pulses (f_s) is also injected into the fiber. When a signal and a control pulse are present together in the fiber, a switched signal appears with frequency f_i, such that

$$f_i = 2f_c - f_s \tag{14.19}$$

Using a filter at the fiber output, the switched signal can be isolated from the control and original signal sequence. Optical bandwidths of the order of 200 nm can be achieved with a FWM fiber switch, in spite of the limitations imposed by the phase matching condition [102]. Such device can also be used as a demultiplexer [103].

One main problem associated with the conventional FWM-based switching is related to the different wavelengths of the original and switched signal sequences. A technique to solve this problem was recently proposed and is illustrated in Fig. 14.19 [21]. The optical signal wave E_s and the control (pump) wave E_p are input into a nonlinear fiber. The fiber output is connected to a polarizer that transmits

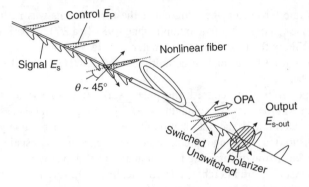

Figure 14.19 Schematic of an OPA fiber switch. (After Ref. [21]; © 2008 IEEE.)

only linearly polarized light along its main axis. Using a polarization controller at the fiber input, the state of polarization of E_s is adjusted such that no light passes the polarizer in the absence of pump light. The SOP of E_p is adjusted to linear polarization and aligned at $\sim 45°$ with respect to the polarizer's main axis.

The configuration of the fiber switch in Fig. 14.19 is similar to that of the conventional optical Kerr shutter, which is based on the XPM effect [69]. However, in contrast to the Kerr shutter, the switch in Fig. 14.19 has the advantage of using optical parametric amplification (OPA), provided by FWM. To achieve the necessary phase matching, the pump wave is tuned to the zero dispersion wavelength, λ_{ZD}, of the fiber. The efficiency of FWM process is maximum when the two waves have the same SOP, whereas it is negligible when the SOPs of the two waves are orthogonal to each other. Therefore, decomposing E_s into two linearly polarized components, parallel

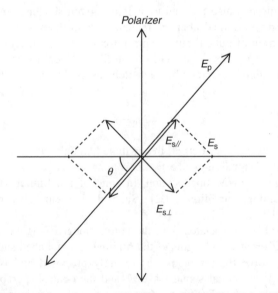

Figure 14.20 Polarization setup of an OPA fiber switch. (After Ref. [21]; © 2008 IEEE.)

$(E_{s\parallel})$ and orthogonal $(E_{s\perp})$ with respect to E_p (see Fig. 14.20), only $E_{s\parallel}$ will be amplified by FWM. The output power at the output of the OPA fiber switch is

$$P_{s\text{-out}} = \frac{1}{2}\Delta P_{s\parallel} \qquad (14.20)$$

where the factor 1/2 is due to the polarizer and $\Delta P_{s\parallel}$ is the power increase of $E_{s\parallel}$ due to parametric amplification. The switching gain is approximately given by [21]

$$G_s \equiv \frac{P_{s\text{-out}}}{P_{s\text{-in}}} = \frac{1}{4}\exp(-\alpha L)\phi_{\text{NL}}^2 \qquad (14.21)$$

where $P_{s\text{-in}}$ is the signal input power, α is the fiber loss coefficient, and

$$\phi_{\text{NL}} = \gamma P_p L_{\text{eff}} \qquad (14.22)$$

is the nonlinear phase shift.

Besides the fact that there is no frequency shift of the signal, the OPA fiber switch of Fig. 14.19 also provided an extinction ratio of more than 30 dB because the unswitched component of E_S was eliminated by the polarizer. Such an extinction ratio could be improved by increasing the switching gain. Considering the ultrafast response of FWM in the fiber, the OPA fiber switch has a potential for application to signal processing at data rates of 1 Tb/s and higher.

PROBLEMS

14.1 Explain the principle of operation of an SPM-based Mamyshev 2R optical regenerator. Assuming that Gaussian pulses are launched into the fiber, derive the condition given by Eq. (14.2).

14.2 Consider an FWM-based wavelength converter composed of a 1 km long single-mode fiber, with a nonlinear parameter $\gamma = 1.2\,\text{W}^{-1}\,\text{km}^{-1}$. Assuming the operation at maximum gain, determine the power of the converted signal for an input signal power $P_s(0) = 5\,\text{mW}$ and a pump power $P_p = 500\,\text{mW}$.

14.3 Describe the principle of operation of an XPM-based wavelength converter and explain the advantages of using Raman gain in this device.

14.4 An XPM-based switch is constructed using a 300 m long HNLF in a Mach–Zehnder interferometer configuration. The fiber has a nonlinear parameter $\gamma = 10\,\text{W}^{-1}\,\text{km}^{-1}$ an effective core area $A_{\text{eff}} = 16\,\mu\text{m}^2$, and an attenuation constant $\alpha = 0.5\,\text{dB/km}$. Determine the required pump power to switch a signal in the 1500 nm window using such an interferometer.

14.5 Explain how XPM and subsequent spectral filtering can be used for demultiplexing a single channel from an OTDM bit stream. Point out the main advantages of using a short HNLF for such a purpose.

14.6 Describe the principle of operation of the OPA fiber switch represented in Fig. 14.19 and derive Eq. (14.21) for its gain.

REFERENCES

1. G. Raybon and P. J. Winzer, *European Conference on Optical Communications* (*ECOC 2007*), Berlin, Germany, 2007, Th4.3.2.

2. M. Saruwatari, *IEEE J. Sel. Top. Quantum Electron.* **6**, 1363 (2000).

3. T. Okuno, M. Onishi, T. Kashiwada, S. Ishikawa, and M. Nishimura, *IEEE J. Sel. Top. Quantum Electron.* **5**, 1385 (1999).

4. P. Ohlen, B. E. Olsson, and D. J. Blumenthal, *IEEE Photon. Technol. Lett.* **12**, 522 (2000).

5. G. Berretini, G. Meloni, A. Bogoni, and L. Poti, *IEEE Photon. Technol. Lett.* **18**, 2439 (2006).

6. J. C. Knight, T. A. Birks, P. S. J. Russel, and M. Atkin, *Opt. Lett.* **21**, 1547 (1996).

7. T. M. Monro, D. J. Richardson, N. G. R. Broderick, and P. J. Bennett, *J. Lightwave Technol. Lett.* **17**, 1093 (1999).

8. K. K. Chow, C. Shu, C. Lin, and A. Bjarklev, *IEEE Photon. Technol. Lett.* **15**, 624 (2005).

9. M. Asobe, T. Kanamori, and K. Kubodera, *IEEE Photon. Technol. Lett.* **4**, 362 (1992).

10. K. Kikuchi, K. Taira, and N. Sugimoto, *Electron. Lett.* **38**, 166 (2002).

11. J. H. Lee, K. Kikuchi, T. Nagashima, et al., *Opt. Express* **13**, 3144 (2005).

12. V. G. Ta'eed, N. J. Baker, L. Fu, et al., *Opt. Express* **15**, 9205 (2007).

13. P. V. Mamyshev, *European Conference on Optical Communications (ECOC98)*, Madrid, Spain, 1998, pp. 475–477.

14. M. Matsumoto, *IEEE Photon. Technol. Lett.* **14**, 319 (2002).

15. T. Yamamoto, E. Yoshida, and M. Nakazawa, *Electron. Lett.* **34**, 1013 (1998).

16. B. E. Olsson, P. Öhlén, L. Rau, and D. J. Blumenthal, *IEEE Photon. Technol. Lett.* **12**, 846 (2000).

17. J. Li, B. E. Olsson, M. Karlsson, and P. A. Andrekson, *IEEE Photon. Technol. Lett.* **15**, 770 (2003).

18. J. H. Lee, T. Tanemura, K. Kikuchi, T. Nagashima, T. Hasegawa, S. Ohara, and N. Sugimoto, *Opt. Lett.* **39**, 1267 (2005).

19. T. Morioka, H. Takara, S. Kawanishi, T. Kitoh, and M. Saruwatari, *Electron. Lett.* **32**, 833 (1996).

20. Y. Wang, C. Yu, T. Luo, L. Yan, Z. Pan, and A. E. Wilner, *J. Lightwave Technol.* **23**, 3331 (2005).

21. S. Watanabe, F. Futami, R. Okabe, R. Ludwig, C. S. Langhorst, B. Huettl, C. Schubert, and H.-G. Weber, *IEEE J. Sel. Top. Quantum Electron.* **14**, 674 (2008).

22. G.-W. Lu, K. Abedin, and T. Miyazaki, *J. Lightwave Technol.* **27**, 409 (2009).

23. M. H. Reeve, A. R. Hunwicks, W. Zhao, S. G. Methley, L. Bickers, and S. Hornung, *Electron. Lett.* **24**, 389 (1988).

24. K. H. Han, E. S. Son, H. Y. Choi, K. W. Lin, and Y. C. Chung, *IEEE Photon. Technol. Lett.* **16**, 2380 (2004).

25. S. S. Wagner and T. Chapuran, *Electron. Lett.* **26**, 696 (1990).

26. S. Kaneko, J. Kani, K. Iwatsuki, A. Ohki, M. Sugo, and S. Kamei, *J. Lightwave Technol.* **24**, 1295 (2006).

27. J. S. Lee, Y. C. Chung, and D. J. DiGiovanni, *IEEE Photon. Technol. Lett.* **5**, 1458 (1993).

28. H. Sanjoh, H. Yasaka, Y. Sakai, K. Sato, H. Ishii, and Y. Yoshikuni, *IEEE Photon. Technol. Lett.* **9**, 818 (1997).

29. L. Boivin, M. Wegmuller, M. C. Nuss, and W. H. Knox, *IEEE Photon. Technol. Lett.* **11**, 466 (1999).

30. K. Tamura, E. Yoshida, and M. Nakazawa, *Electron. Lett.* **32**, 1691 (1996).

31. J. J. Veselka and S. K. Korotky, *IEEE Photon. Technol. Lett.* **10**, 958 (1998).

32. S. Kawanishi, H. Takara, K. Uchiyama, I. Shake, and M. Mori, *Electron. Lett.* **35**, 826 (1999).

33. L. Boivin, S. Taccheo, C. R. Doerr, P. Schiffer, L. W. Stulz, R. Monnard, and W. Lin, *Electron. Lett.* **36**, 335 (2000).

34. L. Boivin and B. C. Collings, *Opt. Fiber Technol.* **7**, 1 (2001).

35. T. Morioka, K. Mori, S. Kawanishi, and M. Saruwatari, *IEEE Photon. Technol. Lett.* **6**, 365 (1994).

36. S. A. E. Lewis, M. J. Guy, J. R. Taylor, and R. Kashyap, *Electron. Lett.* **34**, 1247 (1998).

37. T. Morioka, K. Uchiyama, S. Kawanishi, S. Suzuki, and M. Saruwatari, *Electron. Lett.* **31**, 1064 (1995).

38. S. Kawanishi, H. Takara, K. Uchiyama, I. Shake, O. Kamatani, and H. Takahashi, *Electron. Lett.* **33**, 1716 (1977).

39. K. Takada, M. Abe, T. Shibata, and T. Okamoto, *Electron. Lett.* **38**, 572 (2002).

40. K. Mori, K. Sato, H. Takara, and T. Ohara, *Electron. Lett.* **39**, 544 (2003).

41. R. R. Alfano (Ed.), *The Super Continuum Laser Sources*, Springer, Berlin, 1989.

42. T. Takara, T. Ohara, K. Mori, K. Sato, E. Yamada, K. Jinguji, Y. Inoue, T. Shibata, T. Morioka, and K.-I. Sato, *Electron. Lett.* **36**, 2089 (2000).

43. H. Takara, H. Masuda, K. Mori, K. Sato, Y. Inoue, T. Ohara, A. Mori, M. Kohtoku, Y. Miyamoto, T. Morioka, and S. Kawanishi, *Electron. Lett.* **39**, 382 (2003).

44. H. Takara, T. Ohara, T. Yamamoto, H. Masuda, M. Abe, H. Takahashi, and T. Morioka, *Elecron. Lett.* **41**, 270 (2005).

45. M. Rochette, J. N. Kutz, J. L. Blows, et al., *IEEE Photon. Lett.* **17**, 908 (2005).

46. M. Rochette, L. B. Fu, V. Ta'eed, et al., *IEEE J. Sel. Top. Quantum Electron.* **12**, 736 (2006).

47. G. Raybon, Y. Su, J. Leuthold, R. J. Essiambre, T. Her, C. Joergensen, P. Steinvurzel, K. Dreyer, and K. Feder, *Conference on Optical Fiber Communications (OFC2002), Anaheim, CA,* 2002, PD10-1.

48. K. Igarashi and K. Kikuchi, *IEEE J. Sel. Top. Quantum Electron.* **14**, 551. (2008).

49. T. H. Her, G. Raybon, and C. Headley, *IEEE Photon. Technol. Lett.* **16**, 200 (2004).

50. M. Matsumoto, *J. Lightwave Technol.* **23**, 1472 (2004).

51. T. Her, G. Raybon, and C. Headley, *IEEE Photon. Technol. Lett.* **16**, 200 (2004).

52. T. Miyazaki and F. Kubota, *IEEE Photon. Technol. Lett.* **16**, 1909 (2004).

53. P. Johannisson and M. Karlsson, *IEEE Photon. Technol. Lett.* **17**, 2667 (2005).

54. A. G. Strigler and B. Schmauss, *J. Lightwave Technol.* **24**, 2835 (2006).

55. M. P. Fok and C. Shu, *IEEE J. Sel. Top. Quantum Electron.* **14**, 587 (2008).

56. S. Radic, D. J. Moss, and B. J. Eggleton, *Optical Fiber Telecommunications. VA. Components and Subsystems*, Elsevier Inc., 2008.

57. M. Rochette, J. L. Blows, and B. J. Eggleton, *Opt. Express* **14**, 6414 (2006).

58. J. Suzuki, T. Tanemura, K. Taira, et al., *IEEE Photon. Technol. Lett.* **17**, 423 (2005).

59. M. Daikoku, N. Yoshikane, T. Otani, and H. Tanaka, *J. Lightwave Technol.* **24**, 1142 (2006).

60. T. Yang, C. Shu, and C. Lin, *Opt. Express* **13**, 5409 (2005).

61. K. K. Chow, C. Shu, C. Lin, and A. Bjarklev, *IEEE Photon. Technol. Lett.* **17**, 624 (2005).

62. T. Tanemura, J. Suzuki, K. Katoh, and K. Kikuchi, *IEEE Photon. Technol. Lett.* **17**, 1052 (2005).

63. A. S. Leniham, R. Salem, T. E. Murphy, and G. M. Carter, *IEEE Photon. Technol. Lett.* **18**, 1329 (2006).

64. C. Ito and J. C. Cartledge, *IEEE Photon. Technol. Lett.* **20**, 425 (2008).

65. C. Ito and J. C. Cartledge, *IEEE J. Sel. Top. Quantum Electron.* **14**, 616 (2008).

66. J. Leuthold, C. H. Joyner, B. Mikkelsen, G. Raybon, J. L. Pleumeekers, B. I. Miler, K. Dreyer, and C. A. Burrus, *Electron. Lett.* **36**, 1129 (2000).

67. M. L. Masanovic, V. Lal, J. S. Barton, E. J. Skogen, L. A. Coldren, and D. J. Blumenthal, *IEEE Photon. Technol. Lett.* **15**, 1117 (2003).

68. A. E. Kelly, D. D. Marcenac, and D. Neset, *Electron. Lett.* **33**, 2123 (1997).

69. G. P. Agrawal, *Nonlinear Fiber Optics*, 2nd ed., Academic Press, San Diego, CA, 1995.

70. G. A. Nowak, Y.-H. Kao, T. J. Xia, M. N. Islam, and D. Nolan, *Opt. Lett.* **23**, 936 (1998).

71. J. Yu, X. Zheng, C. Peucheret, A. Clausen, H. Poulsen, and P. Jeppesen, *J. Lightwave Technol.* **18**, 1007 (2000).

72. M. N. Islam and O. Boyraz, *IEEE J. Sel. Top. Quantum Electron.* **8**, 527 (2002).

73. M. Hirano, T. Nakanishi, T. Okuno, and M. Onishi, *IEEE J. Sel. Top. Quantum Electron.* **15**, 103 (2009).

74. M. Takahashi, R. Sugizaki, J. Hiroishi, M. Tadakuma, Y. Taniguchi, and T. Yagi, *J. Lightwave Technol.* **23**, 3615 (2005).

75. T. Okuno, M. Ohmura, and M. Shigematsu, *OFC 2004, Anaheim, CA, 2004*, Paper MF21.

76. O. Aso, S.-I. Arai, T. Yagi, M. Tadakuma, Y. Suzuki, and S. Namiki, *Electron. Lett.* **36**, 709 (2000).

77. M.-C. Ho, M. E. Marhic, K. Y. K. Wong, and L. G. Kazovsky, *J. Lightwave Technol.* **20**, 469 (2002).

78. T. Tanemura and K. Kikuchi, *IEEE Photon. Technol. Lett.* **15**, 1573 (2003).

79. R. Elschner, C. A. Bunge, B. Huttl, A. G. Coca, C. S. Langhorst, R. Ludwig, C. Schubert, and K. Petermann, *IEEE J. Sel. Top. Quantum Electron.* **14**, 666 (2008).

80. J. H. Lee, K. Kikuchi, T. Nagashima, T. Hasegawa, S. Ohara, and N. Sugimoto, *Opt. Express*, **13**, 3144 (2005).

81. J. Yu and P. Jeppese, *IEEE Photon. Technol. Lett.* **13**, 833 (2001).

82. J. Yu, X. Zheng, C. Peucheret, A. Clausen, H. N. Poulsen, and P. Jeppesen, *J. Lightwave Technol.* **18**, 1007 (2000).

83. W. Wang, H. N. Poulsen, L. Rau, H.-F. Chou, J. E. Bowers, and D. J. Blumenthal, *J. Lightwave Technol.* **23**, 1105 (2005).

84. M. Galili, L. K. Oxenlowe, H. C. H. Hansen, A. T. Clausen, and P. Jeppesen, *IEEE J. Sel. Top. Quantum Electron.* **14**, 573 (2008).

85. V. G. Ta'eed, L. B. Fu, M. Pelusi, et al., *Opt. Express* **14**, 10371 (2006).

86. J. H. Lee, T. Nagashima, T. Hasegawa, et al., *Electron. Lett.* **41**, 918 (2005).

87. L. B. Fu, M. Rochette, V. G. Ta'eed, et al., *Opt. Express*, **13**, 7637 (2005).

88. V. W. S. Chan, K. L. Hall, E. Modiano, and K. A. Rauschenbach, *J. Lightwave Technol.* **16**, 2146 (1998).

89. I. W. White, R. V. Penty, and R. E. Epworth, *Electron. Lett.* **24**, 340 (1988).

90. K. J. Blow, N. J. Doran, B. K. Nayar, and B. P. Nelson, *Opt. Lett.* **15**, 248 (1990).

91. M. Jino and T. Matsumoto, *Electron. Lett.* **27**, 75 (1991).

92. A. D. Ellis and D. A. Cleland, *Electron. Lett.* **28**, 405 (1992).

93. H. Bülow and G. Veith, *Electron. Lett.* **29**, 588 (1993).

94. E. Bodtkrer and J. E. Bowers, *J. Lightwave Technol.* **13**, 1809 (1995).

95. D. M. Patrick, A. D. Ellis, and D. M. Spirit, *Electron. Lett.* **29**, 702 (1993).

96. K. Uchiyama, H. Takara, T. Morioka, S. Kawanishi, and M. Saruwarari, *Electron. Lett.* **29**, 1313 (1993).

97. T. Yamamoto, E. Yoshida, and M. Nakazawa, *Electron. Lett.* **34**, 1013 (1998).

98. M. Jinno and T. Matsumoto, *IEEE Photon. Technol. Lett.* **2**, 349 (1990).

99. H. Sotobayashi, C. Sawagushi, Y. Koyamada, and W. Chujo, *Opt. Lett.* **27**, 1555 (2002).

100. J. D. Moores, K. Bergman, H. A. Haus, and E. P. Ippen, *Opt. Lett.* **16**, 138 (1991).

101. J. E. Sharping, M. Fiorentino, P. Kumar, and R. S. Windeler *IEEE Photon. Technol. Lett.* **14**, 77 (2002).

102. M. C. Ho, K. Uesaka, M. Marhic, Y. Akasaka, and L. G. Kazovsky, *J. Opt. Lightwave. Technol.* **19**, 977 (2001).

103. P. A. Andrekson, N. A. Olsson, J. R. Olsson, J. R. Simpson, T. Tanbun-Ek, R. A. Logan, and M. Haner, *Electron. Lett.* **27**, 922 (1991).

INDEX

2-R regeneration, 343–348
3-R regeneration, 348, 349

Acoustic jitter, 168
Acoustic velocity, 275
Acoustic wave, 133, 274, 277
  light scattering at, 274–277
Active fibers, 282–284
Adiabatic pulse compression, 210
Amplification
  Brillouin, 286–293
  distributed, 164–166, 259,
    282–284
  lumped, 163, 164, 259
  parametric, 123–128
  Raman, 258–264
Amplified spontaneous
    emission, 166, 263, 289
Amplitude jitter, 147, 150, 189
Anti-Stokes, 114, 245
ASK, 285, 286, 291, 346, 354
Attenuation, 44, 45, 160
  constant, 23

Background instability, 170, 171, 174
Beat length, 55, 227
BER, 150, 151, 343, 347
Birefringence, 43, 55, 29, 234, 311
  linear, 55, 56, 229–234
  randomly varying, 57–60, 234–236
Bragg condition, 213
Brewster angle, 20

Brillouin
  Leon, 273
Brillouin amplifier, 286–293
  gain, 287–289
  noise, 289, 290
Brillouin distributed sensor, 292, 293
Brillouin filter, 292
Brillouin fiber lasers, 296–300
  multiwavelength, 298
Brillouin gain
  frequency shift, 275
  linewidth, 279
  self-homodyne scheme, 290, 291
Brillouin gain coefficient, 279, 280
  chalcogenide fiber, 279
  tellurite fiber, 279
Brillouin scattering
  spontaneous, 133, 273, 276
  stimulated, 273, 277
Brillouin threshold, 280, 282

CGLE, 174–183, 203, 213
  analytical solutions of, 176–180
  numerical solutions of, 180–183
Cherenkov radiation, 318
Chirp, 50–53, 91, 92, 142
  IXPM-induced, 142–146
  SPM-induced, 87–89
  XPM-induced, 100
Chirp parameter, 51, 92
Chromatic dispersion, 46–54
Coherence length, 117

Nonlinear Effects in Optical Fibers. By Mário F. S. Ferreira.
Copyright © 2011 John Wiley & Sons, Inc. Published 2011 by John Wiley & Sons, Inc.

Complex Ginzburg-Landau equation,
  see CGLE
Correlation length, 57
Crosstalk
  Brillouin-induced, 284
  FWM-induced, 121
  Raman-induced, 255–58
Cutoff condition, 33, 42

DCF, 54
DDF, 210
  pulse propagation in, 211
DGD, 56
Dipole moment, 22
Dispersion
  anomalous, 47
  chromatic, 46–54
  group velocity, 15, 47
  intermodal, 35
  material, 48
  normal, 47
  polarization-mode, 54–60, 234–242
  waveguide, 48
Dispersion compensation, 53, 54
Dispersion length, 50, 75
Dispersion managed soliton, 183–189, 202
  PMD effects, 240
  variational approach, 185–189
Dispersion management, 149, 183
  asymmetric, 150
  FWM suppression with, 123
  symmetric, 145, 150
Dispersion map, 184
Dispersion map strength, 240
Dispersion parameter, 47
Dispersion relation, 32, 95, 104, 275
Distributed amplification, 164–166, 259, 282
DPSK, 150, 354
DWDM, 340, 342

EDFA, 162, 164, 258, 298, 299
Effective area, 44, 68, 72, 249, 313
Effective length, 69, 70, 253, 356
Eigenvalue equation, 41
Eikonal equation, 25, 36
Electrostrictive coefficient, 277
Energy density, 13, 14
Euler-Lagrange equation, 90
Evanescent wave, 36

Extrusion technique, 309
Eye diagram, 120, 150, 347

Fiber
  AllWave, 276
  attenuation, 44, 45
  birefringent, 55, 29, 234, 311
  bismuth-oxide, 296, 314–317, 354, 358
  chalcogenide, 279, 296, 314–317, 344,
    356–358
  dispersion decreasing, 210
  dispersion flattened, 48
  dispersion-shifted, 48
  endlessly single-mode, 311
  graded-index, 36–39
  HG, 256
  highly nonlinear, 305–332
  hollow-core microstructured, 309
  microstructured, 118, 130, 134, 309–313
  modes, 39–42
  multimode, 30
  photonic crystal, 309
  polarization maintaining, 43, 225
  single-mode, 30, 42,43
  standard single-mode, 307
  step-index, 33–36
  tapered, 308
  tellurite, 279, 296, 314, 315
  TrueWave, 276
Fiber Bragg gratings, 213–220
  Bragg resonance, 213
  chirped, 216
  dispersion parameters, 216
  group velocity, 216
  reflection coefficient, 214
Fiber Bragg solitons, 216–220
Fiber lasers
  Brillouin, 296–300
  figure-eight, 201
  soliton, 199–204
  stretched-pulse, 202
  Raman, 264–269
Flux density, 14
FOPA, 123–128, 321
  bandwidth, 127
  gain, 124, 321, 351
  parametric gain, 124
  two-pump, 127, 128
FOPO, 128–131, 321

Frequency filters,
 fixed, 169
 sliding, 170
Frequency shift
 Brillouin, 275
 FWM, 118
 Raman, 250
 XPM-induced, 191
Fourier transform, 39, 49
Fresnel equations, 17–19
FSK, 285, 286
FWM, 111–135, 320–323, 351–354
 amplifier, 123
 coherence length, 117
 crosstalk, 121
 degenerate, 116, 351
 effective phase mismatch, 116
 efficiency, 121
 impact and control of, 118
 mathematical description of, 114, 115
 nonlinear phase conjugation
  with, 131
 phase matching, 113, 115
 spontaneous, 133, 134
 wave mixing, 112

Gain
 Brillouin, 278–280
 parametric, 124, 125, 351
 Raman, 250–252
Gaussian pulse, 48, 51, 87, 91
GAWBS, 133
Geometrical optics, 24, 30
Ghost pulse, 141, 147–149, 152
Ginzburg-Landau equation, see CGLE
Gordon-Haus effect, 155, 166, 167
Grating-fiber compressor, 204–207
Group velocity, 15, 16
Group velocity dispersion (GVD), 16, 47
GVD parameter, 47

Harmonic oscillator, 21, 246
Helmholtz equation, 39, 214
Highly nonlinear fiber, 305–332
 microstructured, 309–314
 non-silica, 314–317
 silica, 306–308
Hollow-core fiber, 309
Idler wave, 114, 116, 320, 351

IFWM, 140, 147–149
 ghost pulse, 148, 149
 suppression, 149
Interaction length, 102, 190
Interferometer,
 Mach-Zehnder, 359
 Sagnac, 211, 360
Intrachannel nonlinear effects, 139–152
 mathematical description, 140–142
 control of, 149–152
Inverse scattering theory, 77
IXPM, 140, 142–146
 phase shift, 142
 suppression, 149
 timing jitter, 143

Jitter
 acoustic, 168
 amplitude, 150
 Gordon-Haus, 166, 167
 PMD-induced, 168
 suppression, 169–176
 IXPM-induced, 145, 150

Kerr coefficient, 67, 68, 306
 dopant dependence, 306, 307
Kerr shutter, 229, 362
Kleinmans symmetry condition, 66

Lagrangian, 90–92, 186
Lorentzian spectrum, 279
Lumped amplification, 163, 164, 259

Mach-Zehnder interferometer, 359
Manakov-PMD equation, 235
Manakov soliton, 236
Maxwell
 James, 9
Maxwell's equations, 9
Maxwellian distribution, 58
Microstructured fiber, 118, 130, 134, 309–313
Mid-span spectral inversion, 131
Mode
 fundamental, 43
 Gaussian approximation, 43
 guided, 30
 radiation, 30
Modulation, 41
Modulation frequency, 15

Modulation instability
   SPM-induced, 94–97
   XPM-induced, 104–106

NALM, 201
NLSE, 70–73
   coupled, 98, 226, 234
   normalized form, 74, 75
   perturbed, 160, 165, 174
   soliton solutions, 75
   numerical solution of, 81
Noise
   Brillouin amplifier, 289, 290
   Raman amplifier, 263, 264
   XPM, 101
NOLM, 201, 360
Nonlinear gain, 174
Nonlinear length, 75
Nonlinear parameter, 72, 306–308, 313
Nonlinear phase conjugation, 131–133
Nonlinear polarization, 63, 226, 247, 248
Nonlinear refractive index, 66, 306, 314
Nonlinear Schrödinger equation,
   see NLSE
Nonlinear wave equation, 66
Nonsoliton radiation, 318, 325
NRZ, 45
Numerical aperture, 35
NZDSF, 103

Optical ray, 24, 26
   meridional, 38
Optical regeneration, 343–349
Optical switching, 358–363
   FWM-based, 361–363
   SPM-induced, 221
   XPM-based 359–361
OTDM, 340, 341, 350, 356, 360

Parametric amplifier, see FOPA
Parametric gain coefficient, 124,
   125, 351
Pauli matrices, 234
Penetration depth, 36
Permittivity, 10, 247, 278
Phase conjugation, 131–133
Phase matching, 115–118
   condition, 113, 118, 320
   SPM-induced, 118

Phase mismatch, 113, 117, 352
   effective, 116, 351
Phase shift,
   nonlinear, 227–229
   IXPM-induced, 142
   SBS-induced, 291
   SPM-induced, 86
   XPM-induced, 99
Phase velocity, 15
Photonic crystal fiber, 309
Photon pair source, 133
Planar waveguides, 30
PMD, 54–60
   first-order compensation, 239
   intrinsic, 55–57
   nonlinear, 236
   pulse broadening induced by, 236
   solitons and, 236–242
   timing jitter induced by, 168
   vector, 58
PMD parameter, 168, 237
Polarization
   induced, 10, 22
   linear, 63
   nonlinear, 63, 227
   third-order, 65
Polarization effects, 152, 225–242
   XPM-induced coupling, 233
Polarization maintaining fibers, 43, 225
Polarization-mode dispersion,
   see PMD
Poynting vector, 13
Principal states of polarization, 58, 236
PSK, 93, 285, 286
Pulse
   chirped Gaussian, 52, 91
   Gaussian, 48, 87
   hyperbolic secant, 91
   super-Gaussian, 88
Pulse broadening
   dispersion-induced, 50
   loss induced, 161
   PMD-induced, 236–242
Pulse compression, 204–220
   factor, 206, 208, 212
   grating-fiber, 204–207
   soliton-effect, 207–210
   fundamental soliton, 210–213
   fiber Bragg gratings for, 214

Pulse shaping, 229
Pulse train generation, 349, 350

Quasi-adiabatic regime, 162, 165

Raman
  Chandrasekhara, 245
Raman amplification, 258–264
  broadband, 259
  gain, 261
  noise, 263
Raman gain coefficient, 249–252
  chalcogenide fiber, 252
  dopant dependence, 251, 252
  polarization dependence, 250
  tellurite fiber, 252
Raman laser, 264–269
  cascaded, 265
  multiwavelength, 265, 266
  tunable, 267
Raman scattering
  intrapulse, 165
  harmonic oscillator model, 246–249
  spontaneous, 245, 253, 263
Raman soliton, 317–319
Raman susceptibility, 248
Raman threshold, 252
Rayleigh length, 68
Rayleigh scattering, 44
Reflection
  coefficient, 18, 19
  total internal, 17, 30, 33
Reflectance, 20
Refractive index, 11
  complex, 23
Relative permittivity, 24
Resonance frequency, 22, 246

Sagnac interferometer, 211, 360
SBS, 273–300
  active fibers, 282–284
  amplifier, 286–293
  applications, 290–293
  ASE noise, 289, 290
  coupled equations, 278, 279, 282
  gain, 287–289
  gain coefficient, 279
  laser, 296–300
  impact on communication
    systems, 284–286

nonlinear phase shift, 291
slow light, 293–296
threshold, 280–282
Scattering,
  Brillouin, 273
  Raman, 245
  Rayleigh, 44
Second-order susceptibility, 65
Self-frequency shift, see SSFS
Sellmeier formula, 24
Sensor, 292, 293
Signal wave
  Brillouin, 287
  FWM, 115, 351
  parametric amplifier, 124, 361
  Raman, 260
Signal processing, 339–363
Signal regeneration, 343–349
Single-mode fiber, 30, 42, 311
Sliding frequency filters, 170
Slow light, 293–296
Slowly varying envelope approximation,
  71, 277
Snell's laws, 17
SNR, 166, 343
Soliton lightwave systems, 155–192
Soliton transmission control,
  168–176
Solitons,
  amplification, 161–164
  arbitrary-amplitude, 178
  birefringence effects on, 234–239
  Bragg, 216–220
  collision, 189–192
  composite pulse, 181, 182
  dark, 80
  dispersion-managed, 183–189
  dissipative, 176–183
  exploding, 182, 183
  fiber Bragg, 216–220
  fiber loss effects, 160, 161
  flat-top, 181
  fundamental, 76, 156
  higher order, 78
  interaction, 157–159
  moving, 181
  order, 78, 207
  path-averaged, 162–168
  period, 79, 207
  perturbation theory, 160

Solitons (*Continued*)
  PMD effects, 237–242
  power, 79, 157, 208
  pulsating, 182
  second-order, 78
  self-frequency shift of, 165, 317–320
  third-order dispersion effect, 165
  transmission control, 168
  WDM systems, 189–192
Soliton-effect compressor, 207–210
Soliton fiber laser, 199–204
  figure-eight, 201
  modeling, 203, 204
  stretched-pulse,202, 203
  the first, 200
Spectrum slicing, 340–343
Split-step Fourier method, 81, 330
SPM, 74–96, 323
  frequency shift, 88
  impact on communication systems, 93
  modulation instability by, 94
  phase shift, 86
  spectral broadening, 88, 89
  spectral narrowing, 92
Squeezing, 133
SRS, 245–269, 323
  amplifier gain, 260–263
  ASE noise, 263, 264
  coupled equations, 249
  gain coefficient, 250
  impact on communication
    systems, 255–258
  laser, 264–269
  threshold, 252–255
SSFS, 165, 166, 317–320
Stokes wave,
  FWM, 114
  higher-order, 255, 282
  SBS, 274
  SRS, 246
Supercontinum, 323–332
  basic physics, 323–330
  modeling, 330–332
Susceptibility
  first order, 64
  linear, 22
  second order, 65
  third order, 65, 339
Synchronous modulators, 173

Tapered fiber, 308
Threshold
  SBS, 280–282
  SRS, 252–255
Timing jitter, 143–145, 166–168
  acoustic, 168
  Gordon-Hauss, 166, 167
  control of, 149, 150, 168–174
  PMD-induced, 168
Third order dispersion, 165, 318, 319
Total internal reflection, 17, 30, 33
Transmission coefficient, 18, 19
Transmittance, 20
Two-photon absorption, 345

Unequally channel spacing, 118–120

V number, 32, 42, 307
Vacuum permeability, 10
Vacuum permitivitty, 10
Variational approach, 89, 144, 160, 185, 231, 232, 240
Velocity
  group, 15, 16
  light, 11
  phase, 15

Walk-off, 102, 360
Wave equation, 11, 66
Wavelength conversion, 350–358
  FWM, 351–354
  Raman enhanced, 355–357
  XPM, 354–358
WDM systems, 189–192, 255–258
  Sources for, 340–343

XPM, 97–106, 227–234, 323
  coupled equations, 98, 100
  impact on communication
    systems, 100
  intrachannel, 142–146
  modulation instability by, 103
  phase shift, 97
  polarization effects, 227–234
  soliton formed by, 106

Zero-dispersion wavelength, 48, 311–313, 319